ALSO BY WALTER ISAACSON

Leonardo da Vinci

The Innovators: How a Group of Hackers, Geniuses, and Geeks Created the Digital Revolution

Steve Jobs

American Sketches

Einstein: His Life and Universe

Benjamin Franklin: An American Life

Kissinger: A Biography

The Wise Men: Six Friends and the World They Made (with Evan Thomas)

Pro and Con

THE CODE BREAKER

Jennifer Doudna, Gene Editing, and the Future of the Human Race

WALTER ISAACSON

Simon & Schuster

NEW YORK LONDON TORONTO SYDNEY NEW DELHI

Simon & Schuster
1230 Avenue of the Americas
New York, NY 10020

First Simon & Schuster hardcover edition March 2021

Manufactured in the United States of America

1 3 5 7 9 10 8 6 4 2

Library of Congress Cataloging-in-Publication Data

ISBN 978-1-9821-1585-2
ISBN 978-1-9821-1587-6 (ebook)

CONTENTS

INTRODUCTION Into the Breach xiii

Part One: The Origins of Life

CHAPTER 1 Hilo 3
CHAPTER 2 The Gene 11
CHAPTER 3 DNA 17
CHAPTER 4 The Education of a Biochemist 29
CHAPTER 5 The Human Genome 37
CHAPTER 6 RNA 43
CHAPTER 7 Twists and Folds 51
CHAPTER 8 Berkeley 63

Part Two: CRISPR

CHAPTER 9 Clustered Repeats 71
CHAPTER 10 The Free Speech Movement Café 79
CHAPTER 11 Jumping In 81
CHAPTER 12 The Yogurt Makers 89
CHAPTER 13 Genentech 97
CHAPTER 14 The Lab 103
CHAPTER 15 Caribou 113
CHAPTER 16 Emmanuelle Charpentier 119
CHAPTER 17 CRISPR-Cas9 129
CHAPTER 18 *Science*, 2012 137
CHAPTER 19 Dueling Presentations 143

Part Three: Gene Editing

CHAPTER 20 A Human Tool 153
CHAPTER 21 The Race 157
CHAPTER 22 Feng Zhang 161
CHAPTER 23 George Church 169
CHAPTER 24 Zhang Tackles CRISPR 175
CHAPTER 25 Doudna Joins the Race 187
CHAPTER 26 Photo Finish 191
CHAPTER 27 Doudna's Final Sprint 197
CHAPTER 28 Forming Companies 203
CHAPTER 29 Mon Amie 215
CHAPTER 30 The Heroes of CRISPR 223
CHAPTER 31 Patents 231

Part Four: CRISPR in Action

CHAPTER 32 Therapies 245
CHAPTER 33 Biohacking 253
CHAPTER 34 DARPA and Anti-CRISPR 259

Part Five: Public Scientist

CHAPTER 35 Rules of the Road 267
CHAPTER 36 Doudna Steps In 283

Part Six: CRISPR Babies

CHAPTER 37 He Jiankui 299
CHAPTER 38 The Hong Kong Summit 315
CHAPTER 39 Acceptance 325

Part Seven: The Moral Questions

CHAPTER 40 Red Lines 335
CHAPTER 41 Thought Experiments 341
CHAPTER 42 Who Should Decide? 355
CHAPTER 43 Doudna's Ethical Journey 367

Part Eight: Dispatches from the Front

CHAPTER 44 Quebec 373
CHAPTER 45 I Learn to Edit 379
CHAPTER 46 Watson Revisited 385
CHAPTER 47 Doudna Pays a Visit 395

Part Nine: Coronavirus

CHAPTER 48 Call to Arms 401
CHAPTER 49 Testing 407
CHAPTER 50 The Berkeley Lab 413
CHAPTER 51 Mammoth and Sherlock 421
CHAPTER 52 Coronavirus Tests 427
CHAPTER 53 Vaccines 435
CHAPTER 54 CRISPR Cures 449
CHAPTER 55 Cold Spring Harbor Virtual 459
CHAPTER 56 The Nobel Prize 469
EPILOGUE 477

Acknowledgments 483
Notes 487
Index 517
Image Credits 535

To the memory of Alice Mayhew and Carolyn Reidy.

What a joy it was to see them smile.

Introduction

Into the Breach

Jennifer Doudna couldn't sleep. Berkeley, the university where she was a superstar for her role in inventing the gene-editing technology known as CRISPR, had just shut down its campus because of the fast-spreading coronavirus pandemic. Against her better judgment, she had driven her son, Andy, a high school senior, to the train station so he could go to Fresno for a robot-building competition. Now, at 2 a.m., she roused her husband and insisted that they retrieve him before the start of the match, when more than twelve hundred kids would be gathering in an indoor convention center. They pulled on their clothes, got in the car, found an open gas station, and made the three-hour drive. Andy, an only child, was not happy to see them, but they convinced him to pack up and come home. As they pulled out of the parking lot, Andy got a text from the team: "Robotics match cancelled! All kids to leave immediately!"[1]

This was the moment, Doudna recalls, that she realized her world, and the world of science, had changed. The government was fumbling its response to COVID, so it was time for professors and graduate students, clutching their test tubes and raising their pipettes high, to rush into the breach. The next day—Friday, March 13, 2020—she led a meeting of her Berkeley colleagues and other scientists in the Bay Area to discuss what roles they might play.

A dozen of them made their way across the abandoned Berkeley campus and converged on the sleek stone-and-glass building that housed her lab. The chairs in the ground-floor conference room were clustered together, so the first thing they did was move them six feet apart. Then they turned on a video system so that fifty other researchers from nearby universities could join by Zoom. As she stood in front of the room to rally them, Doudna displayed an intensity that she usually kept masked by a calm façade. "This is not something that academics typically do," she told them. "We need to step up."[2]

It was fitting that a virus-fighting team would be led by a CRISPR pioneer. The gene-editing tool that Doudna and others developed in 2012 is based on a virus-fighting trick used by bacteria, which have been battling viruses for more than a billion years. In their DNA, bacteria develop clustered repeated sequences, known as CRISPRs, that can remember and then destroy viruses that attack them. In other words, it's an immune system that can adapt itself to fight each new wave of viruses—just what we humans need in an era that has been plagued, as if we were still in the Middle Ages, by repeated viral epidemics.

Always prepared and methodical, Doudna (pronounced DOWD-nuh) presented slides that suggested ways they might take on the coronavirus. She led by listening. Although she had become a science celebrity, people felt comfortable engaging with her. She had mastered the art of being tightly scheduled while still finding the time to connect with people emotionally.

The first team that Doudna assembled was given the job of creating a coronavirus testing lab. One of the leaders she tapped was a postdoc named Jennifer Hamilton who, a few months earlier, had spent a day teaching me to use CRISPR to edit human genes. I was pleased, but also a bit unnerved, to see how easy it was. Even I could do it!

Another team was given the mission of developing new types of coronavirus tests based on CRISPR. It helped that Doudna liked commercial enterprises. Three years earlier, she and two of her graduate students had started a company to use CRISPR as a tool for detecting viral diseases.

In launching an effort to find new tests to detect the coronavirus, Doudna was opening another front in her fierce but fruitful struggle with a cross-country competitor. Feng Zhang, a charming young China-born and Iowa-raised researcher at the Broad Institute of MIT and Harvard, had been her rival in the 2012 race to turn CRISPR into a gene-editing tool, and ever since then they had been locked in an intense competition to make scientific discoveries and form CRISPR-based companies. Now, with the outbreak of the pandemic, they would engage in another race, this one spurred not by the pursuit of patents but by a desire to do good.

Doudna settled on ten projects. She suggested leaders for each and told the others to sort themselves into the teams. They should pair up with someone who would perform the same functions, so that there could be a battlefield promotion system: if any of them were struck by the virus, there would be someone to step in and continue their work. It was the last time they would meet in person. From then on the teams would collaborate by Zoom and Slack.

"I'd like everyone to get started soon," she said. "Really soon."

"Don't worry," one of the participants assured her. "Nobody's got any travel plans."

What none of the participants discussed was a longer-range prospect: using CRISPR to engineer inheritable edits in humans that would make our children, and all of our descendants, less vulnerable to virus infections. These genetic improvements could permanently alter the human race.

"That's in the realm of science fiction," Doudna said dismissively when I raised the topic after the meeting. Yes, I agreed, it's a bit like *Brave New World* or *Gattaca*. But as with any good science fiction, elements have already come true. In November 2018, a young Chinese scientist who had been to some of Doudna's gene-editing conferences used CRISPR to edit embryos and remove a gene that produces a receptor for HIV, the virus that causes AIDS. It led to the birth of twin girls, the world's first "designer babies."

There was an immediate outburst of awe and then shock. Arms flailed, committees convened. After more than three billion years of evolution of life on this planet, one species (us) had developed the talent and temerity to grab control of its own genetic future. There was a sense that we had crossed the threshold into a whole new age, perhaps a brave new world, like when Adam and Eve bit into the apple or Prometheus snatched fire from the gods.

Our newfound ability to make edits to our genes raises some fascinating questions. Should we edit our species to make us less susceptible to deadly viruses? What a wonderful boon that would be! Right? Should we use gene editing to eliminate dreaded disorders, such as Huntington's, sickle-cell anemia, and cystic fibrosis? That sounds good, too. And what about deafness or blindness? Or being short? Or depressed? Hmmm . . . How should we think about that? A few decades from now, if it becomes possible and safe, should we allow parents to enhance the IQ and muscles of their kids? Should we let them decide eye color? Skin color? Height?

Whoa! Let's pause for a moment before we slide all of the way down this slippery slope. What might that do to the diversity of our societies? If we are no longer subject to a random natural lottery when it comes to our endowments, will it weaken our feelings of empathy and acceptance? If these offerings at the genetic supermarket aren't free (and they won't be), will that greatly increase inequality—and indeed encode it permanently in the human race? Given these issues, should such decisions be left solely to individuals, or should society as a whole have some say? Perhaps we should develop some rules.

By "we" I mean *we*. All of us, including you and me. Figuring out if and when to edit our genes will be one of the most consequential questions of the twenty-first century, so I thought it would be useful to understand how it's done. Likewise, recurring waves of virus epidemics make it important to understand the life sciences. There's a joy that springs from fathoming how something works, especially when that something is ourselves. Doudna relished that joy, and so can we. That's what this book is about.

The invention of CRISPR and the plague of COVID will hasten our transition to the third great revolution of modern times. These revolutions arose from the discovery, beginning just over a century ago, of the three fundamental kernels of our existence: the atom, the bit, and the gene.

The first half of the twentieth century, beginning with Albert Einstein's 1905 papers on relativity and quantum theory, featured a revolution driven by physics. In the five decades following his miracle year, his theories led to atom bombs and nuclear power, transistors and spaceships, lasers and radar.

The second half of the twentieth century was an information-technology era, based on the idea that all information could be encoded by binary digits—known as bits—and all logical processes could be performed by circuits with on-off switches. In the 1950s, this led to the development of the microchip, the computer, and the internet. When these three innovations were combined, the digital revolution was born.

Now we have entered a third and even more momentous era, a life-science revolution. Children who study digital coding will be joined by those who study genetic code.

When Doudna was a graduate student in the 1990s, other biologists were racing to map the genes that are coded by our DNA. But she became more interested in DNA's less-celebrated sibling, RNA. It's the molecule that actually does the work in a cell by copying some of the instructions coded by the DNA and using them to build proteins. Her quest to understand RNA led her to that most fundamental question: How did life begin? She studied RNA molecules that could replicate themselves, which raised the possibility that in the stew of chemicals on this planet four billion years ago they started to reproduce even before DNA came into being.

As a biochemist at Berkeley studying the molecules of life, she focused on figuring out their structure. If you're a detective, the most basic clues in a biological whodunit come from discovering how a molecule's twists and folds determine the way it interacts with other molecules. In Doudna's case, that meant studying the structure of RNA. It was an echo of the work Rosalind Franklin had done with DNA, which was used by James Watson and Francis Crick to discover the double-helix structure of DNA in 1953. As it happens, Watson, a complex figure, would weave in and out of Doudna's life.

Doudna's expertise in RNA led to a call from a biologist at Berkeley who was studying the CRISPR system that bacteria developed in their battle against viruses. Like a lot of basic science discoveries, it turned out to have practical applications. Some were rather ordinary, such as protecting the bacteria in yogurt cultures. But in 2012 Doudna and others figured out a more earth-shattering use: how to turn CRISPR into a tool to edit genes.

CRISPR is now being used to treat sickle-cell anemia, cancers, and blindness. And in 2020, Doudna and her teams began exploring how CRISPR could detect and destroy the coronavirus. "CRISPR evolved in bacteria because of their long-running war against viruses," Doudna says. "We humans don't have time to wait for our own cells to evolve natural resistance to this virus, so we have to use our ingenuity to do that. Isn't it fitting that one of the tools is this ancient bacterial immune system called CRISPR? Nature is beautiful that way." Ah, yes. Remember that phrase: Nature is beautiful. That's another theme of this book.

There are other star players in the field of gene editing. Most of them deserve to be the focus of biographies or perhaps even movies. (The elevator pitch: *A Beautiful Mind* meets *Jurassic Park*.) They play important roles in this book, because I want to show that science is a team sport. But I also want to show the impact that a persistent, sharply inquisitive, stubborn, and edgily competitive player can have. With a smile that sometimes (but not always) masks the wariness in her eyes, Jennifer Doudna turned out to be a great central character. She has the instincts to be collaborative, as any scientist must, but ingrained in her character is a competitive streak, which most great innovators have. With her emotions usually carefully controlled, she wears her star status lightly.

Her life story—as a researcher, Nobel Prize winner, and public policy thinker—connects the CRISPR tale to some larger historical threads, including the role of women in science. Her work also illustrates, as Leonardo da Vinci's did, that the key to innovation is connecting a curiosity about basic science to the practical work of devising tools that can be applied to our lives—moving discoveries from lab bench to bedside.

By telling her story, I hope to give an up-close look at how science works. What actually happens in a lab? To what extent do discoveries depend on individual genius, and to what extent has teamwork become more critical? Has the competition for prizes and patents undermined collaboration?

Most of all, I want to convey the importance of *basic* science, meaning quests that are curiosity-driven rather than application-oriented. Curiosity-driven research into the wonders of nature plants the seeds, sometimes in unpredictable ways, for later innovations.[3] Research about surface-state physics eventually led to the transistor and microchip. Likewise, studies of an astonishing method that bacteria use to fight off viruses eventually led to a gene-editing tool and techniques that humans can use in their own struggle against viruses.

It is a story filled with the biggest of questions, from the origins of life to the future of the human race. And it begins with a sixth-grade girl who loved searching for "sleeping grass" and other fascinating phenomena amid the lava rocks of Hawaii, coming home from school one day and finding on her bed a detective tale about the people who discovered what they proclaimed to be, with only a little exaggeration, "the secret of life."

PART ONE

The Origins of Life

The Lord God made a garden in the east, in Eden;
and there he put the man he had made.
Out of the ground the Lord God caused to grow
every tree that is beautiful and good for food;
the tree of life also in the midst of the garden,
and the tree of the knowledge of good and evil.

—Genesis 2:8–9

Jennifer in Hilo

Don Hemmes

Ellen, Jennifer, Sarah, Martin, and Dorothy Doudna

Chapter 1

Hilo

Haole

Had she grown up in any other part of America, Jennifer Doudna might have felt like a regular kid. But in Hilo, an old town in a volcano-studded region of the Big Island of Hawaii, the fact that she was blond, blue-eyed, and lanky made her feel, she later said, "like I was a complete freak." She was teased by the other kids, especially the boys, because unlike them she had hair on her arms. They called her a "haole," a term that, though not quite as bad as it sounds, was often used as a pejorative for non-natives. It imbedded in her a slight crust of wariness just below the surface of what would later become a genial and charming demeanor.[1]

A tale that became part of the family lore involved one of Jennifer's great-grandmothers. She was part of a family of three brothers and three sisters. Their parents could not afford for all six to go to school, so they decided to send the three girls. One became a teacher in Montana and kept a diary that has been handed down over the generations. It is filled with tales of perseverance, broken bones, working in the family store, and other frontier endeavors. "She was crusty and stubborn and had a pioneering spirit," said Jennifer's sister Sarah, the current generation's keeper of the diary.

Jennifer was likewise one of three sisters, but there were no brothers. As the oldest, she was doted on by her father, Martin Doudna, who sometimes referred to his children as "Jennifer and the girls." She was born February 19, 1964, in Washington, D.C., where her father worked as a speechwriter for the Department of Defense. He yearned to be a professor of American literature, so he moved to Ann Arbor with his wife, a community college teacher named Dorothy, and enrolled at the University of Michigan.

When he earned his doctorate, he applied for fifty jobs and got only one offer, from the University of Hawaii at Hilo. So he borrowed $900 from his wife's retirement fund and moved his family there in August 1971, when Jennifer was seven.

Many creative people—including most of those I have chronicled, such as Leonardo da Vinci, Albert Einstein, Henry Kissinger, and Steve Jobs—grew up feeling alienated from their surroundings. That was the case for Doudna as a young blond girl among the Polynesians in Hilo. "I was really, really alone and isolated at school," she says. In the third grade, she felt so ostracized that she had trouble eating. "I had all sorts of digestive problems that I later realized were stress related. Kids would tease me every day." She retreated into books and developed a defensive layer. "There's an internal part of me they'll never touch," she told herself.

Like many others who have felt like an outsider, she developed a wide-ranging curiosity about how we humans fit into creation. "My formative experience was trying to figure out who I was in the world and how to fit in in some way," she later said.[2]

Fortunately, this sense of alienation did not become too ingrained. Life as a schoolkid got better, she developed a genial spirit, and the scar tissue of early childhood began to fade. It would become inflamed only on rare occasions, when some act—an end run on a patent application, a male business colleague being secretive or misleading—scratched deeply enough.

Blossoming

The improvement began halfway through third grade, when her family moved from the heart of Hilo to a new development of cookie-cutter houses that had been carved into a forested slope further up the flanks of the Mauna Loa volcano. She switched from a large school, with sixty kids per grade, to a smaller one with only twenty. They were studying U.S. history, a subject that made her feel more connected. "It was a turning point," she recalled. She thrived so well that by the time she was in fifth grade, her math and science teacher urged that she skip ahead. So her parents moved her into sixth grade.

That year she finally made a close friend, one she kept throughout her life. Lisa Hinkley (now Lisa Twigg-Smith) was from a classic mixed-race Hawaiian family: part Scottish, Danish, Chinese, and Polynesian. She knew how to handle the bullies. "When someone would call me a f—king haole, I would cringe," Doudna recalled. "But when a bully called Lisa names, she would turn and look right at him and give it right back to him. I decided I wanted to be that way." One day in class the students were asked what they wanted to be when they grew up. Lisa proclaimed that she wanted to be a skydiver. "I thought, 'That is so cool.' I couldn't imagine answering that. She was very bold in a way that I wasn't, and I decided to try to be bold as well."

Doudna and Hinkley spent their afternoons riding bikes and hiking through sugarcane fields. The biology was lush and diverse: moss and mushrooms, peach and arenga palms. They found meadows filled with lava rocks covered in ferns. In the lava-flow caves there lived a species of spider with no eyes. How, Doudna wondered, did it come to be? She was also intrigued by a thorny vine called hilahila or "sleeping grass" because its fernlike leaves curl up when touched. "I asked myself," she recalls, "'What causes the leaves to close when you touch them?'"[3]

We all see nature's wonders every day, whether it be a plant that moves or a sunset that reaches with pink fingers into a sky of deep blue. The key to true curiosity is pausing to ponder the causes. What makes a sky blue or a sunset pink or a leaf of sleeping grass curl?

Doudna soon found someone who could help answer such questions. Her parents were friends with a biology professor named Don Hemmes, and they would all go on nature walks together. "We took excursions to Waipio Valley and other sites on the Big Island to look for mushrooms, which was my scientific interest," Hemmes recalls. After photographing the fungi, he would pull out his reference books and show Doudna how to identify them. He also collected microscopic shells from the beach, and he would work with her to categorize them so they could try to figure out how they evolved.

Her father bought her a horse, a chestnut gelding named Mokihana, after a Hawaiian tree with a fragrant fruit. She joined the soccer team, playing halfback, a position that was hard to fill on her team because it required a runner with long legs and lots of stamina. "That's a good analogy to how I've approached my work," she said. "I've looked for opportunities where I can fill a niche where there aren't too many other people with the same skill sets."

Math was her favorite class because working through proofs reminded her of detective work. She also had a happy and passionate high school biology teacher, Marlene Hapai, who was wonderful at communicating the joy of discovery. "She taught us that science was about a process of figuring things out," Doudna says.

Although she began doing well academically, she did not feel that there were high expectations in her small school. "I didn't get the sense that the teachers really expected very much of me," she said. She had an interesting immune response: the lack of challenges made her feel free to take more chances. "I decided you just have to go for it, because what the hell," she recalled. "It made me more willing to take on risks, which is something I later did in science when I chose projects to pursue."

Her father was the one person who pushed her. He saw his oldest daughter as his kindred spirit in the family, the intellectual who was bound for college and an academic career. "I always felt like I was the son that he wanted to have," she says. "I was treated a bit differently than my sisters."

James Watson's The Double Helix

Doudna's father was a voracious reader who would check out a stack of books from the local library each Saturday and finish them by the following weekend. His favorite writers were Emerson and Thoreau, but as Jennifer was growing up he became more aware that the books he assigned to his class were mostly by men. So he added Doris Lessing, Anne Tyler, and Joan Didion to his syllabus.

Often he would bring home a book, either from the library or the local secondhand bookstore, for her to read. And that is how a used paperback copy of James Watson's *The Double Helix* ended up on her bed one day when she was in sixth grade, waiting for her when she got home from school.

She put the book aside, thinking it was a detective tale. When she finally got around to reading it on a rainy Saturday afternoon, she discovered that she was right, in a sense. As she sped through the pages, she became enthralled with what was an intensely personal detective drama, filled with vividly portrayed characters, about ambition and competition in the pursuit of nature's inner truths. "When I finished, my father discussed it with me," she recalls. "He liked the story and especially the very personal side of it—the human side of doing that kind of research."

In the book, Watson dramatized (and overdramatized) how as a twenty-four-year-old bumptious biology student from the American Midwest he ended up at Cambridge University in England, bonded with the biochemist Francis Crick, and together won the race to discover the structure of DNA in 1953. Written in the sparky narrative style of a brash American who has mastered the English after-dinner art of being self-deprecating and boastful at the same time, the book manages to smuggle a large dollop of science into a gossipy narrative about the foibles of famous professors, along with the pleasures of flirting, tennis, lab experiments, and afternoon tea.

In addition to the role of lucky naïf that he concocted as his own persona in the book, Watson's other most interesting character is Rosalind Franklin, a structural biologist and crystallographer whose data he used without her permission. Displaying the casual sexism of the 1950s, Watson refers to her condescendingly as "Rosy," a name she never used, and pokes fun at her severe appearance and chilly personality. Yet he also is generous in his respect for her mastery of the complex science and beautiful art of using X-ray diffraction to discover the structure of molecules.

"I guess I noticed she was treated a bit condescendingly, but what mainly struck me was that a woman could be a great scientist," Doudna says. "It may sound a bit crazy. I guess I must have heard about Marie Curie. But reading the book was the first time I really thought about it, and it was an eye-opener. Women could be scientists."[4]

The book also led Doudna to realize something about nature that was at once both logical and awe-inspiring. There were biological mechanisms that governed living things, including the wondrous phenomena that caught her eye when she hiked through the rainforests. "Growing up in Hawaii, I had always liked hunting with my dad for interesting things in nature, like the 'sleeping grass' that curls up when you touch it," she recalls. "The book made me realize you could also hunt for the reasons why nature worked the way it did."

Doudna's career would be shaped by the insight that is at the core of *The Double Helix*: the shape and structure of a chemical molecule determine what biological role it can play. It is an amazing revelation for those who are interested in uncovering the fundamental secrets of life. It is the way that chemistry—the study of how atoms bond to create molecules—becomes biology.

In a larger sense, her career would also be shaped by the realization that she was right when she first saw *The Double Helix* on her bed and thought that it was one of those detective mysteries that she loved. "I have always loved mystery stories," she noted years later. "Maybe that explains my fascination with science, which is humanity's attempt to understand the longest-running mystery we know: the origin and function of the natural world and our place in it."[5]

Even though her school didn't encourage girls to become scientists, she decided that is what she wanted to do. Driven by a passion to understand how nature works and by a competitive desire to turn discoveries into inventions, she would help make what Watson, with his typical grandiosity cloaked in the pretense of humility, would later tell her was the most important biological advance since the double helix.

SIGNET NON-FICTION • Q3770 • 95c

A NATIONAL BESTSELLER! THE INTENSELY HUMAN STORY BEHIND THE MOST SIGNIFICANT BIOLOGICAL DISCOVERY SINCE DARWIN "AN ENORMOUS SUCCESS...A CLASSIC"
—*The New York Review of Books*

The Double Helix

BY NOBEL PRIZE WINNER
JAMES D. WATSON

"A publishing triumph...
Clearly a great book"
—*John Fischer*

Darwin

Mendel

CHAPTER 2

The Gene

Darwin

The paths that led Watson and Crick to the discovery of DNA's structure were pioneered a century earlier, in the 1850s, when the English naturalist Charles Darwin published *On the Origin of Species* and Gregor Mendel, an underemployed priest in Brno (now part of the Czech Republic), began breeding peas in the garden of his abbey. The beaks of Darwin's finches and the traits of Mendel's peas gave birth to the idea of the gene, an entity inside of living organisms that carries the code of heredity.[1]

Darwin had originally planned to follow the career path of his father and grandfather, who were distinguished doctors. But he found himself horrified by the sight of blood and the screams of a strapped-down child undergoing surgery. So he quit medical school and began studying to become an Anglican parson, another calling for which he was uniquely unsuited. His true passion, ever since he began collecting specimens at age eight, was to be a naturalist. He got his opportunity in 1831 when, at age twenty-two, he was offered the chance to ride as the gentleman collector on a round-the-world voyage of the privately funded brig-sloop HMS *Beagle*.[2]

In 1835, four years into the five-year journey, the *Beagle* explored a dozen or so tiny islands of the Galápagos, off the Pacific coast of South America. There Darwin collected carcasses of what he recorded as finches, blackbirds, grosbeaks, mockingbirds, and wrens. But two years later, after he returned to England, he was informed by the ornithologist John Gould that the birds were, in fact, different species of finches. Darwin began to formulate the theory that they had all evolved from a common ancestor.

He knew that horses and cows near his childhood home in rural England were occasionally born with slight variations, and over the years breeders would select the best to produce herds with more desirable traits. Perhaps nature did the same thing. He called it "natural selection." In certain isolated locales, such as the islands of the Galápagos, he theorized, a few mutations (he used the playful term "sports") would occur in each generation, and a change in conditions might make them more likely to win the competition for scarce food and thus be more likely to reproduce. Suppose a species of finch had a beak suited for eating fruit, but then a drought destroyed the fruit trees; a few random variants with beaks better suited for cracking nuts would thrive. "Under these circumstances, favorable variations would tend to be preserved, and unfavorable ones to be destroyed," he wrote. "The results of this would be the formation of a new species."

Darwin was hesitant to publish his theory because it was so heretical, but competition acted as a spur, as often happens in the history of science. In 1858, Alfred Russel Wallace, a younger naturalist, sent Darwin a draft of a paper that proposed a similar theory. Darwin rushed to get a paper of his own ready for publication, and they agreed that they would present their work on the same day at an upcoming meeting of a prominent scientific society.

Darwin and Wallace had a key trait that is a catalyst for creativity: they had wide-ranging interests and were able to make connections between different disciplines. Both had traveled to exotic places where they observed the variation of species, and both had read "An Essay on the Principle of Population" by Thomas Malthus, an English economist. Malthus argued that the human population was likely to grow faster than the food supply. The resulting overpopulation would lead to famine that would weed out the weaker and poorer people. Darwin and Wallace realized this could be applied to all species and thus lead to a theory of evolution driven by the survival of the fittest. "I happened to read for amusement Malthus on population, and . . . it at once struck me that under these circumstances favorable variations would tend to be preserved and unfavorable ones to be destroyed," Darwin recalled. As the science fiction writer and biochemistry professor Isaac Asimov later noted concerning the genesis of evolutionary theory, "What you needed was someone who studied species, read Malthus, and had the ability to make a cross-connection."[3]

The realization that species evolve through mutations and natural selection left a big question to be answered: What was the mechanism? How could a beneficial variation in the beak of a finch or the neck of a giraffe occur, and then how could it get passed along to future generations? Darwin thought that organisms might have tiny particles that contained hereditary information, and he speculated that the information from a male and female blended together in an embryo. But he soon realized, as did others, that this would mean that any new beneficial trait would be diluted over generations rather than be passed along intact.

Darwin had in his personal library a copy of an obscure scientific journal that contained an article, written in 1866, with the answer. But he never got around to reading it, nor did almost any other scientist at the time.

Mendel

The author was Gregor Mendel, a short, plump monk born in 1822 whose parents were German-speaking farmers in Moravia, then part of the Austrian Empire. He was better at puttering around the garden of the abbey in Brno than being a parish priest; he spoke little Czech and was too shy to be a good pastor. So he decided to become a math and science teacher. Unfortunately, he repeatedly failed his qualifying exams, even after studying at the University of Vienna. His performance on one biology exam was especially dreadful.[4]

With little else to do after his final failure at passing the exams, Mendel retreated to the abbey garden to pursue what had become his obsessive interest in breeding peas. In previous years, he had concentrated on creating purebreds. His plants had seven traits that came in two variations: yellow or green seeds, white or violet flowers, smooth or wrinkled seeds, and so on. By careful selection, he produced purebred vines that had, for example, only violet flowers or only wrinkled seeds.

The following year he experimented with something new: breeding together plants with differing traits, such as those that had white flowers with those that had violet ones. It was a painstaking task that involved snipping off each of the plant's receptors with forceps and using a tiny brush to transfer pollen.

What his experiments showed was momentous, given what Darwin was writing at the time. There was no blending of traits. Tall plants cross-bred with short ones did not produce medium-size offspring, nor did purple-flowered plants cross-bred with white-flowered ones produce some pale mauve hue. Instead, all the offspring of a tall and a short plant were tall. The offspring from purple flowers crossbred with white flowers produced only purple flowers. Mendel called these the dominant traits; the ones that did not prevail he called recessive.

An even bigger discovery came the following summer, when he produced offspring from his hybrids. Although the first generation of hybrids had displayed only the dominant traits (such as all purple flowers or tall stems), the recessive trait reappeared in the next generation. And his records revealed a pattern: in this second generation, the dominant trait was displayed in three out of four cases, with the recessive trait appearing once. When a plant inherited two dominant versions of the gene or a dominant and a recessive version, it would display the dominant trait. But if it happened to get two recessive versions of the gene, it would display that less common trait.

Science advances are propelled by publicity. The quiet friar Mendel, however, seemed to have been born under a vanishing cap. He presented his paper in 1865, in two monthly installments, to forty farmers and plant-breeders of the Natural Science Society in Brno, which later published it in its annual journal. It was rarely cited between then and 1900, at which point it was rediscovered by scientists performing similar experiments.[5]

The findings of Mendel and these subsequent scientists led to the concept of a unit of heredity, what a Danish botanist named Wilhelm Johannsen in 1905 dubbed a "gene." There was, apparently, some molecule that encoded bits of hereditary information. Painstakingly, over many decades, scientists studied living cells to try to determine what molecule that might be.

Watson and Crick with their DNA model, 1953

Chapter 3

DNA

Scientists initially assumed that genes are carried by proteins. After all, proteins do most of the important tasks in organisms. They eventually figured out, however, that it is another common substance in living cells, nucleic acids, that are the workhorses of heredity. These molecules are composed of a sugar, phosphates, and four substances called bases that are strung together in chains. They come in two varieties: ribonucleic acid (RNA) and a similar molecule that lacks one oxygen atom and thus is called deoxyribonucleic acid (DNA). From an evolutionary perspective, both the simplest coronavirus and the most complex human are essentially protein-wrapped packages that contain and seek to replicate the genetic material encoded by their nucleic acids.

The primary discovery that fingered DNA as the repository of genetic information was made in 1944 by the biochemist Oswald Avery and his colleagues at Rockefeller University in New York. They extracted DNA from a strain of bacteria, mixed it with another strain, and showed that the DNA transmitted inheritable transformations.

The next step in solving the mystery of life was figuring out how DNA did it. That required deciphering the clue that is fundamental to all of nature's mysteries. Determining the exact structure of

DNA—how all the atoms fit together and what shape resulted—could explain how it worked. It was a task that required mixing three disciplines that had emerged in the twentieth century: genetics, biochemistry, and structural biology.

James Watson

As a middle-class Chicago boy breezing through public school, James Watson was wickedly smart and cheeky. This ingrained in him a tendency to be intellectually provocative, which would later serve him well as a scientist but less so as a public figure. Throughout his life, his rapid-fire mumbling of unfinished sentences would convey his impatience and inability to filter his impulsive notions. He later said that one of the most important lessons his parents taught him was "Hypocrisy in search of social acceptance erodes your self-respect." He learned it too well. From his childhood into his nineties, he was brutally outspoken in his assertions, both right and wrong, which made him sometimes socially unacceptable but never lacking in self-respect.[1]

His passion growing up was bird-watching, and when he won three war bonds on the radio show *Quiz Kids* he used them to buy a pair of Bausch and Lomb binoculars. He would rise before dawn to go with his father to Jackson Park, spend two hours seeking rare warblers, and then take the trolley to the Lab School, a cauldron of whiz kids.

At the University of Chicago, which he entered at fifteen, he planned to indulge his love of birds, and his aversion to chemistry, by becoming an ornithologist. But in his senior year he read a review of *What Is Life?*, in which the quantum physicist Erwin Schrödinger turned his attention to biology to argue that discovering the molecular structures of a gene would show how it hands down hereditary information through generations. Watson checked the book out of the library the next morning and was thenceforth obsessed with understanding the gene.

With modest grades, he was rejected when he applied to study for a doctorate at Caltech and was not offered a stipend by Harvard.[2] So he went to Indiana University, which had built, partly by recruiting

Jews who were having trouble getting tenure on the East Coast, one of the nation's best genetic departments, starring the future Nobel Prize winner Hermann Muller and the Italian émigré Salvador Luria.

With Luria as his PhD advisor, Watson studied viruses. These tiny packets of genetic material are essentially lifeless on their own, but when they invade a living cell, they hijack its machinery and multiply themselves. The easiest of these viruses to study are the ones that attack bacteria, and they were dubbed (remember the term, for it will reappear when we discuss the discovery of CRISPR) "phages," which was short for "bacteriophages," meaning bacteria-eaters.

Watson joined Luria's international circle of biologists known as the Phage Group. "Luria positively abhorred most chemists, especially the competitive variety out of the jungles of New York City," said Watson. But Luria soon realized that figuring out phages would require chemistry. So he helped Watson get a postdoctoral fellowship to study the subject in Copenhagen.

Bored and unable to understand the mumbling chemist who was supervising his studies, Watson took a break from Copenhagen in the spring of 1951 to attend a meeting in Naples on the molecules found in living cells. Most of the presentations went over his head, but he found himself fascinated by a lecture by Maurice Wilkins, a biochemist at King's College London.

Wilkins specialized in crystallography and X-ray diffraction. In other words, he took a liquid that was saturated with molecules, allowed it to cool, and purified the crystals that formed. Then he tried to figure out the structure of those crystals. If you shine a light on an object from different angles, you can figure out its structure by studying the shadows it casts. X-ray crystallographers do something similar: they shine an X-ray on a crystal from many different angles and record the shadows and diffraction patterns. In the slide that Wilkins showed at the end of his Naples speech, that technique had been used on DNA.

"Suddenly I was excited about chemistry," Watson recalled. "I knew that genes could crystallize; hence they must have a regular structure that could be solved in a straightforward fashion." For the next couple of days, Watson stalked Wilkins with the hope of cadging an invitation to join his lab, but to no avail.

Francis Crick

Instead, Watson was able, in the fall of 1951, to become a postdoctoral student at Cambridge University's Cavendish Laboratory, which was directed by the pioneering crystallographer Sir Lawrence Bragg, who more than thirty years earlier had become, and still is, the youngest person to win a Nobel Prize in science.[3] He and his father, with whom he shared the prize, discovered the basic mathematical law of how crystals diffract X-rays.

At the Cavendish Lab, Watson met Francis Crick, forming one of history's most powerful bonds between two scientists. A biochemical theorist who had served in World War II, Crick had reached the ripe age of thirty-six without having secured his PhD. Nevertheless, he was sure enough of his instincts, and careless enough about Cambridge manners, that he was unable to refrain from correcting his colleagues' sloppy thinking and then crowing about it. As Watson memorably put it in the opening sentence of *The Double Helix*, "I have never seen Francis Crick in a modest mood." It was a line that could likewise have been written of Watson, and they admired each other's immodesty more than their colleagues did. "A youthful arrogance, a ruthlessness, and an impatience with sloppy thinking came naturally to both of us," Crick recalled.

Crick shared Watson's belief that discovering the structure of DNA would provide the key to the mysteries of heredity. Soon they were lunching together on shepherd's pie and talking volubly at the Eagle, a well-worn pub near the labs. Crick had a boisterous laugh and booming voice, which drove Sir Lawrence to distraction. So Watson and Crick were assigned to a pale brick room of their own.

"They were complementary strands, interlocked by irreverence, zaniness, and fiery brilliance," the writer-physician Siddhartha Mukherjee noted. "They despised authority but craved its affirmation. They found the scientific establishment ridiculous and plodding, yet they knew how to insinuate themselves into it. They imagined themselves quintessential outsiders, yet felt most comfortable sitting in the inner quadrangles of Cambridge colleges. They were self-appointed jesters in a court of fools."[4]

The Caltech biochemist Linus Pauling had just rocked the scientific world, and paved the way for his first Nobel Prize, by figuring out the structure of proteins using a combination of X-ray crystallography, his understanding of the quantum mechanics of chemical bonds, and Tinkertoy model building. Over their lunches at the Eagle, Watson and Crick plotted how to use the same tricks to beat Pauling in the race to discover the structure of DNA. They even had the tool shop of the Cavendish Lab cut tin plates and copper wires to represent the atoms and other components for the desktop model they planned to tinker with until they got all the elements and bonds correct.

One obstacle was that they would be treading on the territory of Maurice Wilkins, the King's College London biochemist whose X-ray photograph of a DNA crystal had piqued Watson's interest in Naples. "The English sense of fair play would not allow Francis to move in on Maurice's problem," Watson wrote. "In France, where fair play obviously did not exist, these problems would not have arisen. The States also would not have permitted such a situation to develop."

Wilkins, for his part, seemed in no rush to beat Pauling. He was in an awkward internal struggle, both dramatized and trivialized in Watson's book, with a brilliant new colleague who in 1951 had come to work at King's College London: Rosalind Franklin, a thirty-one-year-old English biochemist who had learned X-ray diffraction techniques while studying in Paris.

She had been lured to King's College with the understanding that she would lead a team studying DNA. Wilkins, who was four years older and already studying DNA, was under the impression that she was coming as a junior colleague who would help him with X-ray diffraction. This resulted in a combustible situation. Within months they were barely speaking to each other. The sexist structure at King's helped keep them apart: there were two faculty lounges, one for men and the other for women, the latter unbearably dingy and the former a venue for elegant lunches.

Franklin was a focused scientist, sensibly dressed. As a result she ran afoul of English academia's fondness for eccentrics and its tendency to look at women through a sexual lens, attitudes apparent in Watson's descriptions of her. "Though her features were strong, she was not unattractive and might have been quite stunning had she taken even a mild interest in clothes," he wrote. "This she did not. There was never lipstick to contrast with her straight black hair, while at the age of thirty-one her dresses showed all the imagination of English bluestocking adolescents."

Franklin refused to share her X-ray diffraction pictures with Wilkins, or anyone else, but in November 1951 she scheduled a lecture to summarize her latest findings. Wilkins invited Watson to take the train down from Cambridge. "She spoke to an audience of about fifteen in a quick, nervous style," he recalled. "There was not a trace of warmth or frivolity in her words. And yet I could not regard her as totally uninteresting. Momentarily I wondered how she would look if she took off her glasses and did something novel with her hair. Then, however, my main concern was her description of the crystalline X-ray diffraction pattern."

Watson briefed Crick the next morning. He had not taken notes, which annoyed Crick, and thus was vague about many key points, particularly the water content that Franklin had found in her DNA samples. Nevertheless, Crick started scribbling diagrams, declaring that Franklin's data indicated a structure of two, three, or four strands twisted in a helix. He thought that, by playing with different models, they might soon discover the answer. Within a week they had what they thought was a solution, even though it meant that some of the atoms were crushed together a little too close: three strands swirled in the middle, and the four bases jutted outward from this backbone.

In a fit of hubris, they invited Wilkins and Franklin to come up to Cambridge and take a look. The two arrived the next morning and, with little small talk, Crick began to display the triple-helix structure. Franklin immediately saw that it was flawed. "You're wrong for the following reasons," she said, her words ripping like those of an exasperated teacher.

She insisted that her pictures of DNA did not show that the molecule was helical. On that point she would turn out to be wrong. But her other two objections were correct: the twisting backbones had to be on the outside, not inside, and the proposed model did not contain enough water. "At this stage the embarrassing fact came out that my recollection of the water content of Rosy's DNA samples could not be right," Watson drily noted. Wilkins, momentarily bonding with Franklin, told her that if they left for the station right away, they could make the 3:40 train back to London, which they did.

Not only were Watson and Crick embarrassed; they were put in a penalty box. Word came down from Sir Lawrence that they were to stop working on DNA. Their model-building components were packed up and sent to Wilkins and Franklin in London.

Adding to Watson's dismay was the news that Linus Pauling was coming over from Caltech to lecture in England, which would likely catalyze his own attempt to solve the structure of DNA. Fortunately, the U.S. State Department came to the rescue. In the weirdness engendered by red-baiting and McCarthyism, Pauling was stopped at the airport in New York and had his passport confiscated because he had been spouting enough pacifist opinions that the FBI thought he might be a threat to the country if allowed to travel. So he never got the chance to discuss the crystallography work done in England, thus helping the U.S. lose the race to figure out DNA.

Watson and Crick were able to monitor some of Pauling's progress through his son Peter, who was a young student in their Cambridge lab. Watson found him amiable and fun. "The conversation could dwell on the comparative virtues of girls from England, the Continent, and California," he recalled. But one day in December 1952, young Pauling wandered into the lab, put his feet up on a desk, and dropped the news that Watson had been dreading. In his hand was a letter from his father in which he mentioned that he had come up with a structure for DNA and was about to publish it.

Linus Pauling's paper arrived in Cambridge in early February. Peter got a copy first and sauntered into the lab to tell Watson and Crick that his father's solution was similar to the one they had tried: a three-chain helix with a backbone in the center. Watson grabbed the paper from Peter's coat pocket and began to read. "At once I felt something was not right," he recalled. "I could not pinpoint the mistake, however, until I looked at the illustrations for several minutes."

Watson realized that some of the atomic connections in Pauling's proposed model would not be stable. As he discussed it with Crick and others in the lab, they became convinced that Pauling had made a big "blooper." They got so excited they quit work early that afternoon to dash off to the Eagle. "The moment its doors opened for the evening, we were there to drink a toast to the Pauling failure," Watson said. "Instead of sherry, I let Francis buy me a whiskey."

"The secret of life"

They knew they could no longer waste time or continue to honor the edict that they defer to Wilkins and Franklin. So Watson took the train down to London one afternoon to see them, carrying his early copy of Pauling's paper. Wilkins was out when he arrived, so he ambled uninvited into the lab of Franklin, who was bending over a light box measuring the latest of her ever-sharper X-ray images of DNA. She gave him an angry look, but he launched into a summary of Pauling's paper.

For a few moments they argued about whether DNA was likely to be a helix, with Franklin still dubious. "Interrupting her harangue, I asserted that the simplest form for any regular polymeric molecule was a helix," Watson recalled. "Rosy by then was hardly able to control her temper, and her voice rose as she told me that the stupidity of my remarks would be obvious if I would stop blubbering and look at her X-ray evidence."

The conversation spiraled downward, with Watson pointing out, correctly but impolitely, that as a good experimentalist Franklin would be more successful if she knew how to collaborate with theorists.

"Suddenly Rosy came from behind the lab bench that separated us and began moving toward me. Fearing that in her hot anger she might strike me, I grabbed up the Pauling manuscript and hastily retreated."

Just as the confrontation climaxed, Wilkins walked by and whisked Watson off to have some tea and calm down. He confided that Franklin had taken some pictures of a wet form of DNA that provided new evidence of its structure. He then went into an adjacent room and retrieved a print of what became known as "photograph 51." Wilkins had gotten hold of the picture validly: he was the PhD advisor of the student who had worked with Franklin to take it. Less proper was showing it to Watson, who recorded some of the key parameters and took them back to Cambridge to share with Crick. The photograph indicated that Franklin had been correct in arguing that the backbone strands of the structure were on the outside, like the strands of a spiral staircase, rather than inside of the molecule, but she was wrong in resisting the possibility that DNA was a helix. "The black cross of reflections which dominated the picture could arise only from a helical structure," Watson immediately saw. A study of Franklin's notes shows that even after Watson's visit she was still many steps away from discerning the DNA structure.[5]

Rosalind Franklin

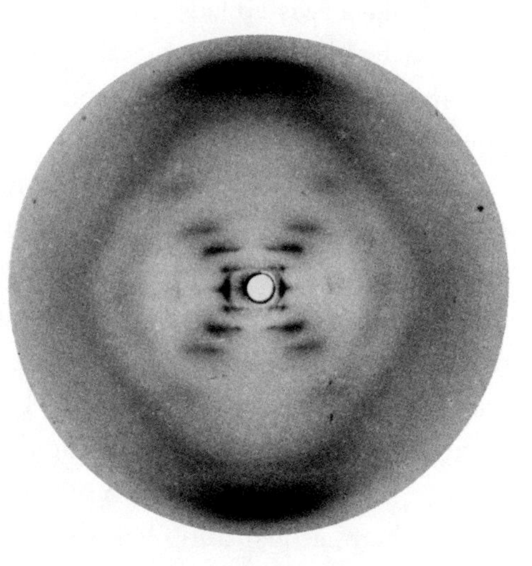

"Photograph 51"

In the unheated train car back to Cambridge, Watson sketched ideas in the margins of his copy of *The Times*. He had to climb over the back gate into his residential college, which had locked up for the night. The next morning, when he went into the Cavendish lab, he encountered Sir Lawrence Bragg, who had demanded that he and Crick steer clear of DNA. But confronted with Watson's excited summary of what he had learned, and hearing of his desire to get back to model-building, Sir Lawrence gave his assent. Watson rushed down the stairs to the machine shop to set them to work on making a new set of components.

Watson and Crick soon got more of Franklin's data. She had submitted to Britain's Medical Research Council a report on her work, and a member of the council shared it with them. Although Watson and Crick had not exactly stolen Franklin's findings, they had appropriated her work without her permission.

By then Watson and Crick had a pretty good idea of DNA's structure. It had two sugar-phosphate strands that twisted and spiraled to form a double-stranded helix. Protruding from these were the four bases in DNA: adenine, thymine, guanine, and cytosine, now commonly known by the letters A, T, G, and C. They came to agree with Franklin that the backbones were on the outside and the bases pointed inward, like a twisted ladder or spiral staircase. As Watson later admitted in a feeble attempt at graciousness, "Her past uncompromising statements on this matter thus reflected first-rate science, not the outpourings of a misguided feminist."

They originally assumed that the bases would each be paired with themselves, for example, a rung that was made up of an adenine bonded to another adenine. But one day Watson, using some cardboard models of bases that he cut out himself, began playing with different pairings. "Suddenly I became aware that an adenine-thymine pair held together by two hydrogen bonds was identical in shape to a guanine-cytosine pair held together by at least two hydrogen bonds." He was lucky to work in a lab of scientists with different specialties; one of them, a quantum chemist, confirmed that adenine would attract thymine and guanine would attract cytosine.

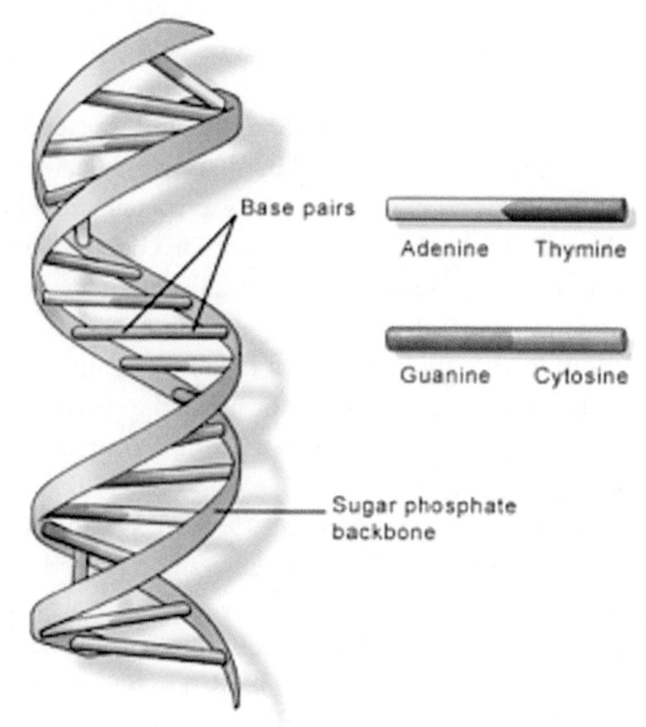

U.S. National Library of Medicine

There was an exciting consequence of this structure: when the two strands split apart, they could perfectly replicate, because any half-rung would attract its natural partner. In other words, such a structure would permit the molecule to replicate itself and pass along the information encoded in its sequences.

Watson returned to the machine shop to prod them to speed up production of the four types of bases for the model. By this point the machinists were infused with his excitement, and they finished soldering the shiny metal plates in a couple of hours. With all the parts now on hand, it took Watson only an hour to arrange them so that the atoms comported with the X-ray data and the laws of chemical bonds.

In Watson's memorable and only slightly hyperbolic phrase in *The Double Helix*, "Francis winged into the Eagle to tell everyone within hearing distance that we had found the secret of life." The solution was too beautiful not to be true. The structure was perfect for the molecule's function. It could carry a code that it could replicate.

Watson and Crick finished their paper on the last weekend of March 1953. It was a mere 975 words, typed by Watson's sister, who was persuaded to do so by his argument that "she was participating in perhaps the most famous event in biology since Darwin's book." Crick wanted to include an expanded section on the implications for heredity, but Watson convinced him that a shorter ending would actually carry more punch. Thus was produced one of the most significant sentences in science: "It has not escaped our notice that the specific pairing we have postulated immediately suggests a possible copying mechanism for the genetic material."

The Nobel Prize was awarded in 1962 to Watson, Crick, and Wilkins. Franklin was not eligible because she had died in 1958, at age thirty-seven, of ovarian cancer, likely caused by her exposure to radiation. If she had survived, the Nobel committee would have faced an awkward situation: each prize can be awarded to only three winners.

Two revolutions coincided in the 1950s. Mathematicians, including Claude Shannon and Alan Turing, showed that all information could be encoded by binary digits, known as bits. This led to a digital revolution powered by circuits with on-off switches that processed information. Simultaneously, Watson and Crick discovered how instructions for building every cell in every form of life were encoded by the four-letter sequences of DNA. Thus was born an information age based on digital coding (0100110111001 . . .) and genetic coding (ACTGGTAGATTACA . . .). The flow of history is accelerated when two rivers converge.

CHAPTER 4

The Education of a Biochemist

Girls do science

Jennifer Doudna would later meet James Watson, work with him on occasion, and be exposed to all of his personal complexity. In some ways he would be like an intellectual godfather, at least until he began saying things that seemed to emanate from the dark side of the Force. (As Chancellor Palpatine said to Anakin Skywalker, "The dark side of the Force is a pathway to many abilities that some consider to be unnatural.")

But her reactions when she first read his book as a sixth-grader were far simpler. It sparked the realization that it was possible to peel back the layers of nature's beauty and discover, as she says, "how and why things worked at the most fundamental and inner level." Life was made up of molecules. The chemical components and structure of these molecules governed what they would do.

The book also sparked the feeling that science could be fun. All of the previous science books she read had "pictures of emotionless men wearing lab coats and glasses." But *The Double Helix* painted a more vibrant picture. "It made me realize that science can be very exciting, like being on a trail of a cool mystery and you're getting a clue here and a clue there. And then you put the pieces together." The tale of Watson

In the lab at Pomona College

and Crick and Franklin was one of competition and collaboration, of letting data dance with theory, and of being in a race with rival labs. All of that resonated with her as a kid, and it would continue to do so throughout her career.[1]

In high school Doudna got a chance to do the standard biology experiments involving DNA, including one that involved breaking apart salmon sperm cells and stirring their gooey contents with a glass rod. She was inspired by an energetic chemistry teacher and by a woman who gave a lecture on the biochemical reasons that cells become cancerous. "It reinforced my realization that women could be scientists."

There was a thread that wove together her childhood curiosity about the eyeless spiders in the lava tubes, the sleeping grass that curled when you touched it, and the human cells that became cancerous: they were all connected to the detective story of the double helix.

She decided that she wanted to study chemistry at college, but like many female scientists of the time, she met resistance. When she explained her college goals to her school's guidance counselor, an older Japanese American man with traditional attitudes, he began to grunt, "No, no, no." She paused and looked at him. "Girls don't do science," he asserted. He discouraged her from even taking the College Board chemistry test. "Do you really know what that is, what that test is for?" he asked her.

"It hurt me," Doudna recalled, but it also stiffened her resolve. "Yes I will do it," she remembers telling herself. "I will show you. If I want to do science, I am going to do it." She applied to Pomona College in California, which had a good program in chemistry and biochemistry, was admitted, and enrolled in the fall of 1981.

Pomona

At first she was unhappy. Having skipped a grade in school, she was now only seventeen. "I was suddenly a small fish in a very big pond," she recalled, "and I doubted I had what it took." She was homesick and, once again, felt out of place. Many of her classmates came from wealthy Southern California families and had their own cars, while she was on a scholarship and worked part time to pay her living expenses. In those days, it was expensive to phone home. "My parents didn't have a lot of money so they told me to call collect, but only once a month."

Having willed herself to major in chemistry, she began to doubt she could handle it. Perhaps her high school counselor had been right. Her general chemistry class had two hundred students, most of whom had gotten a 5 on the AP chemistry test. "It made me question whether I'd set my sights on something that was just not achievable by me," she said. Because of her competitive streak, the field had little appeal if she was going to be just a mediocre student. "I thought, 'I don't want to become a chemist if I'm not going to have a shot at being at the top.'"

She thought about changing her major to French. "I went to talk to my French teacher about that, and she asked what I was majoring in." When Doudna replied that it was chemistry, the teacher told her to stick with it. "She was really insistent. She said 'If you major in chemistry you'll be able to do all sorts of things. If you major in French you will be able to be a French teacher.'"[2]

Her outlook brightened the summer after her freshman year when she got a job working in the lab of her family's friend Don Hemmes, the University of Hawaii biology professor who had taken her on nature walks. He was using electron microscopy to investigate the movement of chemicals inside cells. "Jennifer was fascinated by the ability to look inside cells and study what all the small particles were doing," he recalled.[3]

Hemmes was also studying the evolution of tiny shells. An active scuba diver, he would scoop up samples of the smallest ones, almost microscopic in size, and his students would help him embed them in resin and slice thin sections for analysis under an electron microscope. "He taught us how to use various kinds of chemicals to stain the samples differently, so we could look at shell development," explained Doudna. She kept a lab notebook for the first time.[4]

In chemistry class at college, most of the experiments were

conducted by following a recipe. There was a rigid protocol and a right answer. "The work in Don's lab wasn't like that," she said. "Unlike in class, we didn't know the answer we were supposed to get." It gave her a taste of the thrill of discovery. It also helped her see what it would be like to be part of the community of scientists, making advances and piecing them together to discover the ways that nature worked.

When she returned to Pomona in the fall, she made friends, fit in better, and became more confident in her ability to do chemistry. As part of her work-study program, she had a series of jobs in the college chemistry labs. Most did not engage her because they did not explore how chemistry intersected with biology. But that changed after her junior year, when she got a summer position in the lab of her advisor Sharon Panasenko, a biochemistry professor. "It was more challenging for women biochemists at universities back then, and I admired her not only for being a good scientist but also for being a role model."[5]

Panasenko was studying a topic that aligned with Doudna's interest in the mechanisms of living cells: how some bacteria found in soil are able to communicate so that they can join together when they are starved for nutrients. They form a commune called a "fruiting body." Millions of the bacteria figure out how to aggregate by sending out chemical signals. Panasenko enlisted Doudna to help figure out how those chemical signals worked.

"I have to warn you," Panasenko told her, "that a technician in my lab has been working on growing these bacteria for six months, and he hasn't been able to make it work." Doudna began trying to grow the bacteria in large baking pans rather than the usual Petri dishes. One night she put her preparations in the incubator. "I came in the next day, and when I peeled back the foil on the baking dish that lacked nutrients, I was stunned to see these beautiful structures!" They looked like little footballs. She had succeeded where the other technician had failed. "It was an incredible moment, and it made me think I could do science."

The experiments yielded strong enough results that Panasenko was able to publish a research paper in the *Journal of Bacteriology*, in which she acknowledged Doudna as one of four lab assistants "whose preliminary observations made significant contributions to this project." It was the first time Doudna's name appeared in a scientific journal.[6]

Harvard

When it came time to go to graduate school, she did not initially consider Harvard, despite being the top student in her physical chemistry class. But her father pushed her to apply. "Come on, Dad," she pleaded, "I will never get in." To which he replied, "You certainly won't get in if you don't apply." She did get in, and Harvard even offered her a generous stipend.

She spent part of the summer traveling in Europe on the money she had saved from her work-study program at Pomona. When her trip ended in July 1985, she went right to Harvard so that she could begin working before classes started. Like other universities, Harvard required graduate chemistry students to work each semester in the lab of a different professor. The goal of these rotations was to allow students to learn different techniques and then select a lab for their dissertation research.

Doudna called Roberto Kolter, who was head of the graduate studies program, to ask if she could begin her rotations in his lab. A young Spanish specialist in bacteria, he had a big smile, an elegant sweep of hair, wireless glasses, and a bouncy style of talking. His lab was international, with many of the researchers from Spain or Latin America, and Doudna was struck by how young and politically active they were. "I had been highly influenced by the media's presentation of scientists as old white men, and I thought that's who I would be interacting with at Harvard. That wasn't my experience at all at the Kolter Lab." Her ensuing career, from CRISPR to coronavirus, would reflect the global nature of modern science.

Kolter assigned Doudna to study how bacteria make molecules that are toxic to other bacteria. She was responsible for cloning (making an exact DNA copy of) genes from the bacteria and testing their functions. She thought of a novel way to set up the process, but Kolter declared it wouldn't work. Doudna was stubborn and went ahead with her idea. "I did it my way and got the clone," she told him. He was surprised but supportive. It was a step in overcoming the insecurity that lurked inside her.

Doudna eventually decided to do her dissertation work in the lab of Jack Szostak, an intellectually versatile Harvard biologist who was studying DNA in yeast. A Canadian American of Polish descent, Szostak was one of the young geniuses then in Harvard's Department of Molecular Biology. Even though he was managing a lab, Szostak was still working as a bench scientist, so Doudna got to watch him perform experiments, hear his thought process, and admire the way he took risks. The key aspect of his intellect, she realized, was his ability to make unexpected connections between different fields.

Her experiments gave her a glimpse of how basic science can be turned into applied science. Yeast cells are very efficient at taking up pieces of DNA and integrating them into their genetic makeup. So she worked on a way to make use of this fact. She engineered strands of DNA that ended with a sequence that matched a sequence in the yeast. With a little electric shock, she opened up tiny passageways in the cell wall of the yeast, allowing the DNA that she made to wriggle inside. It then recombined into the yeast's DNA. She had made a tool that could edit the genes of yeast.

Craig Venter and Francis Collins

CHAPTER 5

The Human Genome

James and Rufus Watson

In 1986, when Doudna was working in Jack Szostak's lab, a massive international science collaboration was being hatched.[1] It was called the Human Genome Project, and its goal was to figure out the sequence of the three billion base pairs in our DNA and map the more than twenty thousand genes that these base pairs encode.

One of the many roots of the Human Genome Project involved Doudna's childhood hero James Watson and his son Rufus. The provocative author of *The Double Helix* was the director of Cold Spring Harbor Laboratory, a haven for biomedical research and seminars on a 110-acre wooded campus on the north shore of Long Island. Founded in 1890, it has a history of important research. It was there in the 1940s that Salvador Luria and Max Delbrück led a study group on phages that included the young Watson. But it is also haunted by more controversial ghosts. From 1904 until 1939, under director Charles Davenport, it served as a center for eugenics, producing studies asserting that different races and ethnic groups had genetic differences in such traits as intelligence and criminality.[2] By the end of Watson's tenure as director there from 1968 to 2007, his own pronouncements on race and genetics would revive these ghosts.

In addition to being a research center, Cold Spring Harbor hosts around thirty meetings a year on selected topics. In 1986, Watson decided to launch an annual series titled "The Biology of Genomes." The agenda for the first year's meeting was to plan the Human Genome Project.

On the day the meeting began, Watson made a shocking announcement to the gathered scientists. His son Rufus had broken out of a psychiatric hospital, where he had been committed after trying to break a window and jump to his death from the World Trade Center. He was now missing, and Watson was leaving to help find him.

Born in 1970, Rufus had the lean face, tousled hair, and lopsided grin of his father. He was also very bright. "I was very pleased," Watson says, "because for a while he would go bird-watching with me, and we had some relationship." Bird-watching was something that Watson had done with his own father as a smart, skinny kid in Chicago. But when Rufus was young, he began to show signs of not being able to interact well with people, and in tenth grade at his boarding school, Exeter, he had a psychotic incident and was sent home. A few days later, he went to the top of the World Trade Center with the plan of ending his life. Doctors diagnosed him as schizophrenic. The elder Watson cried. "I had never seen Jim weep before—or ever since in his life," his wife, Elizabeth, says.[3]

Watson missed most of the Cold Spring Harbor genome meeting, while he and Elizabeth joined the hunt for their son. He was finally found wandering in the woods. Watson's science had intersected with real life. The massive international project to map the human genome would no longer be for him an abstract, academic pursuit. It was personal, and it would ingrain in him a belief, bordering on obsession, in the power of genetics to explain human life. Nature, not nurture, made Rufus the way he was, and it also made different groups of people the way they were.

Or so it appeared to Watson, who saw things through glasses filtered by his DNA discovery and his son's condition. "Rufus is as smart as can be, very perceptive, and can be caring but also intense in his anger," Watson says. "My wife and I hoped when he was young we could set up the right environment for him to succeed. But I soon realized that his troubles lay in his genes. That drove me to lead the Human Genome Project. The only way I could understand our son and help him live at a normal level was to decipher the genome."[4]

The race to sequence

When the Human Genome Project was formally launched in 1990, Watson was anointed its first director. All the major players were men. Watson was eventually succeeded by Francis Collins, who in 2009 became the director of the U.S. National Institutes of Health. Among the whiz kids was the charismatic and driven Eric Lander, a breathtakingly brilliant Brooklyn-bred high school math team captain who did a doctoral dissertation on coding theory as a Rhodes Scholar at Oxford and then decided to become a geneticist at MIT. The most controversial player was the wild and abrasive Craig Venter, who had worked in a U.S. Navy field hospital as a draftee during the Tet Offensive of the Vietnam War, had attempted suicide by swimming out to sea, and then became a biochemist and biotech entrepreneur.

The project began as a collaboration, but as with many tales of discovery and innovation it also became a competition. When Venter found different ways to do the sequencing cheaper and faster than everyone else, he broke away to form a private company, Celera, which sought to profit from patenting its discoveries. Watson enlisted Lander to help reorganize the public effort and speed up its work. Lander bruised some egos, but he was able assure that it could keep pace with Venter's private effort.[5]

In early 2000, as the competition became a public spectacle, President Bill Clinton pushed for a truce between Venter and Collins, who had been sniping at each other in the press. Collins had likened Venter's sequencing to "Cliff's Notes" and "*Mad* magazine"; Venter had ridiculed the government project for costing ten times more to do work at a fraction of the speed. "Fix it—make these guys work together," Clinton told his top science advisor. So Collins and Venter met for pizza and beer to see if they could reach an accord on sharing the credit and agreeing to make public, rather than exploiting for private use, what would soon be the world's most important biological data set.

After a few more private meetings, Clinton was able to host Collins and Venter at a White House ceremony to announce the initial results of the Human Genome Project and the agreement to share credit. James Watson hailed the decision. "The events of the past few weeks have shown that those who work for the public good do not necessarily fall behind those driven by personal gain," he said.

I was editor of *Time* then, and we had been working with Venter for weeks to have exclusive access to his story and feature him on the cover. He was an enticing cover boy, because by then he had used his wealth from Celera to become a flashy yacht-owner, competitive surfer, and party-giver. The week that we were closing the story, I got an unexpected phone call from Vice President Al Gore. He pushed me—very hard and persuasively—to put Francis Collins on the cover as well. Venter resisted. He had been forced to share credit with Collins at a press conference, but he did not want to also share a *Time* cover. He eventually agreed, but at the photo session he could not help ragging on Collins for not being able to keep pace with Celera's sequencing. Collins smiled and said nothing.[6]

"Today we are learning the language in which God created life," President Clinton proclaimed at the White House ceremony featuring Venter, Collins, and Watson. The announcement captured the public imagination. The *New York Times* ran a front-page banner headline, "Genetic Code of Human Life Is Cracked by Scientists." The story, written by the distinguished biology journalist Nicholas Wade, began, "In an achievement that represents a pinnacle of human self-knowledge, two rival groups of scientists said today that they had deciphered the hereditary script, the set of instructions that defines the human organism."[7]

Doudna spent time discussing with Szostak, Church, and others at Harvard whether the $3 billion dedicated to the Human Genome

Project was worth it. Church was skeptical at the time, and remains so. "The three billion dollars didn't buy us much," he says. "We didn't discover anything. None of the technologies survived." Having a map of DNA did not, it turned out, lead to most of the grand medical breakthroughs that had been predicted. More than four thousand disease-causing DNA mutations were found. But no cure sprang forth for even the most simple of single-gene disorders, such as Tay-Sachs, sickle cell, or Huntington's. The men who had sequenced DNA taught us how to read the code of life, but the more important step would be learning how to write that code. This would require a different set of tools, ones that would involve the worker-bee molecule that Doudna found more interesting than DNA.

Jack Szostak

CHAPTER 6

RNA

The central dogma

Accomplishing the goal of being able to *write* as well as to *read* human genes required a shift in focus from DNA to its less famous sibling that actually carries out its coded instructions. RNA (ribonucleic acid) is a molecule in living cells that is similar to DNA (deoxyribonucleic acid), but it has one more oxygen atom in its sugar-phosphate backbone and a difference in one of its four bases.

DNA may be the world's most famous molecule, so well-known that it appears on magazine covers and is used as a metaphor for traits that are ingrained in a society or organization. But like many famous siblings, DNA doesn't do much work. It mainly stays at home in the nucleus of our cells, not venturing forth. Its primary activity is protecting the information it encodes and occasionally replicating itself. RNA, on the other hand, actually goes out and does real work. Instead of just sitting at home curating information, it makes real products, such as proteins. Pay attention to it. From CRISPR to COVID, it will be the starring molecule in this book and in Doudna's career.

At the time of the Human Genome Project, RNA was seen as mainly a messenger molecule that carries instructions from the DNA that is nestled in the nucleus of the cells. A small segment of DNA that encodes a gene is transcribed into a snippet of RNA, which then travels to the manufacturing region of the cell. There this "messenger RNA" facilitates the assembly of the proper sequence of amino acids to make a specified protein.

These proteins come in many types. Fibrous proteins, for example, form structures such as bones, tissues, muscles, hair, fingernails, tendons, and skin cells. Membrane proteins relay signals within cells. Above all is the most fascinating type of proteins: enzymes. They serve as catalysts. They spark and accelerate and modulate the chemical reactions in all living things. Almost every action that takes place in a cell needs to be catalyzed by an enzyme. Pay attention to enzymes. They will be RNA's costars and dancing partners in this book.

Francis Crick, five years after co-discovering the structure of DNA, came up with a name for this process of genetic information moving from DNA to RNA to the building of proteins. He dubbed it the "central dogma" of biology. He later conceded that "dogma," which implies an unchanging and unquestioned faith, was a poor choice of words.[1] But the word "central" was apt. Even as the dogma was modified, the process remained central to biology.

Ribozymes

One of the first tweaks to the central dogma came when Thomas Cech and Sidney Altman independently discovered that proteins were not the only molecules in the cell that could be enzymes. In work done in the early 1980s that would win them the Nobel Prize, they made the surprising discovery that some forms of RNA could likewise be enzymes. Specifically, they found that some RNA molecules can split themselves by sparking a chemical reaction. They dubbed these catalytic RNAs "ribozymes," a word conjured up by combining "ribonucleic acid" with "enzyme."[2]

Cech and Altman made this discovery by studying introns. Some parts of DNA sequences do not code instructions for how to make proteins. When these sequences are transcribed into RNA molecules, they clog things up. So they have to be sliced out before the RNA can scurry out on its mission to direct the making of proteins. The cut-and-paste process of slicing out these introns and then splicing the useful bits of RNA back together requires a catalyst, and that role is usually performed by a protein enzyme. But Cech and Altman discovered that there were certain RNA introns that were self-splicing!

This had pretty cool implications. If some RNA molecules could store genetic information and also act as a catalyst to spur chemical reactions, they might be more fundamental to the origins of life than DNA, which cannot naturally replicate themselves without the presence of proteins to serve as a catalyst.[3]

RNA rather than DNA

When Doudna's lab rotation ended in the spring of 1986, she asked Jack Szostak if she could stay on and do her doctoral research under him. Szostak agreed—but he added a caveat. He was no longer going to focus on DNA in yeast. While other biochemists were getting excited about sequencing DNA for the Human Genome Project, he had decided to shift his lab's attention to RNA, which he believed might reveal secrets about the biggest of all biological mysteries: the origins of life.

He was intrigued, he told Doudna, by the discoveries that Cech and Altman had made about how certain RNAs had the catalytic powers of enzymes. His goal was to pin down whether these ribozymes could use this power to replicate. "Did this piece of RNA have the chemical chops to copy itself?" he asked her. He suggested that should be the focus of her PhD dissertation.[4]

She found Szostak's enthusiasm infectious and signed up to be the first graduate student in his lab to work on RNA. "When I was taught biology, we learned about the structure and code of DNA, and we learned about how proteins do all the heavy lifting in cells, and RNA was treated as this dull intermediary, sort of a middle manager," she recalls. "I was quite surprised to find that there was this young genius, Jack Szostak, at Harvard who wanted to focus a hundred percent on RNA because he thought that it was the key to understanding the origin of life."

For both Szostak, who was well established, and Doudna, who wasn't, switching to a focus on RNA was risky. "Instead of following the herd doing DNA," Szostak recalled, "we felt we were pioneering something new, exploring a frontier that was a little bit neglected but we all thought was exciting." This was long before RNA was being considered as a technology to interfere with gene expression or deliver edits to human genes. Szostak and Doudna pursued the subject out of pure curiosity about how nature works.

Szostak had a guiding principle: *Never do something that a thousand other people are doing*. That appealed to Doudna. "It was like when I was on the soccer field and wanted to play a position that the other kids didn't," she says. "I learned from Jack that there was more of a risk but also more of a reward if you ventured into a new area."

By this point she knew that the most important clue for understanding a natural phenomenon was to figure out the structure of the molecules involved. That would require her to learn some of the techniques that Watson and Crick and Franklin used to unravel the structure of DNA. If she and Szostak succeeded, it could be a significant step in answering one of the grandest of all biological questions, perhaps *the* grandest: How did life begin?

The origins of life

Szostak's excitement about discovering how life began taught Doudna a second big lesson, in addition to taking risks by moving into new fields: *Ask big questions*. Even though Szostak liked diving into the details of experiments, he was a grand thinker, someone who was constantly pursuing truly profound inquiries. "Why else would you do science?" he asked Doudna. It was an injunction that became one of her own guiding principles.[5]

There are some truly grand questions that our mortal minds may never be able to answer: How did the universe begin? Why is there something rather than nothing? What is consciousness? Others may be wrestled into submission by the end of this century: Is the universe deterministic? Do we have free will? Of the really big ones, the closest to being solved is how life began.

The central dogma of biology requires the presence of DNA, RNA, and proteins. Because it's unlikely that all three of these sprang forth at the exact same time from the primordial stew, a hypothesis arose in the early 1960s—formulated independently by the ubiquitous Francis Crick and others—that there was a simpler precursor system. Crick's hypothesis was that, early on in the history of earth, RNA was able to replicate itself. That leaves the question of where the first RNA came from. Some speculate it came from outer space. But the simpler answer may be that the early earth contained the chemical building blocks of RNA, and it didn't require anything other than natural random mixing to jostle them together. The year that Doudna joined Szostak's lab, biochemist Walter Gilbert dubbed this hypothesis "the RNA world."[6]

An essential quality of living things is that they have a method for creating more organisms akin to themselves: they can reproduce. Therefore, if you want to make the argument that RNA might be the precursor molecule leading to the origin of life, it would help to show how it can replicate itself. This was the project that Szostak and Doudna embarked upon.[7]

Doudna used many tactics to create an RNA enzyme, or ribozyme, that could stitch together little RNA pieces. Eventually, she and Szostak were able to engineer a ribozyme that could splice together a copy of itself. "This reaction demonstrates the feasibility of RNA-catalyzed RNA replications," she and Szostak wrote in a 1998 paper for *Nature*. The biochemist Richard Lifton later called this paper a "technical tour de force."[8] Doudna became a rising star in the rarefied realm of RNA research. That was still a bit of a biological backwater, but over the next two decades the understanding of how little strands of RNA behaved would become increasingly important, both to the field of gene editing and to the fight against coronaviruses.

As a young PhD student, Doudna mastered the special combination of skills that distinguished Szostak and other great scientists: she was good at doing hands-on experiments and also at asking the big questions. She knew that God was in the details but also in the big picture.

"Jennifer was fantastically good at the bench, because she was fast and sharp and could seemingly get anything to work," Szostak says. "But we talked quite a bit about why the really big questions are the important questions."

Doudna also proved herself a team player, which counted a lot for Szostak, who shared that trait with George Church and some other scientists at the Harvard Medical School campus. This was reflected in the number of coauthors she had on most of her papers. In scientific publications, the first author listed is usually the younger researcher most responsible for the hands-on experiments, and the last is the principal investigator or head of the lab. Those listed in the middle are generally ordered by the contributions they made. On one of the important papers that she helped produce for the journal *Science* in 1989, Doudna's name appears in the middle of the list because she was mentoring a lucky Harvard undergraduate who worked in the lab part time, and she felt that the student should be the featured lead author. During her final year in Szostak's lab, her name was on four academic papers in prestigious journals, all describing aspects of how RNA molecules can replicate themselves.[9]

What also stood out for Szostak was Doudna's willingness, even eagerness, to tackle challenges. That became evident near the end of her tenure in Szostak's lab in 1989. She realized that in order to understand the workings of a self-splicing piece of RNA, she would have to fully discern its structure, atom by atom. "At that time, RNA structure was viewed as so difficult that it was maybe impossible to figure out," Szostak recalled. "Hardly anyone was trying anymore."[10]

Meeting James Watson

The first time that Jennifer Doudna made a presentation at a scientific conference, it was at the Cold Spring Harbor Laboratory, and James Watson was, as usual, sitting in the front row as the host. It was the summer of 1987, and he had organized a seminar to discuss "the evolutionary events that may have given rise to the living organisms that now exist on earth."[11] In other words, how did life begin?

The focus of the conference was on the recent discoveries showing that certain RNA molecules could replicate themselves. Because Szostak was unavailable, an invitation went out to Doudna, then only twenty-three, to present the work that she and he were doing on engineering a self-replicating RNA molecule. When she got the letter signed by Watson addressed to "Dear Ms. Doudna" (she was not yet Dr. Doudna), she not only immediately accepted; she had it framed.

The talk she gave, based on a paper she had written with Szostak, was highly technical. "We describe deletions and substitution mutations in the catalytic and substrate domains of the self-splicing intron," she began. That's the type of sentence that excites research biologists, and Watson was intently taking notes. "I was so incredibly nervous that my palms were sweating," she recalls. But at the end, Watson congratulated her, and Tom Cech, whose work on introns had paved the way for Doudna and Szostak's paper, leaned over and whispered, "Good job."[12]

While at the meeting, Doudna took a walk down Bungtown Road, which wanders through the campus. Along the way, she saw a slightly stooped woman walking toward her. It was the biologist Barbara McClintock, who had been a researcher at Cold Spring Harbor for more than forty years and had recently been awarded the Nobel Prize for her discovery of transposons, known as "jumping genes," that can change their position in a genome. Doudna paused, but was too shy to introduce herself. "I felt like I was in the presence of a goddess," she says, still in awe. "Here's this woman who's so famous and so incredibly influential in science acting so unassuming and walking toward her lab thinking about her next experiment. She was what I wanted to be."

Doudna would stay in touch with Watson, attending many of the Cold Spring Harbor meetings he organized. Over the years, he would evolve into an increasingly controversial character because of his unmoored blurtings about racial genetic differences. Doudna generally refrained from letting his behavior diminish her respect for his scientific achievements. "When I saw him, he often would say things he thought were provocative," she says with a slightly defensive laugh.

"That was his way. You know how it is." Despite his frequent public comments about women's looks, beginning with Rosalind Franklin in *The Double Helix*, he was a good mentor to women. "He was very supportive to a close woman friend of mine who was a postdoc," Doudna says. "That influenced my opinion of him."

CHAPTER 7

Twists and Folds

Structural biology

Ever since she puzzled over the touch-sensitive leaves of the sleeping grass that she found on her walks as a child in Hawaii, Doudna had been passionately curious about the underlying mechanisms of nature. What made the fernlike leaves curl when touched? How did chemical reactions cause biological activity? She learned how to pause, like we all used to do as children, and wonder about how things worked.

The field of biochemistry provided many answers by showing how the chemical molecules in living cells behave. But there was a specialty that looked even deeper into nature: structural biology. Wielding imaging techniques such as X-ray crystallography, which is what Rosalind Franklin used to find evidence of the structure of DNA, structural biologists try to discover the three-dimensional shape of molecules. Linus Pauling worked out the spiral structure of proteins in the early 1950s, which was followed by Watson and Crick's paper on the double-helix structure of DNA.

Doudna realized that she would need to learn more about structural biology if she wanted to truly understand how some RNA molecules could reproduce themselves. "To figure out how these RNA do chemistry," she says, "I needed to know what they looked like." Specifically,

Rising star at Yale

she needed to figure out the folds and twists of the three-dimensional structure of self-splicing RNA. She was aware that such work would be an echo of that done by Franklin on DNA, and the parallel pleased her. "She had a similar kind of question about the chemical structure of a molecule that was at the heart of all of life," Doudna says. "She believed that its structure would provide all sorts of insights."[1]

Doudna also sensed that once you figured out the structure of a ribozyme, it might lead to groundbreaking genetic technologies. The citation for the Nobel Prize that Thomas Cech won with Sidney Altman hinted at what this might be: "A futurist possibility is to correct certain genetic disorders. Such a future use of gene shears will require that we learn more about the molecular mechanisms." *Gene shears*. Yes, the Nobel committee was prescient.

This pursuit meant that it was time to move on from the lab of Jack Szostak, who admitted to not being a visual thinker or expert in structural biology. So in 1991, Doudna considered where she could do her postdoctoral work. There was one obvious choice, the structural biologist who had just shared the Nobel Prize for discovering the catalytic RNA that she and Szostak had been studying: Thomas Cech (pronounced "check") of the University of Colorado in Boulder, who was using X-ray crystallography in order to explore each nook and cranny of the structure of RNA.

Thomas Cech

Doudna already knew Cech. He was the one who whispered "Good job" after her sweaty-palmed lecture at Cold Spring Harbor in the summer of 1987. She had met him again when she took a trip to Colorado that year. "Because we were sort of friendly competitors both racing to make discoveries about the self-splicing introns, I sent him a note," she recalled.

It was a real note, on paper, because email was not yet common. She wrote that she was going to be traveling through Boulder and asked if it would be possible to visit his lab. To her surprise, he quickly got back to her, telephoning one day when she was at work in Szostak's lab.

"Hey, Tom Cech is on the phone for you," the colleague who picked up the phone called out. Her lab mates gave her a curious look, but she just shrugged.

They met in Boulder on a Saturday. Cech had brought his two-year-old daughter to the lab, and he bounced her on his knee as he talked to Doudna, who was completely charmed by both his mind and his fatherly instincts. Their encounter was an example of the mix of competition and collegiality that marks scientific research (and many other endeavors). "I think the reason Tom met with me was that the Szostak Lab was doing work that was potentially competitive but also that there might be opportunities to learn from each other," she says. "And he probably thought it was a way to get some information about what our lab was up to."

After she earned her PhD in 1989, she decided to do her postdoctoral work with Cech. "I realized that if I really wanted to figure out the structure of RNA molecules, my smart move was to go to the very best RNA biochemistry lab," she says. "Who can be better than Tom Cech? This was the lab that had first discovered self-splicing introns."

Tom Griffin

There was one other reason that Doudna decided to go to Boulder for her postdoctoral work. In January 1988 she had married a Harvard Medical School student named Tom Griffin, who was working in a lab next to hers. "He saw in me things I didn't see at the time, including capabilities in science," she says. "He pushed me to be bolder than I would have been."

Griffin, from a military family, loved Colorado. "When we were thinking of where to go when we finished our degrees, he really, really wanted to move to Boulder," says Doudna. "I realized that if we went to Boulder, I could work with Tom Cech." So they moved there in the summer of 1991, and Griffin got a job at a startup biotech company.

At first the marriage worked well enough. Doudna bought a mountain bike, and they would ride along Boulder Creek. She also took up roller-blading and cross-country skiing. But her passion was science, and Griffin didn't have her single-minded focus. Science for him was a nine-to-five endeavor, and he had no aspirations to be an academic researcher. He loved music and books, and he became an early fan of personal computers. Doudna respected his broad range of interests but didn't share them. "I'm someone who's thinking about science all the time," she says. "I'm always focused on what's cooking in the lab, the next experiment, or the bigger question to pursue."

Doudna believes their differences "say something negative about me," though I'm not sure she really believes that, nor do I. People are different in their approaches to their work and passions. She wanted to spend weekends and nights in the lab doing experiments. Not everybody should be that way. But some people should.

After a few years, they decided to go their separate ways and get a divorce. "I was obsessed with what my next experiment was going to be," she says. "He didn't have that same intensity. That just created a critical wedge that was not fixable."

The structure of a ribozyme

Doudna's mission when she arrived at the University of Colorado as a postdoc was to map the intron that Cech had discovered could be a self-splicing piece of RNA, showing all of its atoms, bonds, and shapes. If she succeeded in figuring out its three-dimensional structure, that would help show how its twists and folds could bring the right atoms together to cause chemical reactions and allow the snippet of RNA to replicate itself.

It was a high-risk venture, one that involved going to a region of the playing field where few others wanted to run. At the time there was not much work being done on RNA crystallography, and most people would look at her like she was nuts. But if she succeeded, there would be a huge payoff for science.

During the 1970s, biologists had figured out the structure of a smaller and simpler RNA molecule. But little progress had been made in the twenty years since then because scientists found it difficult to isolate and get images of bigger RNAs. Colleagues told Doudna that getting a good image of a large RNA molecule would, at that time, be a fool's errand. As Cech put it, "If we had asked the National Institutes of Health to fund this project, we would have been laughed out of the room."[2]

The first step was to crystallize the RNA—in other words, convert the liquid RNA molecule into a well-organized solid structure. That was necessary in order to use X-ray crystallography and other imaging techniques to discern its components and shapes.

Helping her was a quiet but cheery graduate student named Jamie Cate. He had been using X-ray crystallography to study the structure of proteins, but when he met Doudna he joined her quest to focus on RNA. "I told him about the project I was working on and he got very interested," she says. "It was really out there. We had no idea what we were going to find." They were pioneering a new field. It was not even clear that RNA molecules would have well-defined structures like proteins do. Unlike Tom Griffin, Cate loved to focus on lab work. He and Doudna would talk every day about how to crystallize the RNA, and soon they were continuing their discussions over coffee and sometimes dinner.

One breakthrough came as a result of the random things that often happen in science: a slight blunder, like the mold that got on Alexander Fleming's Petri dishes and led to the discovery of penicillin. One day a technician was working with Doudna to try to make crystals, and she put the experiment into an incubator that was not working properly. They thought the experiment was spoiled, but when they looked at the samples through a microscope they could see crystals growing. "The crystals had RNA in them and were beautiful," Doudna recalled, "and that was the first breakthrough showing us that to get these crystals we had to elevate the temperature."

Another advance shows the enduring power of being in the same location as other smart people. Tom and Joan Steitz, a husband-and-wife team of Yale biochemists who were studying RNA, were on sabbatical in Boulder for a year. Tom was particularly sociable and liked hanging around the lunchroom of the Cech lab holding a mug of coffee. Doudna mentioned to him one morning that she had been able to get good crystals of the RNA molecule she was researching, but they tended to break down too quickly when they were exposed to X-rays.

Steitz replied that in his Yale lab he had been testing a new technique for cryocooling crystals. They plunged crystals into liquid nitrogen so they would freeze very rapidly. That helped to preserve the structure in the crystals even when they were exposed to X-rays. He arranged for Doudna to fly to Yale and spend time with the researchers in his lab there who were pioneering the technique. It worked beautifully. "At that point we knew that we had crystals that were ordered enough that we would eventually be able to solve the structure," she says.

Yale

Her visit to Tom Steitz's lab at Yale, where innovative techniques and equipment such as cryocoolers were being funded, helped convince Doudna to accept a job there in the fall of 1993 as a tenure-track professor. Not surprisingly, Jamie Cate wanted to accompany her. She contacted the Yale authorities and helped arrange for him to transfer there as a graduate student in her lab. "They required him to retake his qualifying exams," she says, "and as I'm sure you can imagine, he aced them with flying colors."

By using the super-cooling techniques, Doudna and Cate were able to create crystals that diffracted X-rays well. But they were stymied by what is known in crystallography as the "phase problem." X-ray detectors can measure properly only the intensity of a wave but not the phase part of the wave. One way to attack the problem is to introduce a metal ion into a few regions of the crystal. The X-ray diffraction pictures show the position of the metal ions, and that can be used to help calculate the rest of the molecular structure. That had been done with protein molecules, but no one had figured out how to do it with RNA.

Cate solved the problem. He did it by using a molecule called *osmium hexamine*, which has an interesting structure that lends itself to interacting in a few nooks of RNA molecules. As a result, the X-ray diffractions could produce an electron-density map that would provide clues for the structure of an important folded region of the RNA they were studying. They began the process of creating these density maps and then building models of potential structures, just as Watson and Crick had done for DNA.

Her father's farewell

When their work was reaching its climax in the fall of 1995, Doudna got a call from her father. He had been diagnosed with melanoma, and it had metastasized to his brain. He told her that he had only three months to live.

She spent the rest of that fall flying back and forth from New Haven to Hilo, a journey of more than twelve hours. Chunks of time spent at her father's bedside were interspersed with hours on the phone with Cate. Each day, Cate would send her a new electron-density map by fax or email, and they would talk about ways to interpret it. "It was an incredible time of highs and lows and intense emotional swings," she recalls.

Fortunately, her father was genuinely curious about her work, and that made the ordeal less painful. In between periods of pain, he would ask her to explain the latest images she had received. She would walk into his bedroom, and he would be lying there looking at the latest data. Before they could discuss his health, he would begin asking questions. "It would remind me of his scientific curiosity and how he had shared it with me when I was a child," she says.

During a visit that November, which lasted through Thanksgiving, an electron-density map arrived from New Haven that she realized was good enough to nail down the structure of the RNA molecule. She could actually see how the RNA was folded up into an amazing three-dimensional shape. She and Cate had been working on it for more than two years, while countless colleagues had declared what they were doing was impossible, and now the latest data showed that they had triumphed.

Her father was completely bedridden by then and could barely move. But he was lucid. She walked into his bedroom and showed him a color printout she had made from a data file of the latest map.

It looked like a green ribbon that was twisted into a really cool shape. "It looks like green fettuccini," he joked. Then he got serious. "What does it mean?" he asked.

By trying to explain it to him, she was able to clarify her own ideas about what the data meant. They pored over a region on the map that was caused by a cluster of metal ions, and she speculated on the ways that the RNA could be folding around such a cluster. "Maybe there's a core of metals here that helps this RNA to fold up into this type of twist," she suggested.

"Why would that be important?" he asked. She explained that RNA is made up of very few chemicals, so it accomplishes complex tasks based on the different ways it is folded. One of the challenges with RNA is that it's a molecule made of only four chemical building blocks, unlike proteins, which have twenty. "Because there is a lot less chemical complexity to RNA," she says, "the challenge is to think about how does it fold into a unique shape."

The visit clarified how time had deepened her relationship with her father. He took science seriously, and he took her seriously. He was attracted to all the details, but he also sought the bigger picture. She recalled the times she had visited his classroom and seen his excitement at communicating his passions. She also recalled, less happily, the times that she had gotten angry at him because she thought he made snap judgments, some of them prejudiced, about people. Bonds can take different forms, both in chemistry and in life. Sometimes an intellectual bond is the strongest.

When Martin Doudna died a few months later, Jennifer and her mother and sisters went with friends on a hike to scatter his ashes high up in the Waipio Valley near Hilo. The name means "curved water," and the river that winds through its lush wilderness has many gorgeous waterfalls. Among those joining them were Don Hemmes, the biology professor who mentored Jennifer, and her closest childhood friend, Lisa Hinkley Twigg-Smith. "As we released his ashes into the wind," Twigg-Smith recalled, "an endemic hawk known as an 'io, which is associated with the gods, soared overhead."[3]

"It was only after he died that I realized how influential he was in my decision to become a scientist," Doudna says. Among the many gifts that he gave her was a love of the humanities and how it intersects with the sciences. The need for that was becoming clearer to her as research led her into realms that required moral guideposts as well as electron-density maps. "I think my father would have loved to understand CRISPR," Doudna reflected. "He was a humanist, a humanities professor, who also loved science. When I talk about CRISPR's effects on our society, I can hear my father's voice in my head."

Triumph

Her father's death coincided with her first major scientific success. She and Cate, along with their lab colleagues, were able to determine the location of every atom in a self-splicing RNA molecule. Specifically, they showed how the structure of a key domain of the molecule allowed RNA to pack helices together to create its three-dimensional shape. A cluster of metal ions in that domain formed a core around which the structure folded. Just as the double-helix structure of DNA revealed how it could store and transmit genetic information, the structure discovered by Doudna and her team explained how the RNA could be an enzyme and was able to slice, splice, and replicate itself.[4]

When their paper was published, Yale sent out a press release that attracted the notice of a local New Haven television station. After trying to explain what a ribozyme is, the news anchor reported that it had baffled scientists because they had never been able to see its shape. "But now a team led by Yale scientist Jennifer Doudna finally was able to capture a snapshot of the molecule," the anchor proclaimed. The story featured a young, dark-haired Doudna in her lab, showing off a blurry image on her computer screen. "We hope our discovery will provide clues as to how we might be able to modify the ribozyme so that it can repair defective genes," she said. It was a momentous statement, though she didn't think about it much at the time. It would be the beginning of a quest to translate basic science about RNA into a tool that could edit genes.

In another, more sophisticated television report, done by a syndicated science news show, Doudna appeared in a white lab coat using a pipette to put a solution into a test tube. "It's been known for fifteen years that RNA molecules could function like proteins in cells, but nobody knew how that could be, because nobody has really known what RNA molecules look like," she explained. "We have now been able to see how an RNA molecule can form itself into a complicated three-dimensional structure." Asked what the implications could be, she again pointed to what would be her future work: "One possibility is that we might be able to cure or treat people who have genetic defects."[5]

Over the next two decades, many people would contribute to the development of gene-editing technologies. What distinguishes Doudna's tale is that, by the time she entered the field of gene editing, she had already established her reputation and earned distinction in the most basic underlying science: the structure of RNA.

With her husband, Jamie Cate, and son, Andy, in Hawaii, 2003

Chapter 8

Berkeley

Going west

In the article that Doudna and her colleagues wrote on their RNA structure discovery, which was published in *Science* in September 1996, her name is listed last, meaning that she was the principal investigator who headed the lab. Jamie Cate's name is listed first because he did the most important experiments.[1] By then they were more than scientific partners; they had become romantically involved. After her divorce was final, they got married in the summer of 2000 at the Melaka Beach Hotel across the Big Island of Hawaii from Hilo. Two years later, they had their only child, Andrew.

By then, Cate had become an assistant professor at MIT, so they were commuting between New Haven and Cambridge. By train it's less than three hours, but for a new couple even that was tiresome, so they decided to see if they could get appointments in the same town.[2]

Yale tried hard to keep Doudna, promoting her to an important professorship. To resolve what is known as the "two-body problem" in academia, it offered Cate a position as well. However, Tom Steitz, the structural biologist who had shown them the techniques of cryocooling, was there doing the same type of research Cate wanted to do, and he felt that would crimp his chance to flourish. "My direct competitor was there," Cate says. "He's a great guy, but it would be hard to be in the same institution."

Harvard offered Doudna a position in the Department of Chemistry and Chemical Biology, which had just been renamed and was growing. She went there as a visiting professor, and on the first day the dean handed her an offer letter for a permanent position. With Cate at MIT, it seemed to be an ideal arrangement. "I was thinking how great it was that I would end up in Boston, back where I was in graduate school and had such a good time," she said.

It is interesting to imagine how her career would have been different if she had stayed at Harvard. Along with MIT and the jointly managed Broad Institute, the university was a cauldron of biotech research, especially in the field of gene engineering. A decade later, she would find herself in a race to develop CRISPR into a gene-editing tool with various Cambridge-based researchers, including Harvard's George Church and the men who would become her bitter rivals, Feng Zhang and Eric Lander of the Broad Institute.

Then she got a call from the University of California at Berkeley. Her first reaction was to deflect any offer, but when she told Cate, he was shocked. "You should call them back," he said. "Berkeley is nice." When he had been a postdoctoral fellow in Santa Cruz, he had often gone up to the Lawrence Berkeley National Laboratory, which was managed by the university, to do experiments at its cyclotron, a particle accelerator.

When they visited the campus, Doudna was still disinclined to move there. But Cate became more enthusiastic. "I'm more of a western guy," he says. "I found Cambridge to be uptight. My director at the time always came to work in a bow tie. I was happier at the thought of being at Berkeley, where the energy level was great." Doudna liked the fact that Berkeley was a public university, and she was easily persuaded. By the summer of 2002, they had moved.

Their choice of Berkeley is a testament to America's investment in public higher education. Its roots stretch back to when Abraham Lincoln, in the middle of the Civil War, thought public education was important enough that he pushed through the Morrill Land-Grant Act of 1862, which used funds from federal land sales to establish new agriculture and mechanical colleges.

Among those was the College of Agricultural, Mining, and Mechanical Arts near Oakland, California, founded in 1866, which two years later merged with the nearby private College of California. It became the University of California, Berkeley, and grew into one of the world's greatest research and learning institutions. In the 1980s, more than half of Berkeley's funding came from the state. However, since then Berkeley, like most other public universities, has faced reductions. When Doudna arrived, state funding accounted for only 30 percent of Berkeley's budget. In 2018, state funding was cut again, and it amounted to less than 14 percent. As a result, Berkeley's undergraduate tuition for a California resident in 2020 was $14,250 per year, more than triple what it was in 2000. Room, board, and other fees raised the total cost to around $36,264. For an out-of-state student, total costs were around $66,000 a year.

RNA interference

Doudna's study of RNA structure led her to a field that would become unexpectedly relevant later in her career: viruses. Specifically, she was interested in how the RNA in some viruses, such as coronaviruses, allow them to hijack the protein-making machinery of cells. During her first semester at Berkeley, in the fall of 2002, there was an outbreak in China of a virus that caused a severe acute respiratory syndrome (SARS). Many viruses are composed of DNA, but SARS was a coronavirus that instead contained RNA. By the time it died out after eighteen months, it had killed close to eight hundred people around the world. It was officially known as SARS-CoV. In 2020, it had to be renamed SARS-CoV-1.

Doudna also became interested in a phenomenon known as RNA interference. Normally, the genes encoded by the DNA in cells dispatch messenger RNAs to direct the building of a protein. RNA interference does just what the name implies: small molecules find a way to mess with these messenger RNAs.

RNA interference was discovered in the 1990s, partly by researchers who were trying to make petunias more purple by juicing up the flower's color genes. But the process ended up suppressing some of the genes, leading to mottled and speckled petunias. Craig Mello and Andrew Fire coined the term "RNA interference" in a 1998 paper and later won the Nobel Prize when they discovered how the phenomenon works in the nematode, a tiny worm.[3]

RNA interference operates by deploying an enzyme known as "Dicer." Dicer snips a long piece of RNA into short fragments. These little fragments can then embark on a search-and-destroy mission: they seek out a messenger RNA molecule that has matching letters, then they use a scissors-like enzyme to chop it up. The genetic information carried by that messenger RNA is thus silenced.

Doudna set about to discover the molecular structure of Dicer. As she had done with self-splicing RNA introns, she used X-ray crystallography to map its twists and folds, which she hoped would show how it worked. Until then, researchers did not know how Dicer was able to cut RNA into precisely the right letter sequences to silence a specific gene. By studying the Dicer structure, Doudna showed that it acted like a ruler that had a clamp at one end, which it used to grab on to a long RNA strand, and a cleaver at the other end, which it used to slice the segment at just the correct length.

Doudna and her team went on to show how a particular domain of the Dicer enzyme could be replaced in order to create tools that would silence other genes. "Perhaps the most exciting finding of this study is that Dicer can be reengineered," their 2006 paper noted.[4] It was a very useful discovery. It permitted researchers to use RNA interference to turn off a wide variety of genes, both to discover what each gene does and to regulate its activity for medical purposes.

In the age of coronaviruses, there is another role that RNA interference may play. Throughout the history of life on our planet, some organisms (though not humans) have evolved ways to use RNA interference to fight off viruses.[5] As Doudna wrote in a scholarly publication back in 2013, researchers hoped to find ways to use RNA interference to protect humans from infections.[6] Two papers published in *Science* that year gave strong evidence that it might work. The hope then was that drugs based on RNA interference might someday be a good option for treating severe viral infections, including those from new coronaviruses.[7]

Doudna's paper on RNA interference appeared in *Science* in January 2006. A few months later, a paper published in a little-known journal described a different virus-fighting mechanism that exists in nature. It was by an obscure Spanish scientist who discovered the mechanism in microorganisms like bacteria, which have a far longer and even more brutal history fighting viruses than we humans do. At first, the handful of scientists studying this system assumed that it worked through RNA interference. They would soon discover that the phenomenon was even more interesting.

Part Two

CRISPR

The scientist does not study nature because it is useful.
He studies it because he takes pleasure in it,
and he takes pleasure in it because it is beautiful.

—Henri Poincaré, *Science and Method*, 1908

Francisco Mojica

Erik Sontheimer and Luciano Marraffini

CHAPTER 9

Clustered Repeats

Francisco Mojica

When Yoshizumi Ishino was a student at Osaka University in Japan, his PhD research included sequencing a gene in *E. coli* bacteria. It was 1986, and gene sequencing was a laborious process, but he eventually succeeded in determining the 1,038 base pairs of DNA that made up the gene in question. In a long paper on the gene that he published the following year, he noted in the last paragraph an oddity that he did not consider important enough to mention in the paper's abstract. "An unusual structure was found," he wrote. "Five highly homologous sequences of 29 nucleotides were arranged as direct repeats." In other words, he found five segments of DNA that were identical to each other. These repeated sequences, each twenty-nine base pairs long, were sprinkled between normal-looking sequences of DNA, which he called "spacers." Ishino had no idea what these clustered repeats were. In the last line of his paper, he wrote, "The biological significance of these sequences is not known." He didn't pursue the topic.[1]

The first researcher to figure out the function of the repeated sequences was Francisco Mojica, a graduate student at the University of Alicante on the Mediterranean coast of Spain. In 1990, he began working on a PhD dissertation on archaea, which, like bacteria, are single-cell organisms without a nucleus. The archaea he was studying thrive in salt ponds that are ten times saltier than the ocean. He was sequencing regions that he thought might explain its love of salt when he spotted fourteen identical DNA sequences that were repeated at regular intervals. They seemed to be palindromes, meaning they read the same backward and forward.[2]

At first he assumed that he had screwed up the sequencing. "I thought it was a mistake, because sequencing was hard back then," he says with a hearty laugh. But by 1992, when his data kept showing these regularly spaced repeats, Mojica wondered if anyone else had found something similar. Google did not yet exist, nor did online indexes, so he manually sorted through citations for the word "repeat" in a set of *Current Contents*, a printed index of scholarly papers. Because this was in a previous century, when very few publications were online, whenever he found a listing that looked promising, he had to go to the library to find the relevant journal. Eventually he found Ishino's paper.

The *E. coli* bacterium that Ishino studied is a very different organism from Mojica's archaea. So it was surprising that they both had these repeated sequences and spacer segments. This convinced Mojica that the phenomenon must have some important biological purpose. In a paper he published in 1995, he and his thesis advisor dubbed them "tandem repeats," and they guessed, incorrectly, that they might have something to do with cell replication.[3]

After doing two quick postdoctoral stints, one in Salt Lake City and the other at Oxford, Mojica returned in 1997 to the University of Alicante, which was just a few miles from where he was born, and launched a research group to study these mysterious repeated sequences. It was difficult to get funding. "I was told to stop obsessing about repeats, because there were a lot of those type of phenomena in organisms, and mine were probably nothing special," he says.

But he knew that bacteria and archaea have small amounts of genetic material. They cannot afford to waste a lot of it on sequences that have no important function. So he kept trying to figure out the purpose of these clustered repeats. Perhaps they helped shape the

DNA structure or formed loops that proteins could latch on to. Both of those speculations also proved wrong.

The name "CRISPR"

By then, researchers had found these repeated sequences in twenty different species of bacteria and archaea, and many different names for them had sprouted. Mojica became dissatisfied with the name his dissertation advisor had foisted on him, "tandem repeats." The sequences were interspaced, not in tandem. So he renamed them, initially, "short regularly spaced repeats," or SRSR. Though more descriptive, it was an unmemorable name with an unpronounceable acronym.

Mojica had been corresponding with Ruud Jansen of Utrecht University in the Netherlands, who was studying these sequences in tuberculosis bacteria. He had been calling them "direct repeats," but he agreed that they needed to come up with a better name. Mojica was driving home from his lab one evening when he came up with the name CRISPR, for "clustered regularly interspaced short palindromic repeats." Although the clunky phrase was almost impossible to remember, the acronym CRISPR was, indeed, crisp and crispy. It sounded friendly rather than intimidating, though the dropped "e" gave it a futurist sheen. When he got home, he asked his wife what she thought of the name. "It sounds like a great name for a dog," she said. "Crispr, Crispr, come here, pup!" He laughed and decided it would work.

On November 21, 2001, the name was anointed in an email from Jansen in reply to Mojica's suggestion. "Dear Francis," he wrote, "What a great acronym is CRISPR. I feel that every letter that was removed in the alternatives made it less crispy, so I prefer the snappy CRISPR over SRSR and SPIDR."[4]

Jansen formalized the decision in a paper he published in April 2002, which reported his discovery of genes that seemed to be associated with CRISPRs. In most organisms that had CRISPRs, the repeated sequences were flanked by one of these genes, which encoded directions for making an enzyme. He named these "CRISPR-associated," or *Cas*, enzymes.[5]

A virus defense

When Mojica began sequencing the DNA of his salt-loving microbes in 1989, gene sequencing was a slow process. But the Human Genome Project, which was just getting started, eventually spawned new high-speed sequencing methods. By 2003, when Mojica focused on figuring out the role CRISPRs played, the genomes of close to two hundred bacteria had been sequenced (as well as those of humans and mice).

That August, Mojica was on holiday in the beach town of Santa Polo, about twelve miles south of Alicante, staying at the house of his wife's parents. That was not his idea of a good time. "I really do not like sand or being on a beach in the summer when it is hot and crowded with people," he says. "My wife would be lying on the beach getting a suntan, and I would head off and drive up to my lab in Alicante for the day. She had fun on the beach, but I had more fun analyzing sequences from *E. coli* bacteria."[6] Spoken like a dedicated scientist.

What fascinated him were the "spacers," those regions of normal-looking DNA segments that were nestled in between the repeated CRISPR segments. He took the spacer sequences of *E. coli* and ran them through databases. What he found was intriguing: the spacer segments matched sequences that were in viruses that attacked *E. coli*. He found the same thing when he looked at other bacteria with CRISPR sequences; their spacer segments matched those of viruses that attacked that bacteria. "Oh my goodness!" he exclaimed at one point.

One evening, when he was sure about his discovery, he explained it to his wife after he got back to the beach house. "I just discovered something really amazing," he said. "Bacteria have an immune system. They're able to remember what viruses have attacked them in the past." She laughed, admitted she didn't quite understand, but said she believed it must be important because he was so excited. He replied, "In a few years, you'll see this thing that I've just discovered will be written about in newspapers and in history books." That part she did not believe.

What Mojica had stumbled upon was a battlefront in the longest-running, most massive and vicious war on this planet: that between bacteria and the viruses, known as "bacteriophages" or "phages," that attack them. Phages are the largest category of virus in nature. Indeed, phage viruses are by far the most plentiful biological entity on earth. There are 10^{31} of them—a trillion phages for every grain of sand, and more than all organisms (including bacteria) combined. In one milliliter (0.03 ounces) of seawater there can be as many as 900 million of these viruses.[7]

As we humans struggle to fight off novel strains of viruses, it's useful to note that bacteria have been doing this for about three billion years, give or take a few million centuries. Almost from the beginning of life on this planet, there's been an intense arms race between bacteria, which developed elaborate methods of defending against viruses, and the ever-evolving viruses, which sought ways to thwart those defenses.

Mojica found that bacteria with CRISPR spacer sequences seemed to be immune from infection by a virus that had the same sequence. But bacteria without the spacer did get infected. It was a pretty ingenious defense system, but there was something even cooler: it appeared to adapt to new threats. When new viruses came along, the bacteria that survived were able to incorporate some of that virus's DNA and thus create, in its progeny, an acquired immunity to that new virus. Mojica recalls being so overcome by emotion at this realization that he got tears in his eyes.[8] The beauty of nature can sometimes do that to you.

It was an astonishing and elegant discovery, one that would have great repercussions. But Mojica had a ridiculously difficult time getting it published. He submitted a paper to *Nature* in October 2003 entitled "Prokaryotic Repeats Are Involved in an Immunity System." In other words, CRISPR systems were a way that bacteria acquired immunity to viruses. The editors did not even send it out for review. It did not contain, they incorrectly judged, much that wasn't in previous CRISPR papers. They also declared, with more validity, that Mojica had not presented any lab experiments showing how the CRISPR system worked.

Mojica's paper was rejected by two other publications. Finally he was able to get it published in the *Journal of Molecular Evolution*, which was not as prestigious but served to get his findings in a peer-reviewed publication. Even at that journal, Mojica had to pester and prod the slow-moving editors. "I reached out and tried to get in touch with the editors almost every week," he says. "Every week was so terrible, such a nightmare, because I knew we had discovered something really great. And I knew that at some point others would discover it. And I couldn't get them to see how important it was."[9] The journal received the paper in February 2004, did not make a decision until October, and it was not actually published until February 2005, two years after Mojica had come up with his findings.[10]

Mojica says he was driven by his love of the beauties of nature. He had the luxury at Alicante of doing basic research without showing how it might translate into something useful, and he never tried to patent his CRISPR discoveries. "When you work as I do on weird organisms that live in unusual environments, like very salty ponds, your only motivation is curiosity," he says. "It didn't seem likely that our discovery would apply to more normal organisms. But we were wrong."

As is often the case in the history of science, discoveries can have unexpected applications. "When you do curiosity-driven research, you never know what it may someday lead to," Mojica says. "Something that's basic can later have wide consequences." His prediction to his wife that his name would someday be in history books proved to be correct.

Mojica's paper was the beginning of a wave of articles providing evidence that CRISPR was, indeed, an immune system that bacteria adapted whenever they got attacked by a new type of virus. Within a year, Eugene Koonin, a researcher at the U.S. National Center for Biotechnology Information, extended Mojica's theory by showing that the role of the CRISPR-associated enzymes was to grab bits of DNA out of the attacking viruses and insert them into the bacteria's own DNA, sort of like cutting and pasting a mug shot of dangerous viruses.[11] But Koonin and his team got one thing wrong. They speculated that the

CRISPR defense system worked through RNA interference. In other words, they thought that bacteria used the mug shots to find a way to interfere with the messenger RNAs that carry out the instructions encoded by DNA.

Others thought so as well. That is why Jennifer Doudna, Berkeley's leading expert on RNA interference, would end up getting a phone call out of the blue from a colleague who was trying to figure out CRISPR.

Jillian Banfield

CHAPTER 10

The Free Speech Movement Café

Jillian Banfield

In early 2006, shortly after she published her first paper on Dicer, Doudna was in her Berkeley office when she got a call from a Berkeley professor she had heard of but didn't know: Jillian Banfield, a microbiologist who, like Mojica, was interested in tiny organisms found in extreme environments. A gregarious Australian with a wry smile and collaborative nature, Banfield was studying bacteria that her team found in a very salty lake in Australia, a hot geyser in Utah, and the extremely acidic waste draining from a California copper mine into a salt marsh.[1]

When Banfield sequenced the DNA of her bacteria, she kept finding examples of the clustered repeated sequences known as CRISPRs. She was among those who assumed that the CRISPR system worked by using RNA interference. When she typed "RNAi and UC Berkeley" into Google, Doudna's name was the top result, so Banfield gave her a call. "I'm looking for someone at Berkeley," she told Doudna, "who is working on RNA guides, and I did a Google search and your name popped up." They agreed to meet for tea.

Doudna had never heard of CRISPR. In fact, she thought that Banfield was saying "crisper." After hanging up, she did a quick online search and found just a few articles about it. When she got to the point in an article where it said CRISPR stood for "clustered regularly interspaced short palindromic repeats," she decided to wait for Banfield to explain it to her.

They met on a blustery spring day at a stone table in the courtyard of the Free Speech Movement Café, a soup-and-salad hangout at the entrance to Berkeley's undergraduate library. Banfield had printed out the papers by Mojica and Koonin. She realized that, in order to figure out the function of these CRISPR sequences, it made sense to collaborate with a biochemist such as Doudna, who could analyze each component of a mysterious molecule in a laboratory.

When I sat down with the two of them to hear about that meeting, they displayed the same excitement they described feeling back then. They both talked rapidly, especially Banfield, and they finished each other's sentences amid quick laughs. "We are sitting there and drinking tea, and you had a big pile of pages that had all this data of the sequences you had found," Doudna recalled. Banfield, who usually works on her computer and rarely prints out anything, agreed. "I kept showing you the sequences," she recalled. Doudna chimed in, "You were so passionate, and you were talking so fast. You had a lot of data. And I'm thinking, 'She's really, really excited about this.' "[2]

At the café table, Banfield drew a string of diamonds and squares that represented segments of the DNA she had found in her bacteria. The diamonds, she said, all had identical sequences, but the interspersed squares each had unique sequences. "It's like they are diversifying so fast in response to *something*," she told Doudna. "I mean, what was causing these strange clusters of DNA sequences? How did they actually work?"

Until then, CRISPRs had largely been the purview of microbiologists, such as Mojica and Banfield, who studied living organisms. They had come up with elegant theories about CRISPR, some of them correct, but they had not done controlled experiments in test tubes. "At the time, nobody had actually isolated the molecular components of the CRISPR system, tested them in a lab, and figured out their structures," Doudna said. "So the time was right for biochemists and structural biologists like me to jump in."[3]

CHAPTER 11

Jumping In

Blake Wiedenheft

When Banfield asked her to collaborate on CRISPR, Doudna was initially stymied. She had nobody in her lab to work on it.

Then an unusual candidate walked into her office to interview for a postdoctoral position. Blake Wiedenheft, a charismatic and bear-cub-loveable Montanan with an enthusiasm for the outdoors, had spent most of his academic career, when he wasn't taking time off to pursue wilderness adventures, collecting microorganisms from extreme environments, from Kamchatka in Russia to Yellowstone National Park in his backyard, just like Banfield and Mojica. His letters of reference were not stellar, but he was earnest and passionate about switching his interest from the biology of small organisms to the biology of molecules, and when Doudna asked him what he wanted to work on, he said the magic words: "Have you ever heard of CRISPR?"[1]

Wiedenheft was born in Fort Peck, Montana, population 233, an outpost eighty-one miles from the Canadian border and near nothing else. The son of a fisheries biologist for Montana's Wildlife Department, he ran track, skied, wrestled, and played football in high school.

As an undergraduate at Montana State, he majored in biology, but he spent little time in the lab. Instead, he enjoyed going into nearby

Blake Wiedenheft in Kamchatka, Russia

Yellowstone and collecting microorganisms that can survive in the boiling acid springs there. "It made a huge impression on me," he says, "to scoop up sample organisms from an acid hot spring, bring them back in a thermos, grow them in these artificial hot springs we rigged up in the lab, and then take those samples to the microscope and peer through the lens and see something that has never been seen before. That changed how I imagined life."

Montana State was a perfect university for him, because it allowed him to indulge his love of adventure. "I'm always looking for what's over the next peak," he says.[2] When he graduated, he had no plans to become a research scientist. Instead, like his father, he was interested in fish biology, and he signed up to work on a crabbing vessel in the Bering Sea off Alaska, collecting data for government agencies. He then spent a summer teaching science to young students in Ghana, followed by a stint as a ski patroller in Montana. "I was addicted to adventure."

But during his travels, he would find himself rereading his old biology textbooks at night. His college mentor Mark Young was studying the viruses that attacked the bacteria in Yellowstone's boiling acid springs. "Mark's excitement for understanding how these biological machines work was, literally, infectious."[3] After three years of wandering, Wiedenheft decided there were adventures to be found not only outdoors but in labs. He returned to Montana State as a PhD student under Young, and together they studied how these viruses invade bacteria.[4]

Although Wiedenheft was able to sequence the DNA of the viruses, he found himself wanting more. "Once I started peering at the DNA sequences, I realized they were uninformative," he says. "We had to determine structures, because structures, the folds and shapes, are conserved over a longer evolutionary period than the nucleic acid sequences." In other words, the sequence of letters in the DNA did not reveal how it worked; what was important was how it folded and twisted, which would reveal how it interacted with other molecules.[5]

He decided that he needed to learn structural biology, and there was no place better for that than Doudna's lab at Berkeley.

Wiedenheft is too earnest to be insecure, and that came through when he interviewed with Doudna. "I was coming from a small lab in Montana, and I had enough hubris not to be completely intimidated, though I should have been," he recalls. He had a few subject areas he planned to pitch, but when Doudna showed interest in CRISPR, which was his first passion, he became energized. "I just started yammering and tried to sell myself best I could." He went to the whiteboard and mapped out the CRISPR projects being pursued by other researchers, including John van der Oost and Stan Brouns, a team from the Netherlands he had worked with when they came to Yellowstone to collect microorganisms from the hot springs.

He and Doudna brainstormed about opportunities that her lab might pursue, most notably figuring out the functions of the CRISPR-associated (Cas) enzymes. Doudna was struck by his energy and infectious enthusiasm. For his part, Wiedenheft was impressed that Doudna shared his enthusiasm for CRISPR. "She has a knack for seeing around corners to know what the next big thing is," he says.[6]

Wiedenheft threw himself into his work in Doudna's lab with the joyful passion he displayed as an outdoorsman. He was willing to charge headlong into techniques he had never used before. At lunchtime he would go on a hard-core bike ride, then work through the afternoon and evening still wearing his cycling gear, wandering around the lab in his helmet. He once spent forty-eight hours straight on one experiment, sleeping next to it.

Martin Jinek

Wiedenheft's desire to learn structural biology caused him to latch on to, both intellectually and socially, a postdoc who was the Doudna Lab's expert in crystallography. Martin Jinek (YEE-nik) was born in the Silesian town of Třinec in what was then Czechoslovakia. He studied organic chemistry at Cambridge University and did his doctoral work under the Italian biochemist Elena Conti in Heidelberg. This produced, in addition to an agile scientific outlook, a hybrid accent that featured very precisely pronounced phrases repeatedly interspaced with the interjection "basically."[7]

In Conti's lab, Jinek developed a passion for the star molecule of this book, RNA. "It's such a versatile molecule—it can do catalysis, it can fold into 3D structures," he later told Kevin Davies of the *CRISPR Journal.* "At the same time, it's a carrier of information. It's an all-rounder in the world of biomolecules!"[8] His goal was to work in a lab where he could figure out the structure of complexes that combined RNA and enzymes.[9]

Jinek was good at charting his own path. "He was somebody who could work independently, which has always been important in my lab because I'm not a close hands-on advisor," she says. "I like to hire people who have their own creative ideas and want to work under my guidance and as part of my team, but not with daily direction." She arranged to meet Jinek when she went to Heidelberg for a conference, then enticed him to come to Berkeley and sit down with the members of her lab. She felt it was important that people on her team were comfortable with each new hire.

Jinek's initial work in Doudna's lab focused on how RNA interference works. Researchers had described the process in living cells, but Jinek knew that a full explanation required re-creating the process in a test tube. The *in vitro* experiments allowed him to isolate the enzymes that are essential to interfering with the expression of a gene. He also was able to determine the crystal structure of one particular enzyme, thus showing how it is able to cut up the messenger RNA.[10]

Jinek and Wiedenheft, with their very different backgrounds and personalities, became complementary particles. Jinek was a crystallographer who wanted more experience working with living cells, and Wiedenheft was a microbiologist who wanted to learn crystallography. They took an instant liking to each other. Wiedenheft had a much more playful sense of humor than Jinek, but it was so contagious that Jinek soon acquired it. On one trip they took with other lab members to the Argonne National Laboratory near Chicago, they were working in the huge circular building that houses the Advanced Photon Source, a powerful X-ray machine. It is so large that there are tricycles for researchers to use to get around. At 4 a.m., after working all night, Wiedenheft organized a tricycle race around the entire circuit of the building, which he of course won.[11]

Doudna decided that her lab's goal would be to dissect the CRISPR system into its chemical components and study how each worked. She and Wiedenheft decided to focus first on the CRISPR-associated enzymes.

Cas1

Let's pause for a quick refresher course.

Enzymes are a type of protein. Their main function is to act as a catalyst that sparks chemical reactions in the cells of living organisms, from bacteria to humans. There are more than five thousand biochemical reactions that are catalyzed by enzymes. These include breaking down starches and proteins in the digestive system, causing muscles to contract, sending signals between cells, regulating metabolism, and (most important for this discussion) cutting and splicing DNA and RNA.

By 2008, scientists had discovered a handful of enzymes produced by genes that are adjacent to the CRISPR sequences in a bacteria's DNA. These CRISPR-associated (Cas) enzymes enable the system to cut and paste new memories of viruses that attack the bacteria. They also create short segments of RNA, known as CRISPR RNA (crRNA), that can guide a scissors-like enzyme to a dangerous virus and cut up its genetic material. Presto! That's how the wily bacteria create an adaptive immune system!

The notation system for these enzymes was still in flux in 2009, largely because they were being discovered in different labs. Eventually they were standardized into names such as Cas1, Cas9, Cas12, and Cas13.

Doudna and Wiedenheft decided to focus on what became known as Cas1. It's the only Cas enzyme that appears in all bacteria that have CRISPR systems, which indicates that it performs a fundamental function. Cas1 had another advantage for a lab that was using X-ray crystallography to try to discover how the structure of a molecule determines its functions: it was easy to get it to crystallize.[12]

Wiedenheft was able to isolate the Cas1 gene from bacteria and then clone it. Using a vapor diffusion, he was then able to crystallize it. But he was stymied when he tried to figure out the exact crystal structure because he did not have enough experience in using X-ray crystallography.

Doudna drafted Jinek, who had just finished publishing a paper with her on RNA interference,[13] to help Wiedenheft with the crystallography. Together they went to the particle accelerator at the nearby Lawrence Berkeley National Laboratory, and Jinek helped analyze the data in order to build an atomic model of the Cas1 protein. "In the process, I got infected by Blake's enthusiasm," he recalls. "After that, I decided to stay involved with the CRISPR part of Jennifer's lab."[14]

They discovered that Cas1 has a distinct fold, indicating that it is the mechanism that bacteria use to cleave a snippet of DNA from invading viruses and incorporate it into their CRISPR array, thus being the key to the memory-forming stage of the immune system. In June 2009, they published their discovery in a paper that was the Doudna Lab's initial contribution to the CRISPR field. It was the first explanation of a CRISPR mechanism based on a structural analysis of one of its components.[15]

Rodolphe Barrangou

Philippe Horvath

CHAPTER 12

The Yogurt Makers

Basic research and the linear model of innovation

Historians of science and technology, including myself, often write about what is called the "linear model of innovation." It was propagated by Vannevar Bush, an MIT engineering dean who cofounded Raytheon and during World War II headed the U.S. Office of Scientific Research and Development, which oversaw the invention of radar and the atom bomb. In a 1945 report, "Science, the Endless Frontier," Bush argued that basic curiosity-driven science is the seed corn that eventually leads to new technologies and innovations. "New products and new processes do not appear full-grown," he wrote. "They are founded on new principles and new conceptions, which in turn are painstakingly developed by research in the purest realms of science. Basic research is the pacemaker of technological progress."[1] Based on this report, President Harry Truman launched the National Science Foundation, a government agency that provides funding for basic research, mainly at universities.

There is some truth to the linear model. Basic research in quantum theory and surface-state physics of semiconducting materials led to the development of the transistor. But it wasn't quite that simple or linear. The transistor was developed at Bell Labs, the research organization of the American Telephone and Telegraph Company. It employed many basic science theorists, such as William Shockley and John Bardeen. Even Albert Einstein dropped by. But it also threw them together with practical engineers and pole-climbers who knew how to amplify a phone signal. Added to the mix were business development executives who pushed ways to enable long-distance calls across the continent. All of these players informed and prodded each other.

The story of CRISPR at first seems to accord with the linear model. Basic researchers such as Francisco Mojica pursued an oddity of nature out of pure curiosity, and that seeded the ground for applied technologies such as gene editing and tools to fight coronaviruses. However, as with the transistor, it was not simply a one-way linear progression. Instead, there was an iterative dance among basic scientists, practical inventors, and business leaders.

Science can be the parent of invention. But as Matt Ridley points out in his book *How Innovation Works*, sometimes it's a two-way street. "It is just as often the case that invention is the parent of science: techniques and processes are developed that work, but the understanding of them comes later," he writes. "Steam engines led to the understanding of thermodynamics, not the other way round. Powered flight preceded almost all aerodynamics."[2]

The colorful history of CRISPR provides another great tale about this symbiosis between basic and applied science. And it involves yogurt.

Barrangou and Horvath

As Doudna and her team began working on CRISPR, two young food scientists on different continents were studying CRISPR with the goal of improving ways to make yogurt and cheese. Rodolphe Barrangou in North Carolina and Philippe Horvath in France worked for Danisco, a Danish food ingredient company that makes starter cultures, which initiate and control the fermentation of dairy products.

Starter cultures for yogurt and cheese are made from bacteria, and the greatest threats to the $40 billion global market are viruses that can destroy bacteria. So Danisco was willing to spend a lot of money for research into how bacteria defend themselves against these viruses. It had a valuable asset: a historical record of the DNA sequences of bacteria it had used over the years. And that is how Barrangou and Horvath, who first heard of Mojica's research into CRISPR at a conference, became part of the relationship between basic science and business.

Barrangou was born in Paris, which gave him an enthusiasm for food. He also loved science, and in college he decided to combine his passions. He became the only person I've ever encountered who moved from France to North Carolina in order to learn more about food. He enrolled at North Carolina State in Raleigh and got his master's degree in the science of pickle and sauerkraut fermentation. He went on to get his doctorate there, married a food scientist he met in class, and followed her to Madison, Wisconsin, when she went to work at the Oscar Mayer meat company. Madison is also home to a Danisco unit that produces hundreds of megatons of bacteria cultures for fermented dairy products, including yogurt. Barrangou took a job there as a research director in 2005.[3]

Years before, he had become friends with another French food scientist, Philippe Horvath, who was a researcher at a Danisco laboratory in Dangé-Saint-Romain, a town in central France. Horvath was developing tools to identify the viruses that attack different strains of bacteria, and the two began a long-distance collaboration to research CRISPR.

They would talk by phone two or three times a day in French as they plotted their plans. Their method was to use computational biology to study the CRISPR sequences of bacteria in Danisco's vast database, starting with *Streptococcus thermophilus*, the bacteria that is the great workhorse of the dairy culture industry. They compared the bacteria's CRISPR sequences with the DNA of the viruses that attacked them. The beauty of Danisco's historic collection was that there were bacteria strains from every year since the early 1980s, so they could observe the changes that occurred to them over time.

They noticed that bacteria that had been collected soon after a big virus attack had new spacers with sequences from those viruses, indicating that these had been acquired as a way to repel future attacks. Because the immunity was now part of the bacteria's DNA, it was passed down to all future generations of the bacteria. After one specific comparison done in May 2005, they realized they had nailed it. "We saw there was a hundred-percent match between the CRISPR of the bacterial strain and the sequence of the virus that we knew had attacked it," Barrangou recalls. "That was the eureka moment."[4] It was an important confirmation of the thesis put forth by Francisco Mojica and Eugene Koonin.

They then accomplished something very useful: they showed that they could engineer this immunity by devising and adding their own spacers. The French research facility was not approved for genetic engineering, so Barrangou did that part of the experiments in Wisconsin. "I showed that when you add sequences from the virus into the CRISPR locus, the bacteria develops immunity to that virus," he says.[5] In addition, they proved that CRISPR-associated (Cas) enzymes were critical for acquiring new spacers and warding off attacking viruses. "What I did was knock out two Cas genes," Barrangou recalls. "That wasn't easy to do twelve years ago. One of them was Cas9, and we showed when you knock it out you lose the resistance."

They used these discoveries in August 2005 to apply for and get one of the first patents granted for CRISPR-Cas systems. That year Danisco started using CRISPR to vaccinate its bacterial strains.

Barrangou and Horvath produced a paper for the journal *Science*, which was published in March 2007. "That was a great moment in time," Barrangou says. "Here we were, workers at an unknown Danish company, sending a manuscript on a little-known system in an organism that no scientist cares about. Even to get reviewed was amazing. And we got accepted!"[6]

The CRISPR meetings

The article helped kick interest in CRISPR into a higher orbit. Jillian Banfield, the Berkeley biologist who had enlisted Doudna at the Free Speech Movement Café, immediately called Barrangou. They decided to do what pioneers in emerging fields often do: start an annual conference. The first, organized by Banfield and Blake Wiedenheft, met in late July 2008 in Berkeley's Stanley Hall, where Doudna's lab was. Only thirty-five people attended, including Francisco Mojica, who came from Spain to be a featured speaker.

Long-distance collaborations work well in science—and especially in the CRISPR field, as Barrangou and Horvath showed. But physical proximity can spark more powerful reactions; ideas gel when people have tea at places like the Free Speech Movement Café. "Without those CRISPR conferences, the field would not have moved at the speed it has or be as collaborative," Barrangou says. "The camaraderie would never have existed."

The conference rules were loose and trusting. People could talk informally about data they had not yet published, and the other participants would not take advantage of that. "Small meetings, where unpublished data and ideas can be shared and everyone helps everyone, can change the world," Banfield later noted. Among the first accomplishments was standardizing the lingo and names, including adopting a common designation for the CRISPR-associated proteins. Sylvain Moreau, one of the pioneer participants, called the July meeting "our scientific Christmas party."[7]

Sontheimer and Marraffini

The year of the inaugural conference produced a major advance. Luciano Marraffini and his advisor Erik Sontheimer of Northwestern University in Chicago showed that the target of the CRISPR system was DNA. In other words, CRISPR did not work through RNA interference, which had been the general consensus when Banfield first approached Doudna. Instead, the CRISPR system targeted the DNA of the invading virus.[8]

That had a holy-cow implication. As Marraffini and Sontheimer realized, if the CRISPR system was aimed at the DNA of viruses, then it could possibly be turned into a gene-editing tool. That seminal discovery sparked a new level of interest in CRISPR around the world. "It led to the idea that CRISPR could be fundamentally transformative," Sontheimer says. "If it could target and cut DNA, it would allow you to fix the cause of a genetic problem."[9]

There was still a lot to figure out before that could happen. Marraffini and Sontheimer didn't know precisely how the CRISPR enzyme cut the DNA. It could have done so in a way that was incompatible with genetic editing. Nevertheless, they filed a patent application in September 2008 for the use of CRISPR as a DNA-editing tool. It was rejected, and rightly so. Their guess that it could someday be a gene-editing tool was correct, but it was not yet backed up by experimental evidence. "You can't just patent an idea," Sontheimer admits. "You have to actually have invented what you're claiming." They also applied for a grant from the National Institutes of Health to pursue the possibility of a gene-editing tool. That, too, was rejected. But they were on record as being the first to suggest how CRISPR-Cas systems might be used as gene-editing tools.[10]

Sontheimer and Marraffini had studied CRISPR in living cells, such as those of bacteria. So had the other molecular biologists who published papers on CRISPR that year. But a different approach was required in order to determine the essential components of the system: biochemists working with the molecules *in vitro*, in a test tube. By isolating the components in a test tube, biochemists could explain at the molecular level the discoveries made by microbiologists working *in vivo* and by computational geneticists comparing sequencing data *in silico*.

"When you do experiments *in vivo*, you're never completely sure what's causing things," Marraffini concedes. "We cannot look inside a cell and see how things are working." To understand each component fully, you need to take them out of cells and put them into a test tube, where you control precisely what's included. This was Doudna's specialty, and it was what Blake Wiedenheft and Martin Jinek were pursuing in her lab. "Addressing these questions would require us to move beyond genetics research and take a more biochemical approach," she later wrote, "one that would allow us to isolate the component molecules and study their behavior."[11]

But first, Doudna stutter-stepped onto an odd career detour.

Herbert Boyer and Robert A. Swanson

CHAPTER 13

Genentech

Restless

In the fall of 2008, just after this spate of CRISPR papers had been published, Jillian Banfield told Doudna she was worried that the most important discoveries had already been made and that perhaps it was time to "move on." Doudna demurred. "I looked at what had been discovered as being the beginning, not the end, of an exciting journey," she recalls. "I knew there was some kind of adaptive immunity going on and wanted to know how it worked."[1]

Yet at that moment, Doudna was personally planning to move on.

She was forty-four, happily married, with a smart and polite seven-year-old son. Yet despite all of her success, or maybe partly because of it, she was having a mild midlife crisis. "I'd been running an academic research lab for fifteen years, and I started to wonder, 'Is there more?'" she recalls. "I wondered if my work was having an impact in the broader sense."

Despite the excitement of being in the forefront of the emerging field of CRISPR, she was becoming restless with basic science. She was eager to do more applied science and translational research, which aims at turning fundamental scientific knowledge into therapies that enhance human health. Even though there were hints that CRISPR could become a gene-editing tool, which would have great practical value, Doudna was feeling the tug to pursue projects that would have a more immediate impact.

At first she considered going to medical school. "I thought I might like to work with actual patients and be involved in clinical trials," she says. She also considered going to business school. Columbia had an executive MBA program that allowed participants to go to class one weekend a month and do the rest of the work online. The travel to and from Berkeley, and also Hawaii, where her mother was ailing, would be grueling, but she seriously considered it.

Then she ran into a former academic colleague who had joined the San Francisco biotech powerhouse Genentech the year before. The company was a poster child for the innovation and profits that can result when basic science meets patent lawyers meet venture capitalists.

Genentech, Inc.

Genentech was spawned in 1972, when Stanford medical professor Stanley Cohen and biochemist Herbert Boyer of the University of California, San Francisco, attended a conference in Honolulu that dealt with recombinant DNA technology, which was Stanford biochemist Paul Berg's discovery of how to splice pieces of DNA from different organisms to create hybrids. At the conference, Boyer gave a talk about his own discovery of an enzyme that could create these hybrids very efficiently. Cohen then spoke about how to clone thousands of identical copies of a piece of DNA by introducing it into *E. coli* bacteria.

Bored and still a bit hungry after their conference dinner one night, they walked to a New York–style deli, with a neon sign reading "Shalom" rather than the usual "Aloha," in a strip mall near Waikiki Beach. Over pastrami sandwiches, they brainstormed how to combine their discoveries to create a method for engineering and manufacturing new genes. They agreed to work together on the idea, and within four months they had spliced together DNA fragments from different organisms and cloned millions of them, giving birth to the field of biotechnology and launching the genetic engineering revolution.[2]

One of Stanford's alert intellectual property lawyers approached them and, to their surprise, offered to help them file a patent application. In 1974 they did, and it was eventually approved. It had not fully occurred to them that one could patent recombinant DNA processes, which are found in nature. It didn't occur to other scientists either, and many were furious—especially Paul Berg, who had made the original breakthroughs on recombinant DNA. He called the claims "dubious, presumptuous, and hubristic."[3]

In late 1975, a year after the Cohen-Boyer patent application was filed, a struggling young wannabe venture capitalist named Robert Swanson started making unsolicited phone calls to scientists who might be interested in starting a genetic engineering company. Swanson had an unbroken record of failure as a venture capitalist. At the time, he was living in a shared apartment, driving a beat-up Datsun, and surviving on cold-cut sandwiches. But he had read up on recombinant DNA and convinced himself that he had finally found a winning horse. As he went down his list of scientists alphabetically, the first one who agreed to meet him was Boyer. (Berg declined.) Swanson went to his office for what was supposed to be a ten-minute meeting, but he and Boyer ended up spending three hours at a neighborhood bar, where they planned a new type of company that would make medicines out of engineered genes. Each agreed to put in $500 to cover the initial legal fees.[4]

Swanson suggested that they call the company HerBob, a recombination of their first names that sounded like an online dating service or down-market beauty parlor. Boyer wisely rejected that and suggested instead that they call it Genentech, a mash-up of "genetic engineering technology." It began making genetically engineered drugs and, in August 1978, blasted into hypergrowth when it won a bet-the-company race to make a synthetic version of insulin to treat diabetes.

Until then, one pound of insulin required eight thousand pounds of pancreas glands ripped from more than twenty-three thousand pigs or cows. Genentech's success with insulin not only changed the lives of diabetics (and a lot of pigs and cows); it lifted the entire biotechnology industry into orbit. A portrait painting of a smiling Boyer appeared on the cover of *Time* with the headline "The Boom in Genetic Engineering." It came out the same week that Prince Charles of England picked Diana to be his princess, an event that, in those more rarefied times for journalism, received only a secondary mention on the magazine's cover.

Genentech's success led to a memorable front page of the *San Francisco Examiner* in October 1980, when the company became the first biotech company to launch an IPO and become publicly traded. Its stock, trading under the symbol GENE, opened at $35 a share and within an hour was selling at $88. "Genentech Jolts Wall Street," the banner front-page headline blared. Right below it was a picture for a totally separate story: a smiling Paul Berg on the telephone learning the news that he had, on that same day, won the Nobel Prize for his discovery of recombinant DNA.[5]

Detour

By the time Genentech began recruiting Doudna in late 2008, the company was worth close to $100 billion. Her former colleague, who was now working on genetically engineering cancer drugs at Genentech, told her that he was loving his new role. His research was much more focused than when he was an academic, and he was working directly on problems that were going to lead to new therapeutics. "So that got me thinking," Doudna says. "Rather than go back to school, maybe I should just go to a place where I could apply my knowledge."

Her first step was to present a couple of seminars at Genentech describing her work. It was a way for her and the Genentech team to sniff each other out. Among those wooing her was Sue Desmond-Hellmann, chief of product development. They had similar personalities, both eager listeners with quick minds and ready smiles. "When I was being recruited there, she and I sat down in her office and [she] told me she would be my mentor if I came to Genentech," Doudna says.

When Doudna decided to accept the job, she was told she could bring some members of her Berkeley team with her. "We were all preparing for the move," recalls Rachel Haurwitz, one of Doudna's doctoral students who, like most of the others, decided to follow her. "We were figuring out what equipment we were going to take and had begun packing it all up."[6]

But as soon as Doudna began working at Genentech, in January 2009, she realized that she had made a mistake. "I felt very quickly in my gut that I was in the wrong place," she says. "It was a visceral response. Every day and night, I felt I had made the wrong decision." She didn't sleep much. She was upset at home. She had trouble carrying out the most basic functions. Her midlife identity crisis was segueing into a mild mental breakdown. She had always been a very measured person, keeping her insecurities and occasional anxieties under wraps and under control. Until now.[7]

Her turmoil climaxed after only a few weeks. On a rainy night in late January, she found herself lying awake in bed. She got up and went outside in her pajamas. "I sat out in the rain in my backyard, getting soaked, and I thought, 'I'm done,'" she recalls. Her husband found her sitting motionless in the rain and coaxed her back inside. She wondered if she was clinically depressed. She knew she wanted to go back to her research lab at Berkeley, but she feared that door had closed.

Her neighbor Michael Marletta, who was chair of the chemistry department at Berkeley, came to her rescue. She called him the next morning and asked him to come over, which he did. She sent Jamie and their son, Andrew, away so she could have an emotional conversation privately. Marletta was immediately struck by how deeply unhappy she looked, and told her so. "I bet you want to come back to Berkeley," he said.

"I think I may have slammed that door," she replied.

"No, you haven't," he reassured her. "I can help you come back."

Instantly, her mood lifted. That night she could sleep again. "I knew that I was going back to where I was meant to be," she says. She returned to her Berkeley lab at the beginning of March, after only two months away.

From this misstep, she became more aware of her passions and skills—and also her weaknesses. She liked being a research scientist in a lab. She was good at brainstorming with people she trusted. She was not good at navigating a corporate environment where the competition was for power and promotions rather than discoveries. "I didn't have the right skill set or passions to work at a big company." But even though her brief stint at Genentech didn't work out, her desire to tie her research to the creation of practical new tools and companies that could commercialize them would drive the next chapter of her life.

CHAPTER 14

The Lab

Recruiting

There are two components to scientific discovery: doing great research and building a lab that does great research. I once asked Steve Jobs what his best product was, thinking he would say the Macintosh or iPhone. Instead he said that creating great products is important, but what's even more important is creating a team that can continually make such products.

Doudna deeply enjoyed being a bench scientist, a researcher who gets to the lab early, puts on latex gloves and a white coat, and begins working with pipettes and Petri dishes. For the first few years after setting up her lab at Berkeley, she was able to work at the bench half her time. "I didn't want to give that up," she says. "I think I was a pretty good experimenter. That's how my mind works. I can see experiments in my mind, especially when I am working myself." But by 2009, after her return from Genentech, Doudna realized that she had to spend more time cultivating her lab rather than her bacterial cultures.

This transition from player to coach happens in many fields. Writers become editors, engineers become managers. When bench scientists become lab heads their new managerial duties include hiring the

Martin Jinek, Rachel Haurwitz, Blake Wiedenheft,
Kaihong Zhou, and Jennifer Doudna

right young researchers, mentoring them, going over their results, suggesting new experiments, and offering up the insights that come from having been there.

Doudna excelled at these tasks. When considering candidates to be doctoral students or postdoctoral researchers in her lab, she made sure that her other team members believed they would fit in. The goal was to find people who were self-directed yet collegial. As her work on CRISPR ramped up, she found two PhD students with the right mix of eagerness and smarts to become core members of her team alongside Blake Wiedenheft and Martin Jinek.

Rachel Haurwitz

As a young girl growing up in Austin, Texas, Rachel Haurwitz was, in her words, "a science nerd." Like Doudna, she became interested in RNA. She made the molecule a focus of her studies as an undergraduate at Harvard, and then she went to Berkeley to pursue a doctorate. Not surprisingly, she was eager to work in Doudna's lab. She joined in 2008 and was soon swept into the CRISPR orbit of Blake Wiedenheft, attracted by his magnetic personality and joyful enthusiasm for odd bacteria. "When I started working with Blake, I had barely heard of CRISPR, so I read all of the papers that had been published in the field," she recalls. "It took me only about two hours. Neither Blake nor I sensed the tiny tip of the iceberg we were standing on."[1]

Haurwitz was at home studying for her PhD qualifying exam in early 2009 when she heard the news that Doudna had decided to cut short her move to Genentech and return to Berkeley. That was fortunate. Haurwitz had been planning to follow her, but she really wanted to stay at Berkeley and do her dissertation on CRISPR working with Wiedenheft. They shared a love both of biochemistry and of the outdoors; Wiedenheft even helped her develop a new training and eating regimen that got her back into running marathons.

Doudna recognized in Haurwitz something of herself: CRISPR was a risky field because it was so new, and that's what made Haurwitz want to jump in. "She loved the fact that it was a novel field, even though some students would be afraid of that," Doudna says. "So I told her, 'Go for it.'"

After Wiedenheft had worked out the structure of Cas1, he decided to do the same for the five other CRISPR-associated proteins that were in the bacteria he was working on. Four of them were easy. But Cas6* was tough to crack, so he enlisted Haurwitz. "He gave me the problem child," she says.

The source of the difficulty turned out to be that the sequencing of the bacteria's genome had been annotated incorrectly in textbooks and databases. "Blake realized that the reason we were having so much trouble was that they got the start wrong," Haurwitz explains. Once they figured out the problem, they were able to make Cas6 in the lab.[2]

The next step was to figure out what it did and how. "I used the two things the Doudna Lab does," explains Haurwitz: "biochemistry to figure out what its function is, and structural biology to figure out what it looks like." The biochemistry experiments revealed that the role of Cas6 is latching on to the long RNAs made by the CRISPR array and slicing them into the shorter CRISPR RNA snippets, which precisely target the DNA of attacking viruses.

The step after that was deciphering the structure of Cas6, which would explain *how* it operates. "At that point neither Blake nor I had the full set of skills to do structural biology by ourselves," Haurwitz says. "So I tapped on the shoulder of Martin Jinek, sitting at the next bench over, and I asked if he would join the project and help show us how to do this."

They found something unusual. Cas6 binds to RNA in a way that textbooks say should not work: it can find just the right sequence in the RNA that has a structural place for it to bind. "None of the other Cas proteins we had seen could do that," she says. The result was that Cas6 would recognize and cut a very precise place and not mess up other RNA.

In their paper, they called it "an unexpected recognition mechanism."

*At the time, it was generally known as Csy4. It eventually became known as Cas6f.

There was an "RNA hairpin" where the Cas6 could interact with just the right sequence. Once again, the twists and folds of a molecule's shape were the key in discovering how it worked.[3]

Sam Sternberg

In early 2008, Sam Sternberg was accepted into many top PhD programs, including Harvard's and MIT's. He decided to go to Berkeley because he had met Doudna and wanted to work with her on RNA structures. But he ended up deferring his enrollment so that he could finish a scientific paper on the work he had been doing as an undergraduate at Columbia.[4]

During that delay, he was surprised to hear about Doudna's abrupt move to Genentech and even more abrupt rebound. Worried about whether he had made the right choice, he sent her an email asking how committed she was to Berkeley. "I didn't trust myself to ask her in person, because I was too nervous," he admits. Doudna sent back a reassuring reply that she was now sure Berkeley was the right place for her. "It was convincing enough that I decided to go through with my plans to study there."[5]

Haurwitz invited Sternberg to a Passover Seder at the apartment she shared with her boyfriend. Unlike at most other Seders, a main topic of conversation was CRISPR. "I kept asking her to tell me more about the experiments she was doing," he says. She showed him a paper that she was writing about Cas enzymes, and he was hooked. "After that, I made it clear to Jennifer that I didn't want to keep working on RNA interference," he says. "I told her I wanted to work instead on this new CRISPR thing."

After Sternberg heard a talk by Columbia professor Eric Greene about single-molecule fluorescence microscopy, he asked Doudna, very tentatively, if he could try applying that method to one of the CRISPR-Cas proteins. "Oh my gosh, yes," she replied. "Absolutely do that." It was the type of risky approach she liked. Her scientific success had always come from connecting small dots to make big pictures, and she worried that Sternberg was tackling only small CRISPR topics.

After praising him for being bright and talented, she was blunt: "Right now you are punching below your weight. You're not taking the kinds of projects that a student like you is capable of. Why else do we do science? We do it to go after big questions and take on risks. If you don't try things, you're never going to have a breakthrough."[6]

Sternberg was convinced. He had asked whether he could go to Columbia for a week to learn more about the technique. "She not only sent me out there for a week to try it out, she ended up paying for me to spend six whole months there," Sternberg later wrote in the acknowledgments of his PhD dissertation. During his six months back at his alma mater, Sternberg figured out how to use the single-molecule fluorescence method to test the behavior of the CRISPR-associated enzymes.[7] The work resulted in two breakthrough papers—coauthored by Sternberg, Columbia's Eric Greene, Jinek, Wiedenheft, and Doudna—that showed for the first time precisely how the CRISPR system's RNA-guided proteins find the right target sequences of an invading virus.[8]

Sternberg grew especially friendly with Wiedenheft, who became a role model. They got a chance to spend an intense week working together in late 2011, when Wiedenheft was writing a review article on CRISPR for *Nature*.[9] They spent days together sitting side by side at a computer arguing over wording and selecting the illustrations to publish. They bonded more closely when they roomed together at a conference in Vancouver. "That was when my own scientific career began to take off," Sternberg says, "because I began thinking about how I could do something bigger that would bring in Blake."[10]

Sternberg and Wiedenheft and Haurwitz sat in a bay of the lab within a few feet of one another. It became a foxhole for biogeeks. When a big experiment was underway, they would have bets on the outcome. "What are we betting?" Blake would ask, and then himself answer, "We're betting a milkshake." The problem was that the Berkeley area had become too hip, or not yet hip enough, to have milkshake shops that were convenient. Still, they would use the milkshake tally to keep score.

The camaraderie in the lab was not an accident: in hiring, Doudna placed as much emphasis on making sure someone was a good fit as she did assessing their research accomplishments. As we walked through her lab one day, I challenged Doudna about this practice. Might it weed out some brilliant misfits, people who will challenge others or disrupt the group thinking, but in a beneficial way? "I've thought about that a little," she says. "I know some people like creative conflict. But I like having in the lab people who work well together."

Leadership

When Ross Wilson, a newly minted PhD from Ohio State, applied to be a postdoctoral fellow in Doudna's lab, Jinek pulled him aside to give him a word of warning. "You have to be self-sufficient," he told Wilson. "If you're not self-motivated, Jennifer is not going to help you much or do the work for you. At times she will seem disengaged. But if you're a self-starter, she will give you the chance to take risks, provide really smart guidance, and be there when you need her."[11]

Doudna's was the only lab where Wilson interviewed in 2010. He was interested in how RNA interacts with enzymes, and he considered her the world's leading expert. When she accepted him, he cried for joy. "I actually did," he says. "It's the only time in my life I've ever done that."

Jinek's cautionary note, he says, was "one hundred percent accurate," but that made her lab an exciting place to work for a self-driven person. "She definitely doesn't hover over you," says Wilson, who now runs his own Berkeley lab aligned with Doudna's, "but when she goes over your experiments and results with you, there are times when she will lower her voice a bit, look you right in the eye, lean in, and say, 'What if you tried . . . ?'" Then she would describe a new approach, a new experiment, or even a big new idea, usually involving some new way of deploying RNA.

One day, for example, Wilson came to her office to show some results about how two molecules that he had crystallized interacted. "If you can disrupt this interaction based on knowing how it works," she said, "maybe we can make that same disruption inside the cell and see how it changes the behavior of the cell." It pushed Wilson to move beyond the test tube and delve into the inner workings of a living cell. "I would never have thought of doing that," he says, "but it worked."

Most mornings when she is in her lab, Doudna schedules a steady stream of her researchers to come present their most recent results. Her questions tend to be Socratic: Have you thought about adding RNA? Can we image that in living cells? "She has a knack for asking the right critical big questions when you're developing your project," says Jinek. They were designed to get her researchers to look up from the details and see the big picture. Why are you doing this? she will ask. What's the point?

Although she takes a hands-off approach during the early stages of a researcher's project, as it gets close to fruition she engages intensely. "Once something exciting emerges or a real discovery is in the works, she senses when it's a big deal and she gets super involved," says Lucas Harrington, one of her former students. "It comes in pulses." That is when Doudna's competitive juices kick in. She doesn't want another lab to beat hers to a discovery. "She might storm into the lab unexpectedly," Harrington says, "and without raising her voice make it clear what things need to be done and be done quickly."

When her lab produced a new discovery, Doudna was tenacious about getting it published. "I've discovered that the journal editors favor people who are aggressive or pushy," she says. "That's not necessarily my nature, but I have become more aggressive when I feel that journal editors are not appreciating that something we did is really important."

Women in science tend to be shy about promoting themselves, and that has serious costs. A study in 2019 of more than six million articles with women as the principal author showed that they are less likely to use self-promotional terms, such as "novel" and "unique" and "unprecedented," to describe their findings. The trend is especially true for articles in the most prestigious journals, which almost by definition feature research that is groundbreaking. In the highest-impact journals that publish the most important cutting-edge research, women are 21 percent less likely to use positive and self-promotional words in describing their work. Partly as a result, their papers are cited approximately 10 percent less frequently.[12]

Doudna does not fall into that trap. At one point in 2011, for example, she and Wiedenheft, along with her Berkeley colleague Eva Nogales, completed a paper on the array of Cas enzymes dubbed CASCADE. It could home in on an exact spot of DNA in the invading virus and then recruit an enzyme to buzz-saw it into hundreds of pieces. They sent it to one of the most prestigious journals, *Nature*, which accepted it. But the editors said it was not an important enough breakthrough to be a featured "article" in the journal, so they wanted to publish it as a "report," which is a notch down in significance. Most of the team was thrilled to have the paper quickly accepted by such an important publication. But Doudna was upset. She argued strongly that it was a big advance and deserved prominent treatment—writing a letter and soliciting supporting letters—but the editors stuck firm. "Most people are jumping up and down if they get a yes from *Nature*," Wiedenheft says. "Jennifer was jumping up and down because she was mad that it was going to be a report, not an article."[13]

Rachel Haurwitz

CHAPTER 15

Caribou

Bench to bedside

Even though she decided against becoming part of the corporate-science world at Genentech, Doudna retained her desire to translate the basic discoveries about CRISPR into tools that could be useful in medicine. Her opportunity came after Wiedenheft and Haurwitz succeeded in discovering the structure of Cas6.

It was the beginning of a new aspect of her career: looking for ways to turn her CRISPR discoveries into tools that could be useful in medicine. Haurwitz took the idea one more step. If Cas6 could be turned into a medical tool, it could become the basis for a company. "Once we understood how the Cas6 protein worked," she says, "we started to get some ideas on how we might steal it from bacteria and repurpose it for our own uses."[1]

For much of the twentieth century, most new drugs were based on chemical advances. But the launch of Genentech in 1976 shifted the focus of commercialization from chemistry to biotechnology, which involves the manipulation of living cells, often through genetic engineering, to devise new medical treatments. Genentech became the model for commercializing biotech discoveries: scientists and venture capitalists raised capital by divvying up equity stakes, then they entered into agreements with major pharmaceutical companies to license, manufacture, and market some of their discoveries.

Thus did biotech follow the path of digital technology in blurring the lines between academic research and business. This fusion in the digital realm began right after World War II, mainly around Stanford. With prodding from its provost Frederick Terman, Stanford professors were encouraged to turn their discoveries into startups. The companies that sprang out of Stanford included Litton Industries, Varian Associates, and Hewlett-Packard, followed by Sun Microsystems and Google. The process helped turn a valley of apricot orchards into Silicon Valley.

During this period, many other universities, including Harvard and Berkeley, decided it was more appropriate to stick to basic scientific research. Their traditional professors and provosts disdained commercial entanglement. But after envying Stanford's success in the realms of infotech and then biotech, they began to embrace entrepreneurship. Researchers were encouraged to patent their discoveries, partner with venture capitalists, and create businesses. "These companies frequently maintain their links with the universities, working closely with faculty members and postdoctoral candidates on research projects, and sometimes using the university laboratories," Harvard Business School professor Gary Pisano wrote. "In many instances, the founding scientists even retain their faculty posts."[2] This would become Doudna's approach.

Startup

Until then, Doudna had never thought much about commercialization. Money was not then, nor would it be later, a primary motivation in her life. She and Jamie and Andy lived in a spacious but not lavish house in Berkeley, and she never had the desire for a grander one. But she did like the idea of being part of a business, especially one that could have a direct impact on people's health. And unlike Genentech, a startup would have no corporate politics, nor would it drag her away from academia.

Haurwitz likewise felt the allure of business. Although she was good at the lab bench, she realized that she was not cut out to be an academic researcher. So she began taking courses at Berkeley's Haas School of Business. Her favorite was one taught by the venture capitalist Larry Lasky. He split his class into teams of six, half business students and the other half science researchers. Each team built a series of decks for a fictional biotech startup and then spent the semester perfecting how they would pitch it to investors. She also took a class from Jessica Hoover, who had been head of business development at a biotech firm that studied ways to commercialize medical products, including how to secure and license patents.

During Haurwitz's final year in her lab, Doudna asked what she wanted to do next. "Run a biotech company," Haurwitz replied. That would have been an unsurprising response at Stanford, where commercializing research was celebrated, but it was the first time Doudna had heard such an answer at Berkeley, where most PhD students aimed for an academic career.

A few days later, she went into the lab to find Haurwitz. "I've been thinking that maybe we ought to start a company around using Cas6 and some of the other CRISPR enzymes as a tool," she said. With no hesitation, Haurwitz responded, "Of course we should."[3]

So they did. The company was founded in October 2011, and it remained based in Doudna's academic lab for a year while Haurwitz finished her studies. After she got her PhD in the spring of 2012, she became the president and Doudna the chief scientific advisor of the fledgling endeavor.

The idea was that the company, which moved into a low-slung space in a nearby strip mall, would commercialize the patents related to the Cas6 structure and eventually other discoveries to come out of Doudna's lab. Their initial aim was to turn Cas6 into a diagnostic tool that clinics could use to detect the presence of viruses in humans.

The company

By the time Doudna and Haurwitz started their company in 2011, Berkeley had become savvier about encouraging its researchers to be more entrepreneurial. It launched a variety of programs to nurture startups formed by its students and professors. One of them, which was formed in 2000 in partnership with the other University of California campuses in the Bay Area, was the California Institute for Quantitative Biosciences (QB3), which had as its goal "a catalytic partnership between university research and private industry." Doudna and Haurwitz were selected to become participants in QB3's Startup in a Box program, which gave training, legal advice, and banking services to scientist-entrepreneurs who wanted to turn their basic discoveries into commercial ventures.

One day Doudna and Haurwitz took the subway into San Francisco to meet with the lawyer who Startup in a Box enlisted to help them incorporate their new company. When he asked for its name Haurwitz said, "I've been talking to my boyfriend about it, and we think we should call it Caribou." The name is a cut-and-splice mash-up of "Cas" and "ribonucleotides," which are the building blocks of RNA and DNA.

Haurwitz had talents not often found in Silicon Valley entrepreneurs. With her steady personality, she was a naturally good manager. She was down to earth, unflappable, practical, and straightforward. There was no whiff of the combination of ego and insecurity exuded by many startup CEOs. She did not exaggerate or overpromise. That offered many advantages, one of which was that people tended to underestimate her.

On the other hand, she had never been a CEO, so she had some learning to do. That led her to join a local professional development group for young CEOs, the Alliance of Chief Executives, which met for a half-day each month to share problems and solutions. It's hard to imagine Steve Jobs or Mark Zuckerberg joining such a support group, but Haurwitz, like her mentor Doudna, had a self-awareness and humility not usually found among alpha males. Among other things, her Alliance group coached her on how to create a team with different types of expertise.

Today the mere appearance of the word CRISPR in a prospectus is enough to cause venture capitalists to go into heat. But when Doudna and Haurwitz tried to raise money, they had little luck. "At that time, the topic of molecular diagnostics was a turnoff to venture capitalists," Doudna says. "I also feel that there is an anti-female undercurrent, and I was worried that if we took venture money, that Rachel might be pushed out as CEO." None of the venture capitalists they met with was a woman, and this was in 2012. So instead of continuing to seek venture money, they decided to raise what they could from friends and family. Both Doudna and Haurwitz put in their own money.

The triangle

Its bootstrap success may, on the surface, make Caribou Biosciences seem like a poster child for pure free-market capitalism. And there was, nicely, an element of that. But it's important to look deeper and see how, as in so many other companies, from Intel to Google, innovation has been a product of a distinctively American mix of catalysts.

As World War II was ending, the great engineer and public official Vannevar Bush argued that America's innovation engine would require a three-way partnership of government, business, and academia. He was uniquely qualified to envision that triangle, because he had a foot in all three camps. He had been dean of engineering at MIT, a founder of Raytheon, and the chief government science administrator overseeing, among other projects, the building of the atom bomb.[4]

Bush's recommendation was that government should not build big research labs of its own, as it had done with the atomic bomb project, but instead should fund research at universities and corporate labs. This government-business-university partnership produced the great innovations that propelled the U.S. economy in the postwar period, including transistors, microchips, computers, graphical user interfaces, GPS, lasers, the internet, and search engines.

Caribou was an example of this approach. Berkeley, a public university with private philanthropic supporters, housed Doudna's lab and had a partnership with the federally funded Lawrence Berkeley National Laboratory. The amount of federal grants that went from the National Institutes of Health (NIH) to Berkeley to support Doudna's research into CRISPR-Cas systems was $1.3 million.[5] In addition, Caribou itself was able to get a federal grant from the NIH's small business innovation program, which provided $159,000 to the company to create kits to analyze RNA-protein complexes. The program was designed to help innovators turn basic research into commercial products. It kept Caribou alive during the early years, when venture funding was not forthcoming.[6]

There is one other element that is now often added to the academic-government-business triad: philanthropic foundations. In the case of Caribou, that came as a grant from the Bill and Melinda Gates Foundation, which provided $100,000 to fund work on using Cas6 as a tool to diagnose viral infections. "We plan on creating a suite of enzymes that specifically recognize RNA sequences characteristic of viruses including HIV, hepatitis C and influenza," Doudna wrote in her proposal to the foundation. It was a prelude to the funding Doudna would receive from Gates in 2020 to use CRISPR systems to detect coronaviruses.[7]

CHAPTER 16

Emmanuelle Charpentier

The wanderer

Conferences can have consequences. While attending one in Puerto Rico in the spring of 2011, Doudna had a chance meeting with Emmanuelle Charpentier, an itinerant French biologist who had an alluring mix of mystery and Parisian insouciance. She, too, had been studying CRISPR, and she had homed in on the CRISPR-associated enzyme known as Cas9.

Guarded but engaging, Charpentier was a woman of many cities, many labs, many degrees and postdoc programs, but few roots and commitments, ever willing to pack up her pipettes and move, never showing any outward signs of worry or an instinct for competition. This made her much different from Doudna, which is perhaps why they bonded at first, although mainly in a scientific rather than emotional way. They both had warm smiles that made their protective shells almost, but not totally, invisible.

Charpentier grew up in a leafy suburb on the Seine south of Paris. Her father was in charge of the neighborhood park system, and her mother was the administrative nurse in a psychiatric hospital. One day when Charpentier was twelve, she walked past the Pasteur Institute, the Paris research center specializing in infectious diseases. "I

Emmanuelle Charpentier

am going to work there when I grow up," she told her mother. A few years later, when she had to designate a field for her *baccalauréat* exam, which determines a student's course of study in college, she chose life science.[1]

She also was interested in the arts. She took piano lessons from a neighbor who was a concert musician and pursued ballet with the possibility that she might become a professional dancer, continuing her training well into her twenties. "I would like to have been a ballet dancer, but I finally realized that would be too risky as a career," she says. "I was a few centimeters too short and I had a ligament problem that affected the extension of my right leg."[2]

There were lessons from the arts, she would discover, that applied to science. "Methodology is important in both," she says. "You also must know the basics and master the methods. That requires persistence—repeating experiments and repeating them again, perfecting how to prepare the DNA when you clone a gene, and then doing it over and over again. It's part of the training, just like the hard work of a ballet dancer, repeating all day long the same moves and methods." Also like the arts, once a scientist masters the basic routine, she has to combine it with creativity. "You have to be rigorous and disciplined," Charpentier explains, "but also know when to let yourself loose and blend in a creative approach. I found in biological research the right combination of persistence and creativity."

Fulfilling the prediction she made to her mother, she pursued her graduate studies at the Pasteur Institute, where she learned how bacteria can become resistant to antibiotics. She felt at home in the lab. It was a quiet temple for individual persistence and contemplation. She could be creative and independent as she pursued a path toward her own discoveries. "I began to see myself as a scientist and not just as a student," she says. "I wanted to create knowledge, not just learn it."

Charpentier became a postdoctoral pilgrim, enrolling at Rockefeller University in Manhattan in the lab of the microbiologist Elaine Tuomanen, who was studying how the bacteria that cause pneumonia have DNA sequences that can shift, making the bacteria resistant to antibiotics. On the day she arrived, Charpentier found out that Tuomanen was moving, along with her lab and its postdocs, to the St. Jude Children's Research Hospital in Memphis. There Charpentier worked with Rodger Novak, another postdoc in Tuomanen's lab, and he became for a while a romantic companion and then a business partner. While in Memphis, they coauthored with Tuomanen an important study that showed how antibiotics such as penicillin trigger suicidal enzymes in bacteria that dissolve their cell walls.[3]

Charpentier's peripatetic mind and spirit made her ever ready to move to new towns and new topics, and this was hastened by an unpleasant biological discovery she made in Memphis: Mississippi River mosquitoes love French blood. In addition, she wanted to shift her focus from single-cell microbes such as bacteria and learn about genes in mammals, mainly mice. So she switched to a lab at New York University, where she produced a paper on ways to manipulate mouse genes to regulate hair growth. She also did a third postdoc in which she, along with Novak, focused on the role of small RNA molecules in regulating gene expression in *Streptococcus pyogenes*, a bacteria that causes skin infections and strep throat.[4]

After six years in the U.S., she moved back to Europe in 2002 to become the head of a microbiology and genetics lab at the University of Vienna. But once again she became restless. "People in Vienna knew each other a bit too well," she says, which she clearly regarded as a drawback rather than a benefit. "The dynamics got a bit stuck and the structures became inhibiting." So by the time she met Doudna in 2011, she had left behind most of the researchers in her lab to relocate on her own to Umeå, in northern Sweden. Umeå was no Vienna. Four hundred miles north of Stockholm, the town's 1960s-built university consisted of a cluster of modernist buildings on land that had been a grazing ground for reindeer herders. It was best known for its research on trees. "Yes, it was a risky move," Charpentier agrees, "but it gave me a chance to think."

In the years since she entered the Pasteur Institute in 1992, Charpentier had worked in ten institutions in seven cities in five countries.

Her nomadic life reflected the fact, and reinforced the fact, that she resisted bonds. With no spouse or family, she sought out changing environments and adapted to them without any inhibiting personal ties. "I enjoy the freedom of being on my own, of not depending on partnership," she says. She hated the phrase "work-life balance" because it implied that work competes with life. Her work in the lab and her "passion for science," she says, brought her a "happiness that is as fulfilling as any other passion."

Like the organisms she studied, her need to adapt to new environments kept her innovative. "My instinct to keep moving can be destabilizing, but that can be good," she says. "It assures that you never get stuck." Going from one place to another was her way of repeatedly reconsidering her research and forcing herself to start fresh. "The more one moves, the more one learns to analyze as a new situation and see things that others who have been in the system a long time have not identified."

Moving also made her feel like a bit of a foreigner most of the time, the way the young Jennifer Doudna felt as a child in Hawaii. "It's important to know how to be an outsider," Charpentier says. "You're never completely at home, and that can drive you. It can challenge you not to seek being comfortable." As with so many other observant and creative people, she found that a sense of detachment or slight alienation made her better at figuring out the forces at play. That helped her honor the maxim often preached by Louis Pasteur himself: *Be prepared for the unexpected.*

Partly as a result, Charpentier became one of those scientists who could be both focused and distracted. Though impeccably groomed and casually elegant even when riding a bicycle, she also fit the stereotype of an absent-minded professor. When I traveled to see her in Berlin, where she moved after Umeå, she got to my hotel on her bike a few minutes late. It turned out that she had come that morning from a visit to Munich, and when she was leaving the station she realized that she had left her luggage on the train. Somehow she caught up with the train at its terminal, retrieved her luggage, and then biked to my hotel. As we walked to her nearby lab at the Max Planck Institute for

Infectious Diseases, on the grounds of Charité, the venerable teaching hospital in the middle of Berlin, she pushed her bike purposely down a main artery until, after a few blocks, she realized that she had led us in the wrong direction. The next day, when a friend and I took her to see a show at an art museum, she managed to lose her admission ticket between the box office and the main entrance, and when we went to a serene Japanese restaurant for dinner, she left her phone behind. Yet when we were sitting in her lab office or over a multicourse sushi meal, she could speak for hours with super-intense focus.

tracrRNA

In 2009, the year that Charpentier was uprooting from Vienna and moving to Umeå, the CRISPR crowd had coalesced around Cas9 as being the most interesting of the CRISPR-associated enzymes. Researchers had shown that if you deactivated Cas9 in bacteria, the CRISPR system no longer cut up the invading viruses. They had also established the essential role of another part of the complex: CRISPR RNAs, known as crRNAs. These are the small snippets of RNA that contain some genetic coding from a virus that had attacked the bacteria in the past. This crRNA guides the Cas enzymes to attack that virus when it tries to invade again. These two elements are the core of the CRISPR system: a small snippet of RNA that acts as a guide and an enzyme that acts as scissors.

But there was one additional component of the CRISPR-Cas9 system that played an essential role—or, as it turned out, two roles. It was dubbed a "trans-activating CRISPR RNA," or tracrRNA, pronounced "tracer-RNA." Remember this tiny molecule; it will play an outsized role in our tale. That's because science is most often advanced not by great leaps of discovery but by small steps. And disputes in science are often about who made each one of these steps—and how important each really was. This would turn out to be the case for the discoveries involving tracrRNA.

It turns out that tracrRNA performs two important tasks. First, it facilitates the making of the crRNA, the sequence that carries the memory of a virus that previously attacked the bacteria. Then it serves as a handle to latch on to the invading virus so that the crRNA can target the right spot for the Cas9 enzyme to chop.

The process of uncovering these roles of tracrRNA began in 2010, when Charpentier noticed that the molecule kept appearing in her experiments with bacteria. She couldn't figure out its role, but she realized that it was located in the vicinity of the CRISPR spacers, so she speculated that they were connected. She was able to test this by deleting the tracrRNA in some bacteria. The result was that the crRNAs didn't get produced. Researchers had never quite pinned down how the crRNAs were made inside a bacterial cell. Now Charpentier had a hypothesis: it is this tracrRNA that directs the creation of the short crRNAs.

Charpentier was moving to Sweden at the time. When the researchers in her Vienna lab sent her an email saying they had shown that the absence of tracrRNA meant that crRNA wasn't produced, she spent the night drawing up a long plan of experiments for them to do next. "I became obsessed with this tracrRNA," she says. "I am stubborn. It was important for me to follow up. I said 'We have to go for it! I want someone to look at it.'"[5]

The problem was that there was nobody in her Vienna lab who had the time and inclination to pursue the tracrRNA. That's the drawback of being a wandering professor: you leave your students behind, and they move on to other things.

Charpentier considered doing the experiments herself, even though she was in the midst of a move. But she finally found a volunteer in her Vienna lab: a young student from Bulgaria, studying for a master's degree, named Elitza Deltcheva. "Elitza was very dynamic, and she believed in me," Charpentier says. "She understood what was happening, even though she was just a master's student." She even convinced one of the graduate students, Krzysztof Chylinski, to work with her.

Charpentier's little team discovered that the CRISPR-Cas9 system accomplished its viral-defense mission using only three components: tracrRNA, crRNA, and the Cas9 enzyme. The tracrRNA took long strands of RNA and processed them into the small crRNAs that were targeted at specific sequences in an attacking virus. They prepared a paper for *Nature*, which would be published in March 2011, in which Deltcheva got to be the lead author—and the graduate students who had declined to help were lost to history.[6]

A remaining mystery

Charpentier presented the findings at a CRISPR conference in October 2010 in the Netherlands. She was having trouble getting her paper through the editorial process at *Nature*, and it was risky to go public with work before it was published. But she thought that perhaps one of the paper's reviewers would be in the audience and would be convinced to speed up the process.

She was stressed during her presentation because she had not yet figured out what happened to the tracrRNA after it helped to create a crRNA. Was the work of the tracrRNA done by then? Or did the two little RNAs stick together when it came time to guide the Cas protein to cut up an invading virus? One member of the audience asked her directly, "Do the three elements stay together as a complex?" Charpentier tried to deflect the question. "I tried to laugh and to be very confusing on purpose," she says.

That issue—and what Charpentier knew about it—might seem arcane. But it led to a set of disputes that illuminates how CRISPR researchers—and Doudna in particular—can be very competitive about who deserves credit for each small advance. The fact that the tracrRNA did in fact stick around and play an important role in cleavage would later be among the discoveries published in the seminal 2012 paper that Charpentier would write with Doudna. But to Doudna's annoyance, Charpentier would sometimes imply, years later, that she already knew this fact in 2011.

When I press her, Charpentier admits that her 2011 *Nature* paper did not, in fact, describe the full role of the tracrRNA: "It seemed clear to me that the tracrRNA needed to continue to be associated with the crRNA, but there were some details we didn't fully understand, so we didn't put this in the paper." Instead, she made the decision to save writing about the full tracrRNA function until she could find a convincing way to prove it experimentally.

She had studied the CRISPR system in living cells. To get to the next step would require biochemists who could isolate each chemical component in a test tube and figure out precisely how each one works. That is why she wanted to meet Doudna, who was scheduled to speak at the March 2011 conference of the American Society for Microbiology in Puerto Rico. "I knew we were both going to attend," she says, "and I put in my mind that I would find a chance to talk to her."

Puerto Rico, March 2011

When Jennifer Doudna walked into the coffee shop of the hotel in Puerto Rico on the second afternoon of the conference, Emmanuelle Charpentier was at a table in the corner sitting by herself, as she often liked to do, looking far more elegant than the other patrons. Doudna was with her friend John van der Oost, the Dutch CRISPR researcher, who pointed Charpentier out and offered to introduce her. "That would be great," replied Doudna. "I've read her paper."[7]

Doudna found Charpentier to be charming: just a hint of shyness, or feigned shyness, along with an engaging sense of humor and very stylish aura. "I was instantly struck by her intensity but also her sly humor," Doudna says. "I immediately liked her." They chatted for a few minutes and then Charpentier suggested they get together for a more serious discussion. "I've been thinking of contacting you about a collaboration," she said.

The next day they had lunch, followed by a stroll along the cobblestone streets of old San Juan. When the discussion turned to Cas9, Charpentier became excited. "We have to figure out exactly how it works," she urged Doudna. "What's the exact mechanism it uses to cut DNA?"

Charpentier was taken by Doudna's seriousness and attention to detail. "I think it's going to be fun to work with you," she told her. Doudna was similarly moved by Charpentier's intensity. "Somehow, just the way she said that it would be fun to work with me made a chill run down my back," she recalls. The other enticement was that it was just the sort of detective tale that gave Doudna a sense of purpose: the hunt for the key to one of life's basic mysteries.

Right before Doudna left for Puerto Rico, she had a career-counseling conversation with Martin Jinek, the postdoc in her lab who had been working on the structures of Cas1 and Cas6. He was having doubts, which turned out to be unwarranted, about whether he would be successful as an academic researcher, and had thought about becoming an editor at a medical journal instead. But he decided against it. "I think I'm going to be in your lab about one more year," he told her. "What would you like me to work on?" He was especially interested in finding a CRISPR project of his own, he said.

So when Doudna heard Charpentier's pitch, she thought it would be a perfect project for Jinek. "I've got a wonderful biochemist who's also a structural biologist," she told Charpentier.[8] They agreed that they would connect Jinek with the postdoc in Charpentier's lab who had worked on her earlier Cas9 paper, Krzysztof Chylinski, a Polish-born molecular biologist who had stayed in Vienna when she moved to Umeå. Together this foursome would make one of the most important advances in modern science.

CHAPTER 17

CRISPR-Cas9

Success

When Doudna returned to Berkeley, she and Jinek began a series of Skype calls with Charpentier in Umeå and Chylinski in Vienna to plot a strategy for figuring out the mechanisms of CRISPR-Cas9. The collaboration was like a model United Nations: a Berkeley professor from Hawaii, her postdoc from the Czech Republic, a Parisian professor working in Sweden, and her Polish-born postdoc working in Vienna.

"It became a twenty-four-hour operation," Jinek recalls. "I would do an experiment at the end of my day, I would send an email to Vienna, and Krzysztof would read it as soon as he got up in the morning." Then there would be a Skype call, and they would decide what the next step should be. "Krzysztof would execute that experiment during the day and send me the results while I was asleep, so that when I woke up and opened my inbox there would be an update."[1]

At first, Charpentier and Doudna would join the Skype calls only once or twice a month. But the pace picked up in July 2011, when Charpentier and Chylinski flew to Berkeley for the fast-growing annual CRISPR conference. Even though they had bonded over Skype, it was the first time that Jinek had personally met Chylinski, a lanky

Emmanuelle Charpentier, Jennifer Doudna, Martin Jinek, and Krzysztof Chylinski at Berkeley in 2012

researcher with an affable personality and an eagerness to be involved in turning basic research into a tool.[2]

In-person meetings can produce ideas in ways that conference calls and Zoom meetings can't. That had happened in Puerto Rico, and it did so again when the four researchers got together for the first time in Berkeley. There they were able to brainstorm a strategy for figuring out exactly what molecules were necessary for a CRISPR system to cut DNA. Physical meetings are especially useful when a project is in an early phase. "There's nothing like sitting in a room with people and seeing their reactions to things and having a chance to bat around ideas face to face," Doudna says. "That's been a cornerstone to every collaboration that we've had, even those where we are conducting a lot of the work by electronic communication."

Jinek and Chylinski were initially unable to make CRISPR-Cas9 chop up the DNA of a virus in a test tube. They had been trying to make it work with just two components: the Cas9 enzyme and the crRNA. In theory the crRNA would guide the Cas9 enzyme to the virus target, which would then get chopped up. But it didn't work. Something was missing. "It was extremely puzzling to us," Jinek recalls.

This is when the tracrRNA reenters our tale. In her 2011 paper Charpentier showed that tracrRNA was required for producing the crRNA guide. She later said that she suspected it played an even larger, ongoing role, though that possibility had not been part of their initial round of experiments. When those experiments failed, Chylinski decided to throw tracrRNA into his test-tube mix.

It worked: the three-component complex reliably chomped up the target DNA. Jinek immediately told Doudna the news: "Without the tracrRNA, the crRNA guide does not bind to the Cas9 enzyme." After that breakthrough, Doudna and Charpentier became more involved in the daily work. Clearly they were heading to an important discovery: determining the essential components of a CRISPR gene-cutting system.

Night after night, Chylinski and Jinek would ping-pong results back and forth, each adding a tiny bit of the puzzle, with Charpentier and Doudna joining the increasingly frequent strategy calls. They were able to discover the precise mechanisms of each of the three essential components of the CRISPR-Cas9 complex. The crRNA contained a twenty-letter sequence that acted as a set of coordinates to guide the complex to a piece of DNA with a similar sequence. The tracrRNA, which had helped create this crRNA, now had the additional role of acting like a scaffold that held the other components in just the right place when they glommed on to the target DNA. Then the Cas9 enzyme began slicing away.

One evening, right after a key experiment had produced positive results, Doudna was at home cooking spaghetti. The swirls in the boiling water reminded her of the salmon sperm she had studied under a microscope back in high school when learning about DNA, and she started to laugh. Her son, Andy, who was nine, asked her why. "We found this protein, an enzyme called Cas9," she explained. "It can be programmed to find viruses and cut them up. It's so incredible." Andy kept asking how it worked. Over billions of years, she explained, bacteria evolved this totally weird and astonishing way to protect themselves against viruses. And it was adaptable; every time a new virus emerged, it learned how to recognize it and beat it back. He was fascinated. "It was a double joy," she recalled, "a moment of fundamental discovery of something that is so cool, and being able to share it with my son and explaining it in a manner where he can get it." Curiosity can be beautiful that way.[3]

A gene editing tool

This amazing little system, it quickly became clear, had a truly momentous potential application: the crRNA guide could be modified to target any DNA sequence you might wish to cut. It was programmable. It could become an editing tool.

The study of CRISPR would become a vivid example of the call-and-response duet between basic science and translational medicine. At the beginning it was driven by the pure curiosity of microbe-hunters

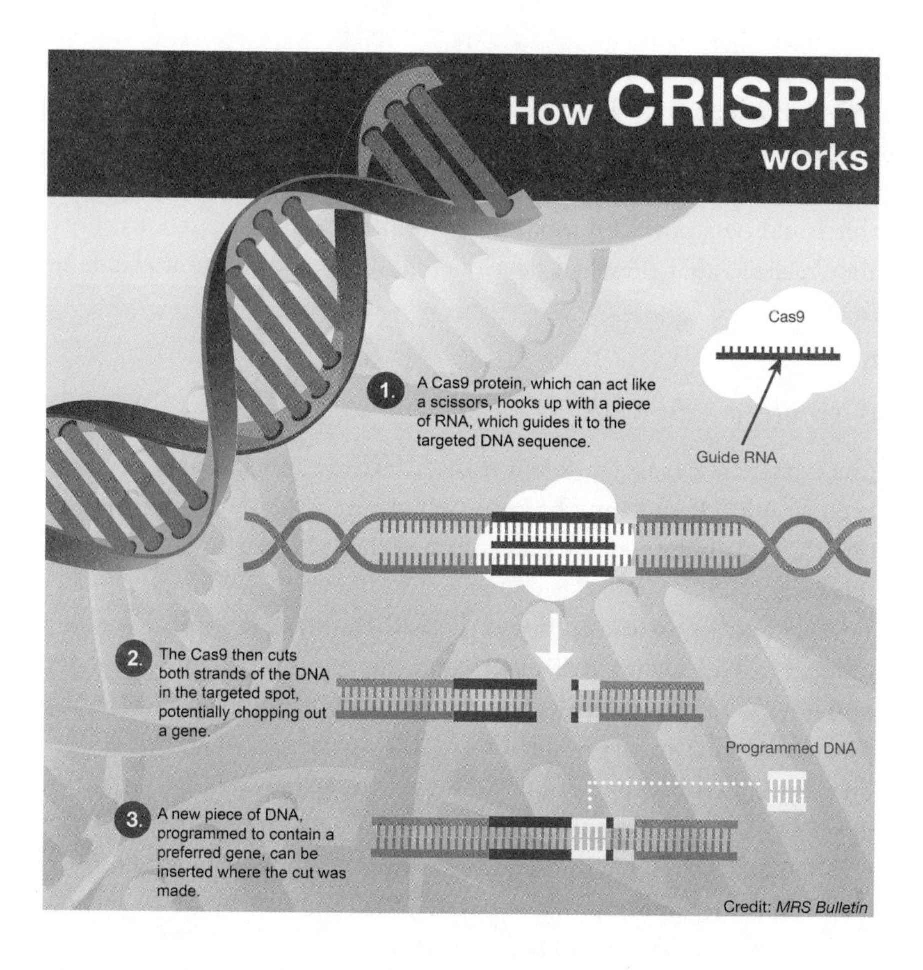

who wanted to explain an oddity they had stumbled upon when sequencing the DNA of offbeat bacteria. Then it was studied in an effort to protect the bacteria in yogurt cultures from attacking viruses. That led to a basic discovery about the fundamental workings of biology. Now a biochemical analysis was pointing the way to the invention of a tool with potential practical uses. "Once we figured out the components of the CRISPR-Cas9 assembly, we realized that we could program it on our own," Doudna says. "In other words, we could add a different crRNA and get it to cut any different DNA sequence we chose."

In the history of science, there are few real eureka moments, but this came pretty close. "It wasn't just some gradual process where it slowly dawned on us," Doudna says. "It was an oh-my-God moment." When Jinek showed Doudna his data demonstrating that you could program Cas9 with different guide RNAs to cut DNA wherever you desired, they actually paused and looked at each other. "Oh my God, this could be a powerful tool for gene editing," she declared. In short, they realized that they had developed a means to rewrite the code of life.[4]

A single-guide RNA

The next step was to figure out if the CRISPR system could be made even simpler. If so, it might become not just a gene-editing tool but one that would be much easier to program and cheaper than existing methods.

One day, Jinek walked down the hall from the lab into Doudna's office. He had been experimenting to determine the minimum requirements for the crRNA that served as a guide and the tracrRNA that clamped it to the target DNA. They were standing at a whiteboard propped in front of her desk, and he was sketching out a diagram of the structure of the two small RNAs. Which parts of the crRNA and tracrRNA, he asked, were essential for cutting up DNA in a test tube? "It appeared that the system had some flexibility as to how long the two RNAs had to be," he says. Each of the little RNAs could be truncated a bit and still function. Doudna had a profound understanding about the structure of RNA and an almost childlike joy in figuring out the ways it worked. As they brainstormed, it became clear to them that they could link the two RNAs together, fusing the tail of one to the head of the other in a way that would keep the combined molecule functional.

Their goal was to engineer a single RNA molecule that would have the guide information on one end and the binding handle on the other. That would create what they ended up calling a "single-guide RNA" (sgRNA). They paused for a moment and looked at each other, then Doudna said, "Wow." As she recalls, "It was one of those moments in science that just comes to you. I had this chill and these little hairs on my neck standing up. In that moment, the two of us realized that this curiosity-driven, fun project had this powerful implication that could change the direction of the project profoundly." It's a fitting scene to imagine: the behavior of a little molecule being able to get the little hairs on Doudna's neck to stand up.

Doudna urged Jinek to begin work right away on fusing these two RNA molecules to work as a single guide for Cas9, and he hastened back down the hall to place an order with a company for the necessary RNA molecules. He also discussed the idea with Chylinski, and they quickly designed a series of experiments. Once they had figured out what parts of the two RNAs could be deleted and how they could be connected, it took only three weeks to make a single-guide RNA that worked.

It was immediately obvious that this single guide would make CRISPR-Cas9 an even more versatile, easy-to-use, and reprogrammable tool for gene editing. What made the single-guide system particularly significant—from both a scientific and an intellectual property standpoint—was that it was an actual human-made invention, not merely a discovery of a natural phenomenon.

So far Doudna's collaboration with Charpentier had produced two significant advances. The first was the discovery that the tracrRNA played an essential role not just in creating the crRNA guide but, more important, holding it together with the Cas9 enzyme and binding it all to the target DNA for the cutting process. The second was the invention of a way to fuse these two RNAs into a single-guide RNA. By studying a phenomenon that evolution had taken a billion or so years to perfect in bacteria, they turned nature's miracle into a tool for humans.

On the day that she and Jinek brainstormed how to engineer a single-guide RNA, Doudna explained the idea to her husband over dinner. Realizing that it would have implications for a possible patent on gene-editing technology, he told her that she needed to have it written up fully in the lab notebook and witnessed. So Jinek went back to the lab that night and wrote a detailed description of their concept.

It was close to 9 p.m., but Sam Sternberg and Rachel Haurwitz were still there. Lab notebooks have witness signature lines at the bottom of each page in order to document important advances, and Jinek asked both of them to sign. Sternberg had never been asked to do that before, so he realized that it was a historic evening.[5]

Chapter 18

Science, 2012

When it came time to write a scholarly paper describing CRISPR-Cas9, Doudna and her teammates used the same round-the-clock collaborative methods they had employed in their experiments. The manuscript was shared in Dropbox, with each of their changes tracked in real time. Jinek and Doudna worked during the day in California, handed things off with a late-night Skype call as dawn was breaking in Europe, and then Charpentier and Chylinski would take the lead for the next twelve hours. Because the sun never set in Umeå during the spring, Charpentier announced she could work any hour of the day. "You can't really sleep much when it's light all the time," she says, "and you're never really tired in those months, so I was on duty at any time."[1]

On June 8, 2012, Doudna hit the Send button on her computer to submit the manuscript to the editors of the journal *Science*. It listed six authors: Martin Jinek, Krzysztof Chylinski, Ines Fonfara, Michael Hauer, Jennifer Doudna, and Emmanuelle Charpentier. An asterisk next to the names of Jinek and Chylinski noted that they had contributed equally. Doudna and Charpentier were listed last because they were the principal investigators leading the labs.[2]

The 3,500-word paper went into great detail on how the crRNA

and the tracrRNA worked to bind the Cas9 protein onto the target DNA. It also showed how the structure of two Cas9 domains determined how each cut one of the DNA strands at a specific location. Finally, it described how they were able to fuse the crRNA and tracrRNA to engineer a single-guide RNA. This system, the authors noted, could be used to edit genes.

When the editors of *Science* received the paper, they were excited. Although many of the activities of CRISPR-Cas9 in living cells had been described before, it was the first time researchers had isolated the essential components of the system and discovered their biochemical mechanisms. In addition, the paper contained a potentially useful invention: the single-guide RNA.

At Doudna's urging, the editors fast-tracked the review process. She knew that other papers on CRISPR-Cas9, including one from a Lithuanian researcher (more on him in a moment), were already circulating, and she wanted to make sure that her team was the first to publish. The editors at *Science* had their own competitive motivation: they didn't want to be scooped by a rival journal. They asked CRISPR pioneer Erik Sontheimer to be one of the reviewers and told him he would have to get his comments back in two days, an unusually fast turnaround. He declined the assignment because he was doing his own work on the topic, but the journal's editors were able to find others to review the paper quickly.

The reviewer comments contained only a few requests for clarification. There was one significant issue that they did not raise. The experiments looked at the CRISPR-Cas9 system of *Streptococcus pyogenes*, a common bacteria that can cause strep throat. Like all bacteria, it is a single-cell organism without a nucleus. But the paper suggested that the CRISPR-Cas9 system could be useful for gene editing in humans. Charpentier thought that would prompt some questions. "I was thinking that the reviewers would ask if there was any evidence that it worked in human cells," she recalls. "But they never raised that, even after the conclusion I wrote saying that it would be an alternative to existing gene-editing methods."[3]

The *Science* editors approved the revisions and formally accepted the paper on Wednesday, June 20, 2012, just as participants were gathering in Berkeley for the annual CRISPR conference. Charpentier had arrived from Umeå and Chylinski from Vienna a few days early so they could be together for the final proofreading and edits. "Krzysztof arrived jet-lagged," Charpentier recalled, "but that was not the case for me because I had been in Umeå where it had been light all the time and I hadn't been sleeping in a rhythm."[4]

They gathered in Doudna's seventh-floor office and watched on her computer as the final PDF files and graphics were uploaded into the journal's online system. "The four of us were sitting in the office watching the status indicators for the uploads," Jinek recalls, "and there was a lot of excitement when the last one reached one hundred percent."

Once the final revisions were submitted, Doudna and Charpentier sat together, just the two of them, in Doudna's office. It had been only fourteen months since they had first met in Puerto Rico. As Charpentier admired the view of the late afternoon sun setting over San Francisco Bay, Doudna spoke of how pleasant it had been collaborating with her. "It was a glorious moment when we finally got to share in person the joy of discovery and also some personal confidences," Doudna recalled. "We got to take a breath and talk about how hard we'd worked together across thousands of miles."

When the talk turned to the future, Charpentier indicated she was interested in returning to a focus on the basic science of microbes rather than making tools for gene editing, and she confided that she was ready to move labs again, probably to the Max Planck Institute in Berlin. Doudna asked, somewhat teasingly, whether she would ever want to settle down, get married, have children. "She said she didn't want that," Doudna recalled. "She said she enjoyed being alone and treasured her private time and was not looking for that kind of companionship."

That evening, Doudna organized a celebratory dinner at Chez Panisse, the Berkeley restaurant where chef Alice Waters pioneered farm-to-table cuisine. Not yet a celebrity outside the rarefied realms of science, Doudna was unable to get a reservation at the fancier downstairs dining room, but she got a long table at the more casual upstairs café. They ordered champagne and toasted what they knew would be a new era in biology. "We felt like we were at the beginning of this intense time when the science was all coming to fruition, and we were thinking about what the implications were," Doudna recalls. Jinek and Chylinski left before dessert. They had to work that night on the slides for the presentation they would make at the conference the next day. On their walk back to the lab, in the last glow of twilight, Chylinski indulged in a cigarette.

Virginijus Šikšnys

Krzysztof Chylinski

Martin Jinek

CHAPTER 19

Dueling Presentations

Virginijus Šikšnys

Virginijus Šikšnys of Vilnius University in Lithuania is a mild-mannered biochemist with wire-rimmed glasses and a shy smile. He studied organic chemistry at Vilnius, got his doctorate at Moscow State University, then returned to his native Lithuania. He became intrigued by CRISPR when he read the 2007 paper by the Danisco yogurt researchers Rodolphe Barrangou and Philippe Horvath showing that CRISPR was a weapon that bacteria acquired in their struggle to fight off viruses.

By February 2012, he had produced a paper, with Barrangou and Horvath as secondary authors, that described how, in a CRISPR system, a Cas9 enzyme was guided by a crRNA to cut up an invading virus. He sent it off to the journal *Cell,* which summarily rejected it. In fact, the journal did not deem the paper interesting enough to send out for peer review. "Even more frustrating, we sent it to *Cell Reports,* which is kind of a sister journal to *Cell,*" Šikšnys says. "They rejected it too."[1]

So his next attempt was to send it to *PNAS*, the publication of the U.S. National Academy of Sciences. One expedited path to be accepted by *PNAS* is for a research paper to be approved by a member of that academy. On May 21, 2012, Barrangou decided to send an abstract of the article to the member who was most familiar with the field: Jennifer Doudna.

Doudna was just finishing her paper with Charpentier, so she recused herself. She read only the abstract, not the full paper. But reading the abstract was enough for her to learn that Šikšnys had discovered many of the mechanisms of how, as the abstract said, "DNA cleavage is executed by Cas9." The abstract also declared that this could lead to a method for editing DNA: "These findings pave the way for engineering of universal programmable RNA-guided DNA endonucleases."[2]

The fact that Doudna subsequently hurried to push her own team's paper into print would cause a small controversy, or at least a few raised eyebrows, among some members of the CRISPR crowd. "You should look at the timing of Jennifer's patent filing and the submission of her paper to *Science*," Barrangou told me. At first glance, it can look suspicious. Doudna got Šikšnys's abstract on May 21, and she and her colleagues filed a patent application on May 25 and submitted their paper to *Science* on June 8.

In fact, the Doudna team's patent application and paper had been in the works well before she got Šikšnys's abstract. Barrangou emphasizes he is not accusing Doudna of doing anything wrong. "It was not improper or even unusual," he says. "It's not like she stole anything. We sent it to her. We can't blame her. This is how science is accelerated, when you know that it's a competitive situation. It gives you an impetus to push the process."[3] As it turned out, Doudna remained friendly with both Barrangou and Šikšnys. Their mix of competition and cooperation were part of a process they all understood.

There was, however, one rival who did question Doudna's haste: Eric Lander, director of the Broad Institute at MIT and Harvard. "She tells the *Science* editors that they have competition, she races the paper in, and *Science* rushes the reviewers," he says. "The whole thing gets done in three weeks, and so she scoops the Lithuanians."[4]

I find Lander's implied criticism of Doudna interesting, even a bit amusing, because he is one of the most cheerfully competitive people I know. The fact that he and Doudna are both very comfortable with being competitive has, I suspect, made their rivalry more intense. But I also think it meant that they understood each other, in the way that the two rivals in C. P. Snow's novel *The Masters* were able to understand each other better than any outsider could. Lander told me over dinner one night that he had the emails Doudna sent to the editors of *Science* that proved she pushed them to hurry her 2012 paper into print after she saw an abstract of Šikšnys's paper. When I ask Doudna about this, she readily agrees that she told the editors of *Science* there was a paper being submitted to a competing journal and requested that the reviewers accelerate their process. "So what?" she says. "Ask Eric if he's ever done that." So the next time I have dinner with Lander, I tell him that Doudna wanted me to ask him that question. He pauses, laughs, and then merrily concedes, "Of course I have. It's how science works. This is completely normal behavior."[5]

Šikšnys presents

Barrangou was one of the organizers of the June 2012 CRISPR conference in Berkeley, the one that Charpentier and Chylinski had flown over to attend, and he invited Šikšnys to present his work there. This set the stage for a face-off between the two teams that were racing to describe the CRISPR-Cas9 mechanisms.

Both Šikšnys and the Doudna-Charpentier team were scheduled to present their work on the afternoon of Thursday, June 21, the day after Doudna uploaded the final version of the *Science* article and went with her colleagues to celebrate at Chez Panisse. Barrangou had decided, even though Šikšnys's work had not yet been accepted for publication, that he should present first, followed immediately by the presentation of the Doudna-Charpentier team.

In the annals of history, the priority had been sealed: the Doudna-Charpentier paper had already been accepted by *Science* and would be published online June 28, while Šikšnys would not get published until September 4. Nevertheless, Barrangou's decision to let Šikšnys be the first to present at the Berkeley conference had the potential to give him a small claim to some of the glory—if his research turned out to match or exceed that of the Doudna-Charpentier team. "I was in charge of the order of speakers," Barrangou says. "I got a request from someone in Jennifer's lab to move their talk to before Virginijus. I rejected that. Virginijus had sent his paper to me first, back in February when we were trying to get it published in *Cell*, and I thought it would be fair for Virginijus to present first."[6]

So just after lunch on Thursday, June 21, Virginijus Šikšnys gave a slide presentation, based on his unpublished paper, in the seventy-eight-seat ground-floor auditorium of Berkeley's new Li Ka Shing Center, where the conference was being held. "We isolated the Cas9-crRNA complex and demonstrated that *in vitro* it generates a double-strand break at specific sites in target DNA molecules," he announced. He went on to say that this system could someday become a gene-editing tool.

There were, however, some gaps in the Šikšnys paper and presentation. Most notably, he spoke of the "Cas9-crRNA complex" and made no mention of the role of tracrRNA in the gene-cutting process. Although he described the tracrRNA role in creating the crRNA, he did not realize that it was necessary for this molecule to stick around in order to bind crRNA and Cas9 onto the DNA site targeted for destruction.[7]

For Doudna, this meant that Šikšnys had failed to discover the essential role played by the tracrRNA. "If you don't know that the tracrRNA is required for DNA cutting," she later said, "there is no way you could implement it as a technology. You haven't defined what the components are to get it to work."

There was competitive tension in the air, and Doudna was intent on making sure that Šikšnys's lapse involving the role of tracrRNA was highlighted. She was seated in the third row of the auditorium, and as soon as Šikšnys finished she raised her hand. Does your data, she asked, show the role of the tracrRNA in the cleaving process?

At first Šikšnys did not engage on the point directly, so Doudna kept pressing him to clarify. He did not try to refute her. "I remember there was a hint of debate in the discussion that followed Jennifer's question, and she was very firm in making her voice heard that the tracrRNA was an essential part that was overlooked in the work that Virginijus presented," says Sam Sternberg. "He did not disagree, but neither was there a full admission that he had missed it." Charpentier was likewise surprised. After all, she had written about part of the tracrRNA role in 2011. "What I don't understand is why Šikšnys, after reading my 2011 paper, did not look further into the role of tracrRNA," she says.[8]

To be fair, Šikšnys deserves a lot of credit, which I hope I've given him, for making many of the biochemical findings at about the same time as Doudna and Charpentier. Perhaps I have put a bit too much focus on the role of the tiny tracrRNA, both because I'm writing the book from Doudna's vantage point and because she emphasized it in many of our interviews. But I actually do think it's important. In explaining the amazing mechanisms of life, little things matter. And very little things matter a lot. Showing precisely the essential role of the two snippets of RNA—the tracrRNA and the crRNA—was key to understanding fully how CRISPR-Cas9 could be a gene-editing tool and how the two RNAs could be fused together to create a simple single guide to the right gene target.

Wow

Immediately after Šikšnys finished, it was time for Doudna and Charpentier to deliver what most attendees by then knew was a set of big breakthroughs. The two sat next to each other in the audience, having decided that the presentation would be made by the postdocs who had done most of the hands-on experiments, Jinek and Chylinski.[9]

When the presentation was about to begin, two Berkeley biology professors walked in with some of their postdocs and students. Doudna had been talking to them about collaborating on getting CRISPR-Cas9 to work in humans, but most of the other participants did not know who they were. Sternberg guessed they were patent lawyers. Their appearance heightened the sense of drama. "I remember people being surprised as a dozen or so unknown people filed in,"

Doudna says. "It was sort of a heads-up that something special was about to happen."

Jinek and Chylinski tried to make their presentation fun. They had prepared the slides so that they could take turns explaining each of the experiments they had done, and they had practiced twice before their appearance. The audience was small, informal, and friendly. Nevertheless, it was very clear they were nervous, especially Jinek. "Martin was very stressed, which made me stressed for him," Doudna says.

There was no need to be nervous. The presentation was a triumph. Sylvain Moineau, a CRISPR pioneer at the University of Laval in Quebec, stood up and said, "Wow!" Others hurriedly emailed and texted their lab colleagues back home.

Barrangou, the Danisco researcher who had been a collaborator on Šikšnys's paper, later said that, as soon as he heard the presentation, he knew that Doudna and Charpentier had taken the field to a whole new level. "Jennifer's paper was clearly so much better than ours," he admits. "It wasn't close. It was the tipping point that moved the CRISPR field from an idiosyncratic interesting microbial-world feature to a technology. So Virginijus and I, we had no hard feelings whatsoever."

One especially informed reaction, a mix of excitement and envy, came from Erik Sontheimer. He had been among the first to predict that CRISPR would become a gene-editing tool. When Jinek and Chylinski finished their presentation, he raised his hand to ask a question: How could the single-guide technology be used for gene editing in eukaryotic cells, meaning ones that had a nucleus? More specifically, would it work in human cells? They suggested it could be adapted, just as many previous molecular technologies had been. After the discussion, Sontheimer, a gentle and old-school type of scientist, turned to Doudna, who was sitting two rows behind him, and mouthed the words "Let's talk." During the next break, they ducked out to meet in a hallway.

"I felt comfortable talking to her because, even though we were going to try to do similar things, I knew she was trustworthy," Sontheimer says. "I told her that I was trying to make CRISPR work in yeast. She said she wanted to keep talking, because adapting CRISPR for eukaryotic cells was going to happen fast."

That evening, Doudna took a walk into downtown Berkeley to eat at a sushi restaurant with three of the researchers who had been, and would continue to be, both colleagues and competitors: Erik Sontheimer and the two men whose paper had just been overshadowed by hers, Rodolphe Barrangou and Virginijus Šikšnys. Rather than being upset that they had been scooped, Barrangou said he realized that they had been bested fairly. In fact, as they were walking down the hill to the restaurant, he asked Doudna whether he and Šikšnys might do well to withdraw their paper that was still pending publication. She smiled. "No, Rodolphe, your paper will be fine," she said. "Don't withdraw it. It makes its own contribution, just like we all try to do."

At the dinner, the four shared where each of their labs might go from there. "It was all very warm, despite the potential for awkwardness," Sontheimer says. "Just a very exciting dinner at a very exciting time when we were all just recognizing how important this was going to be."

The Doudna-Charpentier paper, published online on June 28, 2012, galvanized an entire new field of biotechnology: making CRISPR work in the editing of human genes. "We all knew we were going to be in a big race to do this in human cells," says Sontheimer. "It was an idea whose time had come, and it was going to be a sprint to get there first."

Part Three

Gene Editing

How beauteous mankind is!
O brave new world,
That has such people in't!

—William Shakespeare, *The Tempest*

Chapter 20

A Human Tool

Gene therapies

The road to engineering human genes began in 1972 when Professor Paul Berg of Stanford discovered a way to take a bit of the DNA of a virus found in monkeys and splice it to the DNA of a totally different virus. Presto! He had manufactured what he dubbed "recombinant DNA." Herbert Boyer and Stanley Cohen discovered ways to make these artificial genes more efficiently and then clone millions of copies of them. Thus the science of genetic engineering—and the business of biotechnology—was launched.

It took another fifteen years before scientists began to deliver engineered DNA into the cells of humans. The goal was similar to creating a drug. There was no attempt to change the DNA of the patient; it was not gene *editing*. Instead, gene therapy involved delivering into the patient's cells some DNA that had been engineered to counteract the faulty gene that caused the disease.

The first trial came in 1990 on a four-year-old girl with a genetic mutation that crippled her immune system and left her at risk for infection. Doctors found a way to get functioning copies of the missing gene into the T cells of her blood system. The T cells were removed from her body, given the missing gene, and then reintroduced into her body. This led to a dramatic improvement of her immune system and allowed her to live a healthy life.

The field of gene therapy initially showed modest success, but soon there were setbacks. In 1999, a clinical trial in Philadelphia came to a halt when a young man died due to a massive immune response caused by the virus transporting the therapeutic gene. In the early 2000s, a gene therapy procedure for an immune-deficiency disease inadvertently triggered a cancer-causing gene that led to five patients developing leukemia. Tragedies such as these froze for at least a decade most of the clinical trials, but incremental improvements in gene therapies would lay the groundwork for the more ambitious field of gene editing.

Gene editing

Instead of treating genetic problems through gene therapy, some medical researchers began looking for ways to fix the problems at their source. The goal was to *edit* the flawed sequences of DNA in the relevant cells of the patient. Thus was born the endeavor called gene editing.

Harvard professor Jack Szostak, Doudna's thesis advisor, discovered in the 1980s one of the keys to editing a gene: causing a break in both strands of the DNA double helix, known as a double-strand break. When this happens, neither strand can serve as a template to repair the other. So the genome repairs itself in one of two ways. The first is called "nonhomologous end-joining." ("Homologous" comes from the Greek word for "matching.") In such cases, the DNA is repaired by simply stitching two ends together without trying to find a matching sequence. This can be a sloppy process resulting in unwanted inserts and deletions of genetic material. A more precise process, "homology-directed repair," occurs when the cut DNA finds a suitable replacement template nearby. The cell will usually copy and insert the available homologous sequence where the double-strand breaks occurred.

The invention of gene editing required two steps. First, researchers had to find the right enzyme that could cut a double-strand break in DNA. Then they had to find a guide that would navigate the enzyme to the precise target in the cell's DNA where they wanted to make the cut.

The enzymes that can cut DNA or RNA are called "nucleases." In order to build a system for gene editing, researchers needed a nuclease that could be instructed to cut any sequence that the researchers chose to target. By 2000, they had found a tool to do this. The FokI enzyme, which is found in some soil and pond bacteria, has two domains: one that serves as scissors that can cut DNA and another that serves as a guide telling it where to go. These domains can be separated, and the first can be reprogrammed to go anywhere the researchers want.[1]

Researchers were able to devise proteins that could serve as a guide to get the cutting domain to a targeted DNA sequence. One system, zinc-finger nucleases (ZFNs), came from fusing the cutting domain with a protein that has little fingers shaped by the presence of a zinc ion, which allow it to grasp on to a specified DNA sequence. A similar but even more reliable method, known as TALENs (transcription activator–like effector nucleases), came from fusing the cutting domain with a protein that could guide it to longer DNA sequences.

Just when TALENs were being perfected, CRISPR came along. It was somewhat similar: it had a cutting enzyme, which was Cas9, and a guide that led the enzyme to cut a targeted spot on a DNA strand. But in the CRISPR system, the guide was not a protein but a snippet of RNA. This had a big advantage. With ZFNs and TALENs, you had to construct a new protein guide every time you wanted to target a different genetic sequence to cut; it was difficult and time consuming. But with CRISPR you merely had to fiddle with the genetic sequence of the RNA guide. A good student could do it quickly in a lab.

There was one question, which was either a big one or a trivial one, depending on your perspective and your side in the patent wars that would later erupt. The CRISPR systems worked in bacteria and archaea, which are single-cell organisms that have no nucleus. But that left the question: Would they work in cells that *do* have a nucleus, especially multicell organisms such as plants, animals, you, and me?

As a result, the Doudna-Charpentier paper in June 2012 set off a furious sprint in many labs around the world, including Doudna's, to prove that CRISPR-Cas9 could work in human cells. That triumph was accomplished in five places in about six months. This rather quick success could be taken as evidence, as Doudna and her colleagues would later argue, that making CRISPR-Cas9 work in human cells was an easy and obvious step that was not a separate invention. Or it could be used to argue, as Doudna's competitors have, that it was a major inventive step that came after a fiercely competitive race.

On that question would hang patents and prizes.

CHAPTER 21

The Race

Competition drives discovery. Doudna calls it "the fire that stokes the engine," and it certainly stoked hers. Ever since she was a child, she was not embarrassed to appear ambitious, but she knew how to balance this by being collegial and forthright. She had learned about the importance of competition from reading *The Double Helix*, which describes how the perceived footsteps of Linus Pauling were a catalyst for James Watson and Francis Crick. "Healthy rivalries," she later wrote, "have fueled many of humankind's greatest discoveries."[1]

Scientists are mainly motivated by the joy that comes from understanding nature, but most will admit that they are also driven by the rewards, both psychic and substantive, of being the first to make a discovery: papers published, patents granted, prizes won, and peers impressed. Like any human (is it an evolutionary trait?), they want credit for their accomplishments, payoff for their labor, acclaim from the public, and prize ribbons placed around their necks. That's why they work late into the night, hire publicists and patent attorneys, and even invite writers (like me) into their labs.

Competition gets a bad rap.[2] It's blamed for discouraging collaboration, constricting the sharing of data, and encouraging people to

keep intellectual property proprietary rather than allowing it to be free and open for common use. But the benefits of competition are great. If it hastens the discovery of a way to fix muscular dystrophy, prevent AIDS, or detect cancer, fewer people will die early deaths. To take an example relevant to these days, the Japanese bacteriologist Kitasato Shibasaburō and his Swiss rival Alexandre Yersin both rushed to Hong Kong in 1894 to investigate the pneumonic plague epidemic and, working with different methods, discovered the responsible bacteria within days of each other.

There was one competition in Doudna's life that stands out for becoming heated and then bitter: the race in 2012 to show how CRISPR could edit the genes of humans. It may not be up there with Charles Darwin and Alfred Russel Wallace converging on the idea of evolution or Newton and Leibniz disputing who first figured out calculus. But it is our contemporary counterpart to the race between Pauling and the team of Watson and Crick to discover the structure of DNA.

Doudna entered this competition handicapped by not having a team of collaborators who were experts in working with human cells. Her lab did not specialize in such experiments; its researchers were mainly biochemists comfortable working with molecules in test tubes. So Doudna ended up struggling to keep pace in what turned out to be a six-month frenzied competition.

There were many labs around the world that engaged in this race, but the primary drama—emotionally and personally as well as scientifically—involved three players. All were competitive in their own way, but they were very different in how comfortable they were with their competitiveness:

- Feng Zhang of the Broad Institute of MIT and Harvard. Although as competitive as any star researcher, he was blessed with a cheery sweetness that made him uncomfortable displaying that trait. With deep values imbued by his mother, he had a natural humility that often masked his equally natural ambition. It was as if he had dual cores, one competitive and one beatific, that coexisted quite comfortably. He had a warm smile that rarely left his face except in those moments when the talk turned to competition—or the importance of Doudna's achievements—at which point his lips would continue to smile, but his eyes no longer joined in. He tended to be shy of the limelight, but he was pushed by his mentor Eric Lander, the brilliant and sparky mathematician-turned-scientist who directed the Broad Institute, to compete for credit as well as for discoveries.

- George Church of Harvard, Doudna's longtime friend, who considered himself, at least for a while, to be Zhang's mentor and academic advisor. Both on the surface and as deep as my eye can discern, he was the least competitive of them all. A Santa-bearded vegan who wants to use genetic engineering to bring back the woolly mammoth, he was driven by a playful and earnest curiosity.
- And finally there was Doudna, who was not only competitive but also comfortable with her competitiveness. It was one of the reasons a certain coolness developed between her and Charpentier, who expressed some amusement and a bit of disdain for Doudna's drive for credit. "She is sometimes stressed about credit, which made her seem insecure or not fully grateful for her success," Charpentier says. "I am French and not as worked up, so I was always telling her, 'Surf on the good wave.'" But when pressed, Charpentier admits that the competitiveness that Doudna exhibits is the force that drives most scientific pioneers, and thus science itself. "If it were not for competitive people like Jennifer, our world would not be as good," she says. "Because what drives people to do good things is recognition."[3]

CHAPTER 22

Feng Zhang

Des Moines

When I first approached Feng Zhang to ask if I could spend some time with him, I was nervous. I had told him that I was doing a book focused on Jennifer Doudna, his rival, and I thought he would be put off, perhaps would even push back.

Instead, when I visited him at his lab at the Broad Institute near MIT, with its high windows offering views of the Charles River and the spires of Harvard, he was exceedingly gracious, as he was at our subsequent conversations, lunches, and dinners. I could not tell whether his geniality was genuine or arose from an assessment that it would lead to his being portrayed better in my book. But the more time I spent with him, the more I became convinced that it was the former.

Zhang's journey, which is worthy of a book of its own, is one of those classic immigration tales that has made America great. He was born in 1981 in Shijiazhuang, an industrial city of 4.3 million people southwest of Beijing. His mother taught computer science, his father was a university administrator. The streets of the city were festooned with China's customary banners of exhortations, most notably those touting the patriotic duty to study science. Zhang was sold. "I grew up

playing with robot kits and fascinated by anything to do with science," he recalls.[1]

In 1991, when Zhang was ten, his mother came to the United States as a visiting scholar at the University of Dubuque, a gem nestled in an architecturally rich Iowa city along the Mississippi River. One day she visited a local school, where she marveled at the computer lab and the lack of emphasis on rote memorization. Like any loving parent, she imagined it through the eyes of her child. "She thought I would enjoy being in such a lab and school, so she decided to stay and bring me over," Zhang recalls. She got a job at a paper company in Des Moines and with her H-1B visa was able to bring her son to America the next year.

His father soon followed, but he never learned English well, so Zhang's mother became the driving force in the family. She was the one who pioneered the path to America, got a job, made friends at work, and volunteered to set up computers at local charities. Because of her, and because of the hospitality gene ingrained in heartland towns, the family always had invitations to neighbors' houses for Thanksgiving and other holidays.

"My mother always told me to keep my head down and not be arrogant," Zhang says. She bestowed upon him the gift of easygoing humility, which he wore lightly. But she also instilled in him an ambition to be innovative and never passive. "She pushed me to make things, even on a computer, rather than play with things that other people had made." Years later, as I was writing this book, Zhang's mother had moved in part time with him and his wife in Boston to help take care of their two young kids. As he talks about her while picking at a hamburger in a Cambridge seafood restaurant, Zhang lowers his head and pauses for a moment. "I'm sure going to miss her when she's gone," he says in a very soft voice.

At first Zhang seemed likely to follow the path of so many supersmart kids in the 1990s and become a computer geek. When he got his first computer (a PC, not a Mac) at age twelve, he learned to take it apart and use the components to build other computers. He also became a wizard at using open-source Linux operating system software.

So his mother sent him to computer camp and, just to make sure he was wired for success, debate camp as well. It was the type of enhancement that privileged parents can do even without gene editing.

Instead of pursuing computer science, however, Zhang became a forerunner of what will, I think, soon be common among aspiring geeks: his interests shifted from digital tech to biotech. Computer code was something his parents and their generation did. He became more interested in genetic code.

Zhang's path to biology began with his Des Moines middle school's Gifted and Talented Program, which included a Saturday enrichment class in molecular biology.[2] "Until then, I didn't know much about biology and didn't find it interesting, because in seventh grade all they did was give you a tray with a frog and tell you to dissect it and identify the heart," he recalls. "It was all memorization and not very challenging." In the Saturday enrichment class, the focus was on DNA and how RNA carried out its instructions, with an emphasis on the role played in this process by enzymes, those protein molecules that act as catalysts to spark actions in a cell. "My teacher loved enzymes," Zhang says. "He told me that whenever you face a tough question in biology, just say 'Enzymes.' It's the correct answer to most questions in biology."

They did a lot of hands-on experiments, including one that transformed bacteria to make them resistant to antibiotics. They also watched the 1993 movie *Jurassic Park*, in which scientists bring dinosaurs back from extinction by combining their DNA with that of frogs. "I was excited to discover that animals could be a programmable system," he says. "That meant human genetic coding could be programmable as well." It was more exciting than Linux.

With his corn-fed eagerness to learn and discover, Zhang became an example of the impact that gifted and talented programs can have on turning American kids into world-class scientists. The U.S. Department of Education had just published, in 1993, a study called "A Case for Developing America's Talent," which led to funding for local school districts "to challenge our top performing students to greater heights." Those were the days when people took very seriously, even if it meant spending tax dollars, the aim of creating a world-class education system, one that would keep America the world leader in innovation. In Des Moines, this included a program called STING (Science/Technology Investigations: The Next Generation), which tapped a small group of talented and motivated students to do original projects and work at local hospitals or research institutions.

Zhang's Saturday teacher helped him get selected to spend his afternoons and free time at the gene therapy lab of Methodist Hospital in Des Moines. As a high school student, he worked under a psychologically intense but very personable molecular biologist named John Levy, who explained over tea each day the work he was doing and assigned Zhang to increasingly more sophisticated experiments. On some days Zhang would arrive right after school and work until eight in the evening. "My dear mother would drive each day to pick me up and then sit in the parking lot until I was finished," he says.

His first major experiment involved a fundamental tool in molecular biology: a gene from jellyfish that produces green fluorescent protein, which glows when exposed to ultraviolet light and thus can be used as a marker in cell experiments. Levy first made sure Zhang understood its fundamental natural purpose. Sketching on a piece of paper as he sipped tea, he explained why a jellyfish might need that fluorescent protein as it moved up and down layers of the ocean during different phases of its life cycle. "He drew it in a way that you could just picture the jellyfish and the ocean and nature's wonders."

Levy "held my hand," Zhang recalls, "as I did my first experiment." It involved putting the gene for green fluorescent protein into human melanoma (skin cancer) cells. It was a simple but exciting example of genetic engineering: he had inserted a gene from one organism (a jellyfish) into the cells of another (a human), and he could see the proof of his success when the bluish-green glow emanated from the manipulated cells. "I was so excited that I began to shout, 'It's glowing!'" He had reengineered a human gene.

Zhang spent the next few months studying whether the green fluorescent protein, which absorbs ultraviolet light when it glows, could protect the human cell's DNA from the damage that can be caused by exposure. It worked. "I was using the jellyfish's GFP as a sunscreen to protect human DNA from ultraviolet light damage," he says.

The second science project he did with Levy was to deconstruct HIV, the virus that causes AIDS, and examine how each of the components worked. Part of the goal of the Des Moines enrichment programs was to help students do projects to compete in the Intel Science Search, a national competition. Zhang's virus experiment won him third place, which carried a hefty $50,000 prize. He used it to help pay his tuition when he got into Harvard in 2000.

Harvard and Stanford

Zhang was at Harvard at the same time as Mark Zuckerberg, and it's interesting to speculate on which of them will end up having the most impact on the world. It's a proxy for the larger question, which future historians will answer, of whether the digital revolution or the life-science revolution will end up being the more important.

Majoring in both chemistry and physics, Zhang initially did research with Don Wiley, a crystallographer who was a master at determining the structure of complex molecules. "I don't understand anything in biology unless I know what it looks like," he liked to say, a credo worthy of all structural biologists, from Watson and Crick to Doudna. But in November of Zhang's sophomore year, Wiley mysteriously disappeared one night while attending a conference at St. Jude's Children's Hospital in Memphis, leaving his rental car on a bridge. His body was later found in the river.

That year Zhang also had to help a close friend in his class who was spiraling into major depression. The friend would be sitting in their room studying, and then suddenly he would get hit by an anxiety or depressive attack and not be able to get up or move. "I had heard of depression, but I thought it was like having a bad day and you had to barrel through," Zhang says. "Growing up in my family, I mistakenly thought that psychiatric disease was when someone just wasn't being strong enough." Zhang would sit with his friend to help him avoid suicide. (The student took time off and recovered.) The experience caused Zhang to turn his attention to researching treatments for mental illness.

So when he went to Stanford for graduate school, he asked to join the lab of Karl Deisseroth, a psychiatrist and neuroscientist who was developing ways to make the workings of the brain and its nerve cells, known as neurons, more visible. Along with another graduate student, they pioneered the field of optogenetics, which uses light to stimulate neurons in the brain. That allowed them to map different circuits in the brain and gain insights about how they functioned or malfunctioned.

Zhang focused on inserting light-sensitive proteins into the neurons—an echo of his high school work inserting green fluorescent protein into skin cells. His method was to use viruses as a delivery mechanism. For one demonstration, he inserted these proteins, which become activated when light hits them, into the part of a mouse brain that controls its movement. By using light pulses, the researchers could trigger the neurons and cause the mice to walk in circles.[3]

Zhang faced a challenge. It was difficult to insert the gene for the light-sensitive proteins into the exact right location of the DNA of the brain cell. Indeed, the entire field of genetic engineering was hampered by the lack of simple molecular tools for cutting and pasting desired genes into strands of DNA inside a cell. So after he got his doctorate in 2009, Zhang took a postdoc position at Harvard and began researching the gene-editing tools that were available at the time, such as TALENs.

At Harvard, Zhang focused on ways to make TALENs more versatile so that they could be programmed to target different gene sequences.[4] It was difficult; TALENs are hard to engineer and re-engineer. Fortunately, he was working in the most exciting lab at Harvard Medical School, which was run by a professor who was beloved for embracing new ideas, sometimes wildly, and who fostered a jovial atmosphere that encouraged exploration: Doudna's longtime friend, the avuncular and bushy-bearded George Church, one of the contemporary legends of biology and a scientific celebrity. He became for Zhang, as he did for almost all of his students, a loving and beloved mentor—until the day Church believed that Zhang had betrayed him.

Chapter 23

George Church

Tall and gangly, George Church looks like, and actually is, both a gentle giant and a mad scientist. He is one of those iconic characters who is equally charismatic on Stephen Colbert's TV show and in his bustling Boston lab amid a gaggle of adoring researchers. Always calm and genial, he has the amused demeanor of a time traveler who is eager to get back to the future. With his wild-man beard and halo of hair, he looks like a cross between Charles Darwin and a woolly mammoth, an extinct species that he wants, perhaps out of a vague sense of kinship, to resurrect using CRISPR.[1]

Although he is personable and charming, Church has the literalness often found in successful scientists and geeks. At one point we were discussing some decision that Doudna had made, and I asked him whether he thought it had been necessary. "Necessary?" he replied. "Nothing is necessary. Even breathing is not necessary. You can even stop breathing if you really want to." When I joked that he had taken me too literally, he remarked that one reason he is a good scientist, and also thought of as a bit of a madman, is that he questions the necessity of any premise. He then wandered off into a discourse on free will (which he doesn't believe humans have) until I was able to get him back on track talking about his career.

Born in 1954, he grew up in the marshy exurbs of Clearwater, on Florida's Gulf Coast near Tampa, where his mother went through three husbands. As a result, George had many last names and different schools, which made him feel, he says, "like a real outsider." His birth father had been a pilot at nearby MacDill Air Force Base and a barefoot water-ski champion who was in the Water Ski Hall of Fame. "But he couldn't hold a job, and my mother moved on," Church explains.

The young Church was fascinated by science. In those days when parents were less overprotective, his mother let him roam alone in the marshes and mudflats near Tampa Bay, hunting for snakes and insects. He would crawl through the high swamp grass collecting specimens. One day he found an odd caterpillar that looked like a "submarine with legs" and put it in his jar. The next day he discovered, to his astonishment, that it had transformed into a dragonfly, a metamorphosis that is truly one of nature's thrilling everyday miracles. "That helped set me on my path to be a biologist," he says.

When he came home in the evening, mud on his boots, he would dive into the books his mother provided, including a set of *Collier's Encyclopedia* and a twenty-five-volume series of vibrantly illustrated nature books from Time-Life. Because he was mildly dyslexic, he had trouble reading but could absorb information from pictures. "It made me a more visual person. I could imagine 3-D objects, and by visualizing the structure I could understand how things worked."

When George was nine, his mother married a physician named Gaylord Church, who adopted George and gave him a permanent surname. His new stepfather had a bulging medical bag that George loved to rummage through. He was particularly fascinated by the hypodermic needle, which his stepfather used liberally to administer painkillers and feel-good hormones to his patients and to himself. He taught George how to use the instruments and would sometimes take him on house calls. At a Harvard Square pub over a soybean burger, Church chuckles as he recalls this odd childhood. "My father would let me give his women patients hormone shots, and they loved him for it," he says, "and he let me give him shots of Demerol. I later realized he was addicted to painkillers."

Using the ingredients in his stepfather's medicine bag, Church began to perform experiments. One involved thyroid hormones that his stepfather supplied to grateful patients who complained of fatigue or depression. At age thirteen, Church put some hormones in the water of a group of tadpoles, leaving another group in untreated water. The first group grew faster. "It was my first true biology experiment, with a control set and all," he recalls.

When his mother drove him in her Buick up to the 1964 World's Fair in New York, he became tantalized by the future. It made him feel impatient about being stranded in the present. "I wanted to get to the future, I felt that's where I belonged, and that's when I realized that it was something I had to help create," he says. As the science writer Ben Mezrich noted of Church, "Later in life, he would return to this moment as the instant when he first started to think of himself as a sort of time traveler. Deep down, he started to believe that he was from the far future, and had somehow been left in the past. It was his task in life to try to get back, to try to shift the world to where he had once been."[2]

Bored in his backwater high school, Church soon became a handful, especially to his stepfather, who had initially indulged him. "He decided he wanted me to go away," Church says, "and my mother realized it was a great opportunity, because he would pay for boarding school." So he was packed off to Phillips Academy in Andover, Massachusetts, America's oldest prep school. The idyllic quads with their Georgian buildings were almost as wondrous as the marshlands of his childhood. He taught himself computer coding, maxed out on all the chemistry courses, and then was given a key to the chemistry lab so he could explore on his own. Among his many triumphs: making flytrap plants grow huge by spiking their water with hormones.

He went on to Duke, where he earned two undergraduate degrees in two years and then skipped ahead into a PhD program. There he stumbled. He became so involved in the lab research of his advisor, which included using crystallography to figure out the three-dimensional structure of different RNA molecules, that he stopped going to classes. After failing two of them, he got a letter from the dean coldly informing him, "You are no longer a candidate for the Doctor of Philosophy degree in the department of Biochemistry at Duke University." He kept the letter as a source of pride, the way others keep their framed diplomas.

He had already been a coauthor of five important papers and was able to talk his way into Harvard Medical School. "It's a mystery why Harvard would accept me after flunking out of Duke," he said in an oral history. "Usually, it's the other way around."[3] There he worked with Nobel laureate Walter Gilbert to develop methods for sequencing DNA, and he was at the initial 1984 retreat sponsored by the Department of Energy that led to the launch of the Human Genome Project. But in a preview of their later disputes, he clashed with Eric Lander, who rejected Church's method for streamlining the sequencing tasks by clonally amplifying the DNA.

Church became a quirky popular celebrity in 2008, when the *New York Times* science writer Nicholas Wade interviewed him about the possibility of using his genetic engineering tools to regenerate the extinct woolly mammoth from frozen hairs found in the Arctic. Not surprisingly, the idea had a playful appeal to Church, born of his days juicing up tadpoles with hormones. He became a public face of the effort, still underway, to take the skin cell from a modern elephant, convert it to its embryonic state, and then modify the genes until they match those sequenced from the woolly mammoth.[4]

When Jennifer Doudna was a PhD student at Harvard in the late 1980s, she admired Church's unconventional style and thinking. "He was a new professor, tall and gangly and already had his big beard, and he was quite the maverick," she says. "He was not afraid of being different, and I liked that." Church recalls being impressed by Doudna's demeanor. "She did stellar work, especially on the structure of RNA," he says. "We shared that esoteric interest."

During the 1980s, Church worked to create new gene-sequencing methods. He became prolific not only as a researcher but as a founder of companies to commercialize the work coming out of his lab. Later he focused on finding new tools for gene editing. So when Doudna and Charpentier's *Science* article describing CRISPR-Cas9 went online in June 2012, Church decided to try to get it to work in humans.

He did the polite thing and sent both of them an email. "I was collegial and tried to find out who was working in the field to see if they would mind if I did so," he recalls. An early riser, he dispatched it just after 4 a.m. one day:

> Jennifer and Emmanuelle,
>
> Just a quick note to say how inspiring and helpful is your CRISPR paper in *Science.*
>
> My group is trying to apply some of the lessons from your study to genome engineering in human stem cells. I'm sure that you have received similar appreciative comments from other labs.
>
> I look forward to staying in contact as things progress.
>
> Best wishes, George

Later that day, Doudna wrote back:

> Hi George,
>
> Thanks for your message. We will be very interested to hear how your experiments progress. And yes, there is a lot of interest in Cas9 at the moment—we are hopeful that it will turn out to be useful for genome editing and regulation in various cell types.
>
> All the best, Jennifer

They followed up with some phone conversations, and Doudna told him that she was likewise working on trying to get CRISPR to work in human cells. It was characteristic of the way Church did science: collegially, with a greater inclination toward cooperation and openness than competition and secrecy. "It was very typical of George," Doudna says. "He is incapable of being devious." The best way to get a person to trust you is for you to trust them. Doudna is a guarded person, but she was always open with Church.

There was one person Church did not think of contacting: Feng Zhang. The reason was, he says, that he had no idea that his former doctoral student was working on CRISPR. "If I had known Feng was working on it, I would have asked him about it," Church says. "But he was very secretive when he suddenly hopped on CRISPR."[5]

CHAPTER 24

Zhang Tackles CRISPR

Stealth mode

After completing his postdoctoral work in Church's Harvard Medical School lab in Boston, Zhang had moved across the Charles River to the Broad Institute in Cambridge. Ensconced in state-of-the-art lab buildings on the edge of MIT's campus, the Broad was founded in 2004 by the irrepressible Eric Lander with funding (eventually $800 million) from Eli and Edythe Broad. Its mission was to advance the treatment of diseases using the knowledge spawned by the Human Genome Project, on which Lander had been the most prolific gene-sequencer.

A mathematician turned biologist, Lander envisioned the Broad as a place where different disciplines would work together. This required a new type of institution, one that fully integrated biology, chemistry, mathematics, computer science, engineering, and medicine. Lander also forged something even more difficult: a collaboration between MIT and Harvard. By 2020, the Broad community included more than three thousand scientists and engineers. It thrived because Lander is a joyful and intensely committed mentor, cheerleader, and fundraiser for wave after wave of young scientists who gravitated to the Broad. He also is able to connect science to public policy and social

Luciano Marraffini

good; for example, he is spearheading a movement called "Count Me In" that encourages cancer patients to anonymously share their medical information and DNA sequences in a public database that any researcher can access.

When Feng Zhang moved to the Broad in January 2011, he continued the research he had been doing in Church's lab on using TALENs for gene editing. But each new editing project required building new TALENs. "That would sometimes take up to three months," he says. "I began looking for a better way."

That better way would turn out to be CRISPR. A few weeks after his arrival at the Broad, Zhang attended a seminar by a Harvard microbiologist who was studying a species of bacteria. He happened to mention, in passing, that they contained CRISPR sequences with enzymes that could cut the DNA of invading viruses. Zhang had barely heard of CRISPR, but ever since his seventh-grade enrichment class he had learned to perk up at the mention of enzymes. He was particularly interested in those enzymes, known as nucleases, that cut DNA. So he did what any of us would do: he googled CRISPR.

The next day he flew to Miami for a conference on how genes get expressed, but instead of sitting through all the talks he stayed in his hotel room reading the dozen or so major scientific papers on CRISPR that he found online. In particular, he was struck by the one published that previous November by the two yogurt researchers at Danisco, Rodolphe Barrangou and Philippe Horvath, which showed that the CRISPR-Cas systems can cut a double-stranded DNA at a specific target.[1] "The minute I read that paper, I thought this was pretty amazing," Zhang says.

Zhang had a protégé and friend who was still a graduate student in Church's lab: Le Cong, a Beijing-born geek with big glasses whose childhood love of electronics had, like Zhang's, given way to a passion for biology. Also like Zhang, Cong was interested in genetic engineering because he hoped to alleviate the suffering that came from mental disorders, such as schizophrenia and bipolar disease.

Immediately after reading the CRISPR papers in his Miami hotel room, Zhang emailed Cong and suggested that they work together to see if it could become a gene-editing tool in humans, perhaps one that was better than the TALENs they had been using. "Take a look at this," Zhang wrote, including a link to the Barrangou-Horvath paper. "Maybe we can test in mammalian system." Cong agreed, replying, "It should be very cool." A couple of days later, Zhang sent another email. Cong was still a student in Church's lab, and Zhang wanted to make sure that he kept the idea secret, even from his advisor. "Hey let's keep this confidential," he wrote.[2] Although Cong formally remained one of Church's graduate students at Harvard, he followed that injunction and did not tell Church that he was going to work on CRISPR when he moved to the Broad with Zhang.

Zhang's office, hallways, conference rooms, and lab areas have multiple whiteboards, poised to accommodate any spontaneous insights that may strike. It's part of the atmosphere at the Broad. Whiteboarding is like a sport, the way foosball is in less rarefied offices. On one of Zhang's well-used whiteboards, he and Le Cong began listing what they would have to do to get CRISPR-Cas systems to penetrate into the nucleus of human cells. Then they would pull late-nighters in the lab, subsisting on ramen noodles.[3]

Even before they started experimenting, Zhang filed a "Confidential Memorandum of Invention" to the Broad Institute, dated February 13, 2011. "The key concept of this invention is based on the CRISPR found in many microbial organisms," it read. The system, he explained, used snippets of RNA to guide an enzyme to make cuts in DNA at targeted spots. If it could be made to work in humans, Zhang noted, it would be a much more versatile gene-editing tool than ZFNs and TALENs. His memo, which was never publicly shared, concluded by stating that the "invention could be useful for genome modification of microbes, cells, plants, animals."[4]

Zhang's memo did not, despite its title, describe an actual invention. He had just begun to sketch out a research plan, and he had done no experiments nor devised any techniques that reduced his concept to practice. The memo was merely a stake in the ground, the type of thing that researchers sometimes file in case they end up being successful inventing something and need evidence (as would indeed be the case) that they had been working on the idea for a long time.

Zhang seemed to sense from the outset that the race to turn CRISPR into a human gene–editing tool would turn out to be very competitive. He kept his plans secret. He did not share his memorandum of invention, nor did he mention CRISPR in a video he made at the end of 2011 that described the research projects he had been working on. But he began to document each of his experiments and discoveries on dated and witnessed notebook pages.

In this competition to adapt CRISPR into a gene-editing tool in humans, Zhang and Doudna came into the arena from different routes. Zhang had never worked on CRISPR. People in that field would later refer to him as a latecomer and interloper, one who jumped on CRISPR after others had pioneered the field. Instead, his specialty was gene editing, and for him CRISPR was simply another method to get to the same goal, along the lines of ZFNs and TALENs, though much better. For her part, Doudna and her team had never worked on gene editing in living cells. Their focus for five years had been on figuring out the components of CRISPR. As a result, Zhang would end up having some difficulty in sorting out the essential molecules in a CRISPR-Cas9 system, while Doudna's difficulty would be figuring out how to get the system into the nucleus of a human cell.

By early 2012—before Doudna and Charpentier went online in June with their *Science* paper showing the essential three components of the CRISPR-Cas9 system—Zhang had made no documented progress. He and a group of colleagues from the Broad filed an application for funding to pursue gene-editing experiments. "We will engineer the CRISPR system to target Cas enzymes to multiple specific targets in the mammalian genome," Zhang wrote in the application. But he made no claim that he had already accomplished any of the major steps to this goal. Indeed, the grant application indicated that work on mammalian cells was not expected to begin until a few months later.[5]

Also, Zhang had not yet figured out the full role of the pesky tracrRNA. Recall that Charpentier's 2011 paper and the work done by

Šikšnys in 2012 described the work that this molecule did in creating a guide RNA, known as crRNA, that navigates an enzyme to the correct DNA location to cut. However, one of the discoveries reported by Doudna and Charpentier in their 2012 paper was that the tracrRNA has another important role: it needs to stick around in order for the CRISPR system to do the actual cutting of the target RNA. Zhang's grant application indicated that he had not yet discovered this; it spoke only about "a tracrRNA element to facilitate the processing of guide RNAs." One of the illustrations showed only the crRNA and not the tracrRNA being part of the complex with Cas9 to do the cutting. This may seem like a small thing. But it's over such small discoveries, or lack of them, that battles for historic credit are waged.[6]

Marraffini helps

If things had worked out differently, Feng Zhang and Luciano Marraffini might have been a collaboration story as inspiring as that of Doudna and Charpentier. Zhang's tale was pretty wonderful on its own: the eager and competitive Chinese immigrant whiz kid who is nurtured in Iowa and whose unrelenting curiosity makes him a star at Stanford, Harvard, and MIT. But it would have double-stranded nicely with the story of Marraffini, an immigrant from Argentina who in early 2012 collaborated with Zhang.

Marraffini loved studying bacteria, and as a doctoral student at the University of Chicago he became interested in the newly discovered phenomenon of CRISPR. Because his wife had a job as a translator in the court system of Chicago, he wanted to stay in that city, so he got a postdoc position in the lab of Erik Sontheimer at Northwestern University. Sontheimer was then studying RNA interference, as Doudna had done, but he and Marraffini soon realized that the CRISPR system worked in a more powerful manner. That is how they made their important discovery in 2008, that it works by chopping up the DNA of invading viruses.[7]

Marraffini met Doudna the following year when she came to Chicago for a conference. He made a point of sitting at the table next to her. "I really wanted to meet her because of her work on RNA structure, which was extremely hard," he says. "It's one thing to crystallize proteins, but it's far harder to crystallize RNA, and that impressed me." She had just begun working on CRISPR, and they discussed the possibility that he might join her lab. But there was not the right opening for him, so in 2010 he moved to Rockefeller University in Manhattan, where he set up a lab studying CRISPR in bacteria.

At the very beginning of 2012, he got an email from Zhang, whom he didn't know. "Happy new year!" Zhang wrote. "My name is Feng Zhang and I am a researcher at MIT. I read many of your papers on the CRISPR system with great interest, and I was wondering if you would be interested in collaborating to develop the CRISPR system for applications in mammalian cells."[8]

Marraffini did a Google search to learn who Zhang was, as he was still unknown to most of the CRISPR research community. Zhang had sent his email at around 10 p.m., and Marraffini answered about an hour later. "I will be very interested in a collaboration," he wrote, adding that he had been working on a "minimal" system—in other words, one that had been stripped down to just the essential molecules. They agreed to talk by phone the next day. It seemed that this would be the beginning of a beautiful friendship.

Marraffini got the impression that Zhang was stymied and was trying out a variety of Cas proteins. "He was testing not only Cas9 but all the different CRISPR systems, including Cas1, Cas2, Cas3, and Cas10," Marraffini says. "Nothing was working. He was doing things like a chicken without a head." So Marraffini, at least by his own recollection, became the one to push him to focus on Cas9. "I was very sure of Cas9. I was an expert in the field. I realized the other enzymes were going to be too difficult."

After their phone call, Marraffini sent Zhang a list of things they should do. The very first item was to quit including any enzymes other than Cas9.[9] He also sent, by regular mail, a printout of the bacteria's entire CRISPR sequence, covering multiple pages (ATGG-TAGAAACACTAAATTA . . .). When Marraffini told me the tale, he got up from his desk and printed out pages of the sequence for me. "With all of this data," he told me, "I made Feng realize that he had to use Cas9, and I gave him a roadmap, which he followed."

For a while they collaborated by splitting up the tasks. Zhang would come up with ideas that he hoped would work in humans. Then Marraffini, who specialized in microbes, would test to see if the idea worked in bacteria, an easier experiment. One important case involved adding a nuclear location signal (NLS) that was necessary to get the CRISPR-Cas9 into the nucleus of a human cell. Zhang devised ways to add different nuclear location signals to Cas9, and then Marraffini tested them to see if they worked in bacteria. "If you add an NLS and it stops working in bacteria, then you know it also won't work in humans," he explains.

Marraffini believed that they had a fruitful collaboration going, based on mutual respect, that could lead, if they were successful, to being coauthors on the resulting paper and co-inventors on what might be a lucrative set of patents. For a while, that would indeed be the case.

When did he know it?

The work that Zhang did in early 2012 with Marraffini would not lead to any published results until the beginning of 2013. That would later raise a multimillion-dollar question for prize jurors, patent examiners, and historical chroniclers judging The Great CRISPR Race: What did Zhang know and do before Doudna and Charpentier published their CRISPR-Cas9 *Science* paper online in June 2012?

One person who later reconstructed that history was Eric Lander, Zhang's mentor at the Broad. In a controversial article titled "The Heroes of CRISPR," which I will discuss a little later in this book, Lander would tout Zhang's importance. "By mid-2012," he wrote, Zhang "had a robust three-component system consisting of Cas9 from either *S. pyogenes* or *S. thermophilus*, tracrRNA, and a CRISPR array. Targeting sixteen sites in the human and mouse genomes, he showed that it was possible to mutate genes with high efficiency and accuracy."[10]

Lander offered no proof of this assertion, and Zhang had yet to publish any evidence that he had nailed down experimentally the precise role of all the CRISPR-Cas9 components. "We held back," Zhang says. "I did not realize there was competition."

But then, in June of that year, the Doudna-Charpentier paper was published online. Zhang read it when he got one of the regular email alerts sent out by *Science* magazine, and it prodded him to get moving. "That's when I realized that we have to wrap this up and get published," he says. "I thought to myself, 'We don't want to be scooped on the gene-editing part of this.' That was the bar for me: showing that you could use this for editing in human cells."

Zhang bristles slightly when I ask if he was building on the Charpentier-Doudna discoveries. He had, he insists, been striving for more than a year to turn CRISPR into a gene-editing tool. "I don't look at it as taking the torch from them," he says. He was working in the living cells of mice and humans, not just in a test tube. "Theirs was not a gene-editing paper. It was a biochemistry experiment in a test tube."[11]

To Zhang, "a biochemistry experiment in a test tube" was meant as a disparagement. "Showing that CRISPR-Cas9 cleaves DNA in a test tube is not an advance in terms of gene editing," he says. "In gene editing, you have to know whether or not it cleaves in cells. I always worked directly in cells. Not *in vitro*. Because the environment in cells is different than in the biochemistry environment."

Doudna makes the reverse argument, saying that some of the most important advances in biology come when the molecular components are isolated in a test tube. "What Feng was doing was using the entire Cas9 system, with all the genes and CRISPR array that were a part of it, and expressing that in cells," she says. "They weren't doing biochemistry, so they didn't actually know what the individual components were. They didn't know until our paper came out what was necessary."

They are both right. Cellular biology and biochemistry complement each other. That has been true for many of the important discoveries in genetics, most notably CRISPR, and the need to combine the two approaches was a basis for the collaboration between Charpentier and Doudna.

Zhang insists that his gene-editing ideas were already in hand by the time he read the Doudna-Charpentier paper. He presented notebook pages describing experiments in which he used the three components of a CRISPR-Cas9 system—the crRNA, tracrRNA, and Cas9 enzyme—to make edits in a human cell.[12]

There is, however, evidence that in June 2012 he still had a long way to go. A graduate student from China named Shuailiang Lin worked in Zhang's lab on the CRISPR project for nine months, and he would be listed as a coauthor on the paper Zhang eventually produced. In June 2012, when Lin was about to return to China, he prepared a slide show entitled "Summary of CRISPR Work during Oct. 2011–June 2012." It indicates that Zhang's attempts at gene editing thus far were inconclusive or a failure. "No modification seen," one slide reports. Another shows a different approach and declares, "CRISPR 2.0 fail to induce genome modification." And the final summary slide declares, "Maybe Csn1 [what Cas9 was then called] protein is too big, we tried several methods to target it into nucleus but all failed. . . . Maybe other factors need to be identified." In other words, according to Lin's presentation, Zhang's lab had not been able to get a CRISPR system to cut in human cells by June 2012.[13]

When Zhang was embroiled with Doudna in a patent battle three years later, Shuailiang Lin expanded on his slideshow information in an email to Doudna. "Feng is not only unfair to me, but also to the science history," Lin wrote. "The 15-page declaration of his and Le Cong's luciferase data is mis- and overstated. . . . We did not work it out before seeing your paper, it's really a pity."[14]

The Broad Institute dismissed Lin's email as disingenuous, alleging it was sent in hope of getting a job in Doudna's lab. "There are numerous other examples," the Broad said in a statement, "that make clear that beginning in 2011, Zhang and other members of his lab were actively and successfully engineering a unique CRISPR/Cas9 eukaryotic genome-editing system prior to and independent of what was later published [by Charpentier and Doudna]."[15]

One of Zhang's notebook pages records experiments from the

spring of 2012 that he claims document that he was able to produce results that showed the CRISPR-Cas9 system made edits in human cells. But as is often the case with scientific experiments, the data were open to interpretation. They did not clearly prove that Zhang had succeeded in editing the cells because some of the results indicated otherwise. Dana Carroll, a biochemist at the University of Utah, examined Zhang's notebook pages as an expert witness on behalf of Doudna and her colleagues. He says that Zhang left out some of the conflicting or inconclusive data contained in his notebooks. "Feng cherry-picked the data," he concludes. "They even had data that indicated an editing effect when Cas9 was not included."[16]

There is one other aspect of Zhang's work in early 2012 that seems to have fallen short. It goes back to that issue of the role of the tracrRNA. If you will recall, Charpentier discovered in her 2011 paper that tracrRNA is needed for the creation of the crRNA that serves as a guide for the Cas9 enzyme. But it was not until the Doudna-Charpentier June 2012 paper that it was clear the tracrRNA has the more important role of being part of the binding mechanism that allows Cas9 to cut DNA in the targeted location.

In his January 2012 grant application, Zhang did not describe the full role of the tracrRNA. Likewise, in his notebook pages and declaration describing the work he had done before June 2012, there is no evidence that he appreciated the role the tracrRNA plays in cleaving the targeted DNA. One of the relevant pages, Carroll says, "includes a rather detailed recipe of the components included, and that list does not have anything that suggests that a tracrRNA was included." Zhang's lack of understanding of the role of the tracrRNA, Doudna and her supporters would later say, was the main reason that his experiments were not working well before June 2012.[17]

Zhang himself, in the paper that he and his colleagues eventually published in January 2013, seemed to acknowledge that a full understanding of the role of the tracrRNA did not come until he saw what Doudna and Charpentier had published. He noted that it had "previously been shown" that the tracrRNA was needed to cleave DNA, and he footnoted the Doudna-Charpentier paper at that point. "The reason that Feng knew those two RNAs were required was based on reading our paper," Doudna says. "If you look at Feng's 2013 article, we are cited and we're cited for that reason."

When I ask Zhang about this, he says that he included the footnote as a standard practice, because the Doudna-Charpentier paper was the first to publish on the full role of tracrRNA. But he and the Broad Institute say that he was already experimenting with systems that linked the tracrRNA to the crRNA.[18]

These are murky claims to sort out. For what it's worth, my own assessment is that Zhang was working on using CRISPR for human gene editing beginning in 2011, and by mid-2012 he was focusing on the Cas9 system and showing some success, but not a lot, in getting it to work. However, there is no clear evidence, and certainly no published evidence, that he had fully sorted out the precise components that were essential or that he appreciated the ongoing role of the tracrRNA until after reading the Doudna-Charpentier paper of June 2012.

Zhang was open about one thing he learned from the Doudna-Charpentier paper: the possibility of fusing the crRNA and the tracrRNA into a single-guide RNA that could be programmed to target a desired DNA sequence. "We adapted a chimeric crRNA-tracrRNA hybrid design recently validated in vitro," he later wrote, with a footnote citing the Doudna-Charpentier paper. Marraffini, who was still working with Zhang in June 2012, agrees: "Feng and I began using a single-guide RNA only after we saw Jennifer's paper."

As Zhang points out, the creation of the single guide was a useful but not totally essential invention. The CRISPR-Cas9 system can work with the tracrRNA and crRNA remaining separate rather than fused into a simpler molecule, as Doudna and Charpentier's team had done. The single guide simplifies the system and allows it to be delivered more easily into human cells, but it's not what enables the system to work.[19]

Chapter 25

Doudna Joins the Race

"We were not genome editors"

It was surprising that Jennifer Doudna was even a contender in the race to make CRISPR-Cas9 work in humans. She had never experimented with human cells, nor had she ever engineered gene-editing tools such as TALENs. That was also true of her primary researcher, Martin Jinek. "I had a lab full of biochemists and people doing crystallography and that sort of thing," she says. "Whether it was creating cultured human cells or even those of nematode worms, that was not the kind of science my lab was expert in." So it was a testament to her willingness to take risks that she jumped into what she knew would be a crowded race to take their discoveries about CRISPR-Cas9 and turn it into a tool that would work in human cells.

Doudna realized, correctly, that using CRISPR to edit human genes was the next breakthrough waiting to happen. She assumed that other researchers, including Eric Sontheimer and probably people at the Broad, were racing to do it, and she felt a competitive urgency. "After our June paper, I knew we had to speed up, and it wasn't clear that our collaborators had the same commitment," she recalls. "That was a frustration to me. I'm competitive." So she pushed Jinek to work more aggressively. "You need to make this your absolute priority," she

repeatedly told him, "because if Cas9 is a robust technology for human genome editing then the world changes." Jinek worried that it would be difficult. "We were not genome editors, unlike some of the labs that pioneered the method," he says, "so we had to reinvent what others had already done."[1]

Alexandra East

At first, Doudna later admitted, she suffered "many frustrations" in her quest to make CRISPR-Cas9 work in human cells.[2] But as the fall semester of 2012 began—and Zhang was racing to finish his own experiments—she got a lucky break. A new graduate student named Alexandra East, who had experience working with human cells, joined the lab. What made her arrival especially interesting was where she came from: she had received her training and honed her gene-editing skills as a technician at the Broad Institute, working with Feng Zhang and others.

East was able to grow the necessary human cells and then began testing ways to get Cas9 into the nucleus. When she started getting the data from her experiments, she was not sure that they showed evidence of gene editing. Sometimes biology experiments do not have clear results. But Doudna, who had a far better eye for assessing results, saw the experiments as successful. "When she showed me the data, it was immediately clear to me that she had beautiful evidence of genome editing by Cas9 in the human cells," Doudna says. "This is a classic difference between a student who is in training and someone like me who's been doing this for a while. I knew what I was looking for, and when I saw the data she had, it just clicked and I thought, 'Yes, she's got it.' Whereas she was unsure and thought she might have to do the experiments again, I was saying, 'Oh my gosh, this is huge! This is so exciting!'"[3]

To Doudna, this was evidence that getting CRISPR-Cas9 to edit in a human cell was not a difficult leap or a major new invention: "It was very well known how you could tag proteins with nuclear localization signals to get them to go into the nucleus, which is what we did with Cas9. It was also well known how to change the codon usage in a gene so it would be expressed well in mammalian cells versus bacteria, and we did that as well." So she did not feel that it was a great inventive step, even though she was racing to be the first to do it. It merely required adapting methods that others had used in the past, such as with TALENs, to get enzymes into the nucleus of a cell. East had been able to do it in a few months. "It was easy once you knew the components," Doudna says. "A first-year grad student was able to do it."

Doudna felt it was important to publish something as soon as possible. She realized—correctly, as it turned out—that if other labs became the first to show that CRISPR-Cas9 could be ported to human cells, they would claim that to be a major discovery. So she pushed East to firm up her data through repeated experiments. In the meantime, Jinek worked on ways to turn the single-guide RNA that they had devised in test tubes into a guide that could get Cas9 to the right target in a human cell. It was not easy. The single-guide RNA that he had engineered was not, it turned out, quite long enough to work most efficiently on human DNA.

CHAPTER 26

Photo Finish

Zhang's final lap

When Feng Zhang began to test the idea of using a single-guide RNA, he discovered that the version described in the Doudna-Charpentier paper of June 2012 worked poorly in human cells. So he made a longer version of the single-guide RNA that included a hairpin turn. That made the single guide more efficient.[1]

Zhang's modification showed one difference between doing something in a test tube, like Doudna's team, and doing it in human cells. "Jennifer was probably convinced by the biochemical results that the RNA didn't need that extra chunk," he says. "She thought the short single guide that Jinek had engineered was sufficient, because it worked in a test tube. I knew that biochemistry does not always predict what will actually happen in living cells."

Zhang also did other things to improve the CRISPR-Cas9 system and optimize it so that it would work in human cells. It's sometimes hard to get a large molecule through the membrane surrounding a cell nucleus. Zhang used a technique that involved tagging the Cas9 enzyme with a nuclear localization sequence, which grants a protein access to the otherwise impenetrable cell nucleus.

In addition, he used a well-known technique called "codon optimization" to make the CRISPR-Cas9 system work in human cells. Codons are the three-letter snippets of DNA that provide instructions for the specific arrangement of amino acids, which are the building blocks used to make proteins. A variety of codons can code for the same amino acid. In different organisms, one or another of these alternative codons may work more efficiently. When trying to move a gene-expression system from one organism to another, such as from bacteria to a human, codon optimization switches the codon sequence to the one that works best.

On October 5, 2012, Zhang sent his paper to the editors of *Science*, who accepted it on December 12. Among the authors were Shuailiang Lin, the postdoc who said that Zhang was making little progress until after the Doudna-Charpentier paper appeared, and Luciano Marraffini, who had helped Zhang focus on Cas9 but would later be dropped from his main patent application. After describing their experiments and results, their paper concluded with one of those significant final sentences: "The ability to carry out multiplex genome editing in mammalian cells enables powerful applications across basic science, biotechnology, and medicine."[2]

Zhang vs. Church

For twenty-five years, George Church had been working on various methods to engineer genes. He had trained Feng Zhang and was still nominally the academic advisor of Zhang's lead coauthor, Le Cong. But until the late fall of 2012, he hadn't been told—or thought he hadn't been told—by either of them that they had been working for more than a year on turning CRISPR into a human gene–editing tool.

It was not until November of that year, when Church went to the Broad Institute to give a talk, that he found out that Zhang had submitted a paper to *Science* on using CRISPR-Cas9 in human cells. That was a shock, because Church had just submitted a paper to the same journal on the same topic. He was furious and felt betrayed. He had previously published papers on gene editing with Zhang, and he didn't realize that his former student now considered him a rival rather than a collaborator. "I guess Feng didn't get the full culture of my lab," Church says. "Or maybe he just felt the stakes were so high so he didn't tell me." Although Le Cong had moved to the Broad to work with Zhang, he was still a graduate student at Harvard and Church was still officially his advisor. "It was upsetting and seemed to me a breach of protocol that my own student was doing something he knew would interest me but he kept from me," Church says.

Church raised the issue with the Harvard Medical School's dean for graduate studies, who agreed that it was improper. Eric Lander then accused Church of bullying Le Cong. "I didn't want to make a federal case out of it," Church says. "I didn't think I was bullying him, but Eric did. So I backed off." [3]

In order to sort this out, I shuttled back and forth between the various contending parties, finding myself constantly reminded that memory can be an unreliable guide to history. Zhang insists that he did, in fact, tell Church that he was working on CRISPR in August 2012, when they drove together to the San Francisco airport from a cutting-edge conference, known as Science Foo Camp, held on the Google campus an hour away. Church has narcolepsy, and he admits he could have dropped off to sleep while Zhang was talking. But even if that happened, it does not, at least in Church's opinion, get Zhang off the hook for failing to communicate his plans, since he surely would have noticed that he was getting no response from Church.

Over dinner one night, I ask Lander his view of the dispute. Church's narcolepsy issue is "nonsense," he insists, and he accuses Church of starting his own work on CRISPR only after Zhang told him he was embarked on that task. When I ask Church about this, I think that I can detect his placid face tightening beneath his beard. "That is absurd," he replies. "If my students had told me that they wanted to establish their own name in this, I would have backed off. I had a lot else I could have done."

The quarrel so unsettled Le Cong, who is shy and polite, that he subsequently avoided doing much more work in the CRISPR field. When I tracked him down at Stanford Medical School, where he is focusing his research on immunology and neuroscience, he had just returned from his honeymoon. He told me that he thought he had behaved properly when he withheld from Church the details of what he was doing in Zhang's lab. "The two labs were independent research groups at two institutions," he says. "The principal investigators [Zhang and Church] were responsible for sharing information or materials. This is what we were taught as entering PhD students in our Responsible Conduct of Research class."[4]

When I tell him Cong's version of the story, Church chuckles. He teaches an ethics course at Harvard, and he agrees that the behavior of Zhang and Cong was not unethical. "It was within the norms of science." It did, however, violate the norms he tried to cultivate in his own lab. History would have been a little different, he says, if Zhang and Cong had stayed working for him rather than moving to the Broad. "If they had stayed in my lab, where there was a culture of open behavior, I would have made sure that their relationship with Jennifer was much more collaborative, and there wouldn't have been all the patent battles."

Ingrained in Church's character are instincts that promote reconciliation. Zhang, likewise, avoids conflict. He uses his disarming smile as an effective shield to avoid confrontation. "When one of our grandchildren was born, Feng sent us a colorful play-mat with the alphabet on it," Church says. "He also invites me to his workshops each year. We all move on." Zhang feels likewise. "We hug when we see each other."[5]

Church succeeds

Church and Zhang ended up in a virtual tie in showing how CRISPR-Cas9 could be engineered for use in human cells. Church submitted his paper to *Science* on October 26, three weeks after Zhang sent his. After dealing with referee comments, they were both accepted by the editors on the same date, December 12, and were published online simultaneously on January 3, 2013.

Like Zhang, Church created a version of Cas9 that was codon-optimized and had a nuclear localization sequence. Drawing on (and crediting more generously than Zhang did) the Doudna-Charpentier paper of June 2012, Church also synthesized a single-guide RNA. His version was longer than the one Zhang devised and ended up working even better. In addition, Church provided templates for the homology-directed repair of the DNA after CRISPR-Cas9 had made its double-strand break.

Though their papers differ somewhat, they both came to the same historic conclusion. "Our results establish an RNA-guided editing tool," Church's paper declared.[6]

The editor at *Science* was surprised, and a bit suspicious, that the journal had received two papers on the same topic from researchers who were supposed to be colleagues and collaborators. Was he being gamed? "The editor felt as if Feng and I were doing some kind of double dipping, doing two papers when we should have submitted one," Church recalls. "He required a letter from me saying these papers were actually done without knowledge of each other."

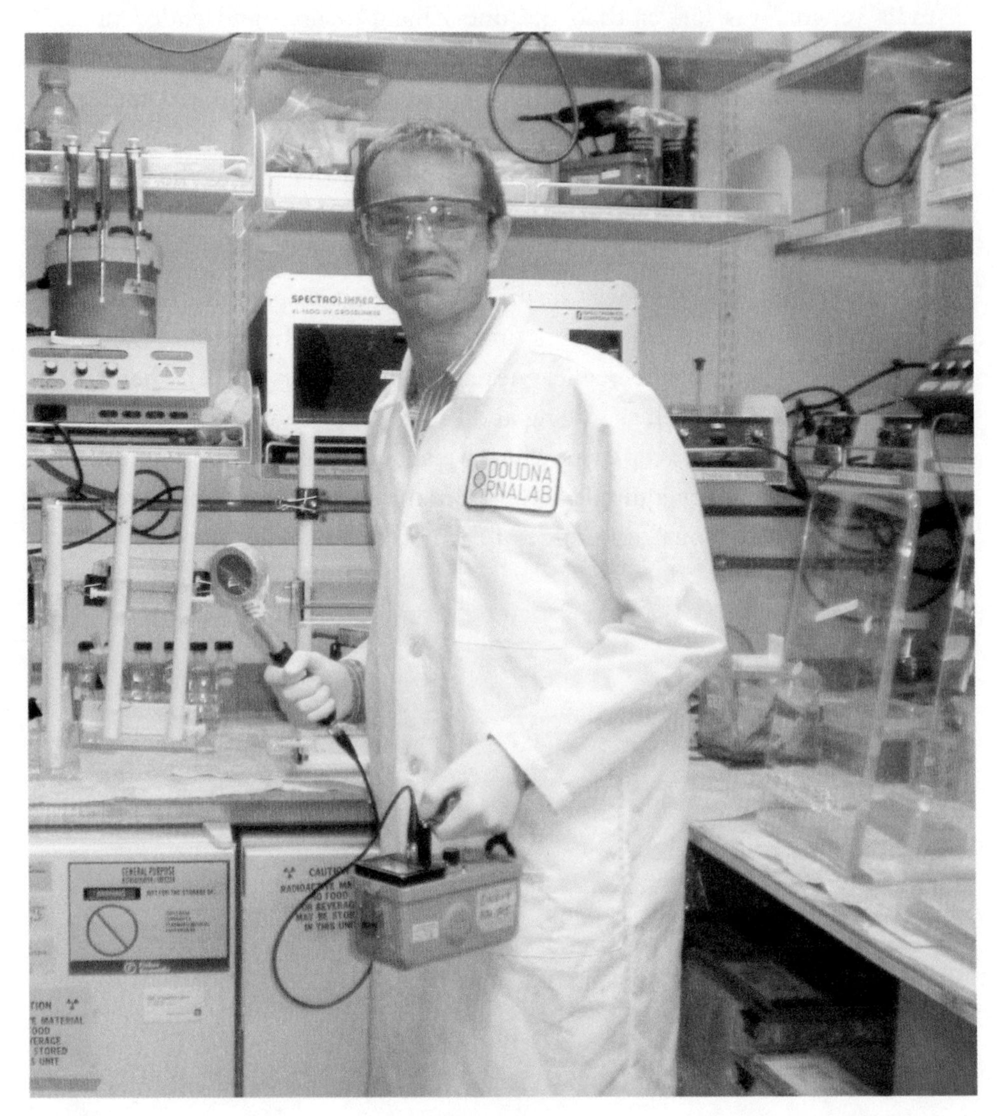

Martin Jinek

CHAPTER 27

Doudna's Final Sprint

In November 2012, Doudna and her team were pushing hard to pin down the results of their experiments so they could win the race to publish on the use of CRISPR-Cas9 in humans. She didn't know that Church had just submitted a paper to *Science*, and she had barely heard of Feng Zhang, who also had. Then she got a phone call from a colleague. "I hope you're sitting down," the caller said. "CRISPR is turning out to be absolutely spectacular in George Church's hands."[1]

Doudna already knew from Church's email that he was working on CRISPR, and when she heard about his progress in making it work in humans, she gave him a call. He was gracious and explained the experiments he had done and the paper he had submitted. By then, Church had learned about Zhang's work, and he told Doudna that it was slated for publication as well.

Church agreed to send Doudna a copy of his manuscript as soon as the editors at *Science* accepted it. When she received it in early December, she was deflated. Jinek was still doing experiments in her lab, and the data they had were not as extensive as those of Church.

"Should I still go ahead and try to publish my work anyway?" she asked Church. He said yes. "He was very supportive of our work and of us publishing," she says. "I thought he behaved as a great colleague." Whatever experimental data she produced, Church told Doudna, would add to the accumulation of evidence, especially on how best to tailor the RNA guide.

"I felt it was important to keep pushing with our experiments, even if others were already doing the same work," Doudna later told me, "because that would show how easy it was to use Cas9 for human genome editing. It showed that you didn't have to have special expertise to use the technology, and I felt that that was important for people to know." Publishing their work would also help her stake a claim that she had demonstrated CRISPR-Cas9 could work in human cells at approximately the same time as competing labs had.

That meant she needed to get her paper published quickly. So she called a colleague at Berkeley who had recently started an open-access electronic journal, *eLife*, that published papers after less review time than traditional journals such as *Science* and *Nature*. "I talked to him, described the data, and sent him a title," Doudna says. "He said it sounded interesting and he would get it reviewed quickly."

Jinek, however, was reluctant to rush their paper into print. "He's a real perfectionist, and he wanted to have a lot more data, a bigger story," she recalls. "He felt what we had wasn't worth publishing." They had many heated discussions, including one in the Berkeley quad in front of their lab in Stanley Hall.

"Martin, we have to publish this, even if it's not quite the story we wish we could tell," Doudna said. "We have to put out the best story that we can, with the data that we have, because we don't have any more time. These other papers are coming out and we have to publish."

"If we publish this work, we're going to look like amateurs in the genome editing field," Jinek shot back.

"But Martin, we are amateurs, and it's okay," she replied. "I don't think people are going to think badly of us. If we had six more months, we could do a lot more, but I think you will understand better as time goes by that it's incredibly important for us to publish this right now."[2]

Doudna recalls that she "put her foot down," and after a bit more discussion they came to an agreement: Jinek would put together the data and figures for the experiments, but Doudna would have to write the paper.

At the time, she was working on revising a second edition of a textbook on molecular biology she had written with two colleagues.[3] "We hadn't been entirely happy with the first edition, so we rented a house in Carmel to have a two-day powwow on how to revise it," she says. As a result, she found herself in mid-December in Carmel, where it was wickedly cold, in a house that had no working heat. The owners said they would call a repair person, but they couldn't get anyone out there right away. So Doudna and her coauthors huddled around the fireplace as they worked late into the night revising their textbook.

After everyone went to bed at 11 p.m., Doudna stayed up to prepare her CRISPR paper for *eLife*. "I was exhausted and cold and I realized that I had to write the paper then or it wouldn't get written," she says. "So I sat up for three hours in bed, pinching myself to stay awake, and typed out the text of a draft." She sent it off to Jinek, who kept coming back with suggestions. "I didn't tell my textbook coauthors or editors about any of this, so you can imagine the scene where I'm in this freezing cold house trying to talk about this textbook but I'm totally distracted because I knew I had to get the paper written, and Martin kept coming back with revisions." Finally, she cut Jinek off and declared the paper finished. On December 15, she emailed it to *eLife*.

A few days later, she and her husband, Jamie, and their son, Andy, left for a ski vacation in Utah. She spent a lot of her time in their room at the lodge as she negotiated little fixes with Jinek and pushed the *eLife* editor to speed up the reviewing process. Every morning she would check the *Science* magazine website to see if the Church or Zhang paper had been published. The main scholar who was doing the peer review of her paper was in Germany,[4] and Doudna was prodding him by email almost daily.

She was also on the phone with her former collaborator Emmanuelle Charpentier, who was in Umeå, where it was now dark all day. "I was trying to manage my relationship with her, and I didn't want her to feel that we had somehow cut her out of that story, but the reality was that she hadn't participated in the science for the *eLife* paper,"

Doudna says. "So we acknowledged her, but in the end she wasn't a coauthor." Doudna sent her a draft of the manuscript, hoping she would not be upset. "I'm fine," Charpentier responded, without much elaboration. There was a certain frostiness. What Doudna did not quite understand was that, even though Charpentier had not wanted to collaborate on the effort to edit human cells, she felt a little proprietary about the CRISPR-Cas9 system. After all, she was the one who had brought Doudna in on that work when they met in Puerto Rico.[5]

When the peer-review coordinator in Germany finally got back with comments, he asked for a few additional experiments. "A few of the mutated targets must be sequenced, just to demonstrate that the expected types of mutations are present," he wrote. Doudna was able to brush him back. Doing the suggested experiments would "require analyses of close to a hundred clones," she replied, which would "be better performed as part of a larger study."[6]

She prevailed, and on January 3, 2013, *eLife* accepted her paper. But she couldn't celebrate. The evening before, she had received, out of the blue, a happy-new-year email that did not portend a happy new year:

From: Feng Zhang
Sent: Wednesday, January 02, 2013 7:36 PM
To: Jennifer Doudna
Subject: CRISPR
Attachments: CRISPR manuscript.pdf

Dear Dr. Doudna,

Greetings from Boston and happy new year!

I am an assistant professor at MIT and have been working on developing applications based on the CRISPR system. I met you briefly during my graduate school interview at Berkeley back in 2004 and have been very inspired by your work since then. Our group in collaboration with Luciano Marraffini at Rockefeller recently completed a set of studies applying the type II CRISPR system to carry out genome editing of mammalian cells. The study was recently accepted by *Science* and it will be publishing online tomorrow. I have attached a copy of our paper for your review. The Cas9 system is very powerful and I would love to talk with you sometime. I am sure we have a lot of synergy and perhaps there are things that would be good to collaborate on in the future!

Very best wishes, Feng
Feng Zhang, Ph.D.
Core Member, Broad Institute of MIT and Harvard

If Jinek had been less balky, I later ask Doudna, might her paper have been published sooner? Might she have been able to tie, or even beat, Zhang and Church, even though her team had finished their experiments after them? "It would have been tough," Doudna says. "I don't think so. We were still doing experiments right up until the very last minute because Martin, rightfully, wanted to make sure that the data included in the paper had been replicated three times. I wish it had been possible to submit earlier, but it probably wasn't."

Their paper did not have an extended version of the guide RNA, which both Zhang and Church showed worked better in human cells. Unlike Church's paper, theirs also did not include templates for homology-directed repair that would create more reliable DNA edits. However, it did show that a lab specializing in biochemistry could quickly move CRISPR-Cas9 from a test tube to human cells. "We show here that Cas9 can be expressed and localized to the nucleus of human cells," Doudna wrote. "These results demonstrate the feasibility of RNA-programmed genome editing in human cells."[7]

Some great discoveries and inventions—such as Einstein's theories of relativity and the creation of the transistor at Bell Labs—are singular advances. Others—such as the invention of the microchip and the application of CRISPR to editing human cells—were accomplished by many groups at around the same time.

On the same day that Doudna's paper appeared in *eLife*, January 29, 2013, a fourth paper was published online showing that

CRISPR-Cas9 worked in human cells. It was by a South Korean researcher, Jin-Soo Kim, who had been corresponding with Doudna and credited her June 2012 paper for laying the ground for his own work. "Your *Science* paper prompted us to start this project," he had written in a July email.[8] A fifth paper published that day, by Keith Joung of Harvard, showed that CRISPR-Cas9 could genetically engineer the embryos of zebrafish.[9]

Even though Doudna had been beaten by a few weeks by Zhang and Church, the fact that five different papers on CRISPR-Cas9 editing in animal cells all appeared in January 2013 reinforced the argument that this discovery was inevitable after it had been shown that it could work in a test tube. Whether that was a difficult step, as Zhang contends, or an obvious step, as Doudna claims, the idea of using an easily programmed RNA molecule to target specific genes and change them was, for humanity, a momentous step into a new age.

CHAPTER 28

Forming Companies

Square dances

In December 2012, a few weeks before the multiple papers on CRISPR gene editing were due to be published, Doudna arranged for one of her business associates, Andy May, to meet with George Church at his Harvard lab. An Oxford-educated molecular biologist, May was the scientific advisor at Caribou Biosciences, the biotech company that Doudna had started with Rachel Haurwitz in 2011, and he wanted to explore the business potential for using CRISPR-based gene editing as a medical technology.

Doudna was giving a seminar in San Francisco when May tried to reach her to report on the meeting. "Can we talk later tonight?" she texted back.

"Yes, but I really need to talk to you," he responded.

When she reached him, she was driving back to Berkeley. He began by saying, "Are you sitting down?"

"Yes, of course, I'm driving home," she replied.

"Well, I hope you don't drive off the road," he said, "because I had this incredible meeting with George who says this will be the most amazing discovery. He's changing his entire gene-editing focus to CRISPR."[1]

Rodger Novak, Jennifer Doudna, and Emmanuelle Charpentier

The excitement over the potential of CRISPR provoked all of the major players to begin square dancing, forming groups and swapping partners in the quest to create companies that would commercialize CRISPR for medical applications. Doudna and May decided, at first, to launch a company with Church and, if they could corral them, some of the other CRISPR pioneers. So in January 2013, Haurwitz accompanied May back to Boston for another meeting with Church.

Church's bushy beard and cultivated eccentricities continued to make him a scientific celebrity, and on the day of the meeting that caused him to be distracted. In an interview with the German magazine *Spiegel*, he had offhandedly speculated about the possibility of resurrecting a Neanderthal by implanting its DNA in the egg of a volunteer surrogate mother. Not surprisingly (except perhaps to him), his phone rang nonstop as tabloid reporters jumped on the story.[2] But he finally focused on his meeting, and within an hour they had a plan. They would try to enlist Emmanuelle Charpentier and Feng Zhang, along with a few top venture capitalists, into a grand consortium to commercialize CRISPR.

Charpentier, in the meantime, was working on a potential startup of her own. Earlier in 2012, she had contacted Rodger Novak, her onetime boyfriend and longtime scientific partner, whom she had befriended when they were researchers at Rockefeller University and in Memphis. They had remained close personal friends, and he had by then joined the pharmaceutical company Sanofi in Paris.

"What do you think about CRISPR?" she asked him.

"What are you talking about?" he replied.

But once he studied her data and consulted with some of his colleagues at Sanofi, he realized it would make sense to launch a business around it. So he called a close friend who was a venture capitalist, Shaun Foy, and they decided to discuss the prospect by going on a surfing trip (even though neither of them knew how to surf) off the northern part of Vancouver Island. A month later, after he had done some more due diligence, Foy called Novak and said they needed to launch a company as soon as possible. "You have to quit your job," he told Novak, who eventually did.[3]

In the hope of getting all of the main players to coalesce, a brunch meeting was scheduled in February 2013 at The Blue Room, a once trendy restaurant with zinc-topped tables nestled in a renovated brick factory near MIT. It was located in Kendall Square in Cambridge, an epicenter of institutions that turn basic science into profitable applications: corporate research centers such as those of Novartis and Biogen and Microsoft, nonprofit institutes such as the Broad and the Whitehead, and a few federal funding agencies such as the National Transportation Systems Center.

Invited to the brunch were Doudna, Charpentier, Church, and Zhang. At the last minute Zhang canceled, but Church urged that they forge ahead without him. "We need to start a company because there is so much we can do with this," he said. "It's so powerful."

"How big do you think it is?" Doudna asked him.

"Well, Jennifer, all that I can tell you is that there is a tidal wave coming," he replied.[4]

Doudna wanted to work with Charpentier, even though they had been drifting apart scientifically. "I spent many hours on the phone with her trying to convince her to come along as a cofounder of what I was doing with George," Doudna says. "But she really did not want to work with some of the folks in Boston. I think she didn't trust them, and in the end I think she was right. But I didn't see that at the time. I was trying to give people the benefit of the doubt."

Church was not as eager to have Charpentier on board. "I became somewhat wary about joining forces with her," he says. "One of the reasons we didn't go in with her was because her boyfriend wanted to be CEO. We just felt that was a nonstarter. You needed to have a process by which you pick the CEO. I was willing to go with it. I tend to be accommodating. But Jennifer laid out the reasons against it, and I said, 'Yeah, you're right.'" (In fact, by then, Novak and Charpentier were no longer romantically involved.)[5]

Andy May had the same negative reaction when Doudna arranged for him to meet with Novak and Foy. "They came in pretty heavy-handed," May says of Charpentier's two business partners. "Their initial approach was that we should get out of the way and let them take care of it."[6]

To be fair, both Novak and Foy had been involved in businesses and knew what they were doing. So along with Charpentier, they broke off discussions with the Doudna-Church group and instead founded a company of their own, CRISPR Therapeutics, initially based in Switzerland but later also in Cambridge, Massachusetts. "It was extremely easy to access money then, especially if you were called CRISPR," Novak says.[7]

For a while in 2013, it seemed as if Doudna and Zhang, despite their rivalry, might become business allies or partners. After he missed the February 2013 brunch at The Blue Room, Zhang sent Doudna an email asking if she might like to collaborate on topics related to the brain, which had long been one of his interests. "I remember sitting at my desk here at my kitchen in Berkeley, seeing him on Skype," she says.

He came out to San Francisco for a conference that spring and met Doudna at the Claremont Hotel in Berkeley. "I went to see her because I thought it was important to have some common alliance around the intellectual property so that you could make this a clean field for people to practice," Zhang says. His idea was that Berkeley's intellectual property and potential patents would be put into a pool with the Broad's, which would make it easy for users to license the CRISPR-Cas9 system. Zhang thought Doudna liked the idea, so Lander phoned her to see if they could establish the framework for such a patent pool. "The next day Eric told me that my trip was productive," Zhang says, "and he thought we had cemented the alliance."

But Doudna had qualms. "I just didn't get a good feeling from Feng," she recalls. "He was not forthright. He was being cagey about when they had actually filed for patents. It didn't sit well with me."

So she decided to give an exclusive license of her intellectual property, which Berkeley managed in coordination with Charpentier, to her existing firm, Caribou Biosciences, and not do an alliance with the Broad. Zhang says that he thinks Doudna "has difficulty trusting people," so she relied too heavily on her former student and Caribou cofounder Haurwitz. "Rachel is a nice person and smart, but not the right person to be the CEO of such a company," he says. "Someone much more seasoned in terms of being able to develop the technology is really important."

The decision not to pool the CRISPR-Cas9 intellectual property would pave the way for an epic patent battle. It also would end up hampering the easy and widespread licensing of the technology. "I think in retrospect, if I had to do it over again, I would have licensed it differently," Doudna says. "When you have a platform technology like CRISPR, it's probably a better idea to license it in a way that offers it as broadly as possible." She had no expertise with intellectual property, and she was at a university that didn't have much either. "It was kind of like the blind leading the blind," she says.

Editas Medicine

Although she did not want to put her intellectual property into a pool with the Broad's, Doudna was still open to becoming a partner in a CRISPR-focused company that would license both her potential patents and those of the Broad. So throughout the spring and summer of 2013, she traveled many times to Boston to dance with a rotating cast of investors and scientists, including Church and Zhang, who were trying to put together companies.

On one trip in early June, she went jogging one evening along the Charles River by Harvard, remembering her days there studying RNA under Jack Szostak. Back then she never thought that her research would lead to commercial ventures. It was not part of the ethos at Harvard. Now Harvard had changed, and so had she. If she wanted to have a direct impact on people, she realized that forming companies would be the best way to translate the basic science of CRISPR into clinical applications.

As the negotiations dragged on through the summer, the stress of figuring out how to form a company began to wear her down. So did flying between San Francisco and Boston every few weeks. Particularly difficult was having to choose between working with Charpentier or with Church and Zhang. "I couldn't tell what the right decision was," she admits. "A couple of people in Berkeley, colleagues that I trusted and had started companies in the past, were telling me to definitely work with the people in Boston, because they were better at business."

Until then, she had rarely gotten sick. But now, in the summer of 2013, she found herself being hit with waves of pain and fever. Her joints locked up in the morning, and sometimes she could barely move. She went to a few doctors, who speculated that she might have a rare virus or perhaps an autoimmune condition.

The problems receded after a month, but then they recurred on a trip to Disneyland with her son late in the summer. "It was just the two of us, and each morning I'd wake up in our hotel and everything was hurting," she recalls. "I didn't want to wake up Andy, so I would go in the bathroom, close the door, and get on the phone with these people in Boston." The stress of the situation, she realized, was affecting her physically.[8]

Nevertheless, she was able to reach an agreement with the Boston-men by the end of the summer. A group coalesced with Doudna, Zhang, and Church at its core. Some Boston-based investment firms—Third Rock Ventures, Polaris Partners, and Flagship Ventures—provided commitments for more than $40 million in initial funding. The group decided to have five scientific founders, so they added two top Harvard biologists who had been working on CRISPR, Keith Joung and David Liu. "It seemed like the five of us were pretty much a dream team," says Church. Their board included representatives from each of the three major investment firms along with some distinguished scientists. There was general consensus about most members, but Church did end up vetoing the selection of Eric Lander.

In September 2013, Gengine, Inc. was founded. Two months later, it changed its name to Editas Medicine. "We have the ability to essentially target any gene," said Kevin Bitterman, a principal at Polaris Partners who served as the interim president for the first few months.

"And we have in our crosshairs any diseases with a genetic component. We can go in and fix the error."[9]

Doudna quits

After only a few months, Doudna's discomfort and stress began to resurface. She sensed that her partners, especially Zhang, were doing things behind her back, and her qualms worsened at a January 2014 medical conference hosted in San Francisco by J.P. Morgan. Zhang came out from Boston with some of the management team from Editas, and they invited Doudna to a couple of meetings with potential investors. She got bad vibes as soon as she walked in. "I could immediately tell from Feng's behavior and body language that something had changed," she says. "He wasn't collegial anymore."

As she watched from a corner, the men at the meeting clustered around Zhang and treated him as the principal. He was introduced as "the inventor" of CRISPR gene editing. Doudna was treated as a secondary player, one of the scientific advisors. "I was being cut out," she says. "There were things involving the intellectual property and I wasn't being kept informed. There was something afoot."

Then she was hit with a surprising piece of news, one that made her understand why she had the queasy sense that Zhang was keeping her in the dark. On April 15, 2014, she received an email from a reporter asking for her reaction to the news that Zhang and the Broad had just been granted a patent for the use of CRISPR-Cas9 as an editing tool. Doudna and Charpentier still had a patent application pending, but Zhang and the Broad, who had put in their own application later, had paid to have their decision fast-tracked. Suddenly it became clear, to Doudna at least, that Zhang and Lander were trying to relegate her and Charpentier to minor players—both in history and in any commercial use of CRISPR-Cas9.

It dawned on Doudna that this was why Zhang and many of the others folks at Editas had seemed secretive with her. The finance people in Boston had been positioning Zhang as the inventor. "They've known about this for months," she said to herself, "and now this patent has been issued and they're trying to completely cut me out and stab me in the back."

It wasn't just Zhang, she felt. It was the gang of men who dominated the biotech and finance world of Boston. "All the Boston people were all so interconnected," she says. "Eric Lander was on an advisory board for Third Rock Ventures, and there was equity going back to the Broad from Editas, and there's licensing agreements that can make them tons of money as long as Feng is seen as the inventor." The episode made her physically ill.

In addition, she was exhausted. She had been flying to Boston once a month for meetings at Editas. "It was brutal. I'd buy an economy class ticket, sit straight up for five hours, and then get in at seven in the morning. I'd go to the United Club, take a shower, change my clothes, go to Editas, have our meetings, and then I'd often go to Church's lab to talk about science. Then I would jump on a six p.m. flight back to California."

So she decided to quit.

She talked to a lawyer about how to extract herself from the agreement she had signed. It took a little time, but by June they had drafted an email to the CEO of Editas saying that she was resigning. They finalized the text over the phone when she was at a meeting in Germany. "Okay, it's ready to go," the lawyer told her after they wrestled with a few final changes. It was evening in Germany and afternoon in Boston when she hit the Send button. "I wondered how many minutes it would be until my phone would ring," she says. "It was less than five minutes, and it was the Editas CEO calling."

"No, no, you can't go, you can't leave," he said. "What's wrong? Why are you doing this?"

"You know what you did to me," she replied. "I'm done. I'm not going to work with people I can't trust, people who stab you in the back. You stabbed me in the back."

The Editas CEO denied being involved in Zhang's patent filings. "Look," Doudna replied, "you may be right or you may be wrong, but either way I can't be part of this company anymore. I'm done."

"What about all your stock?" he asked.

"I don't care," she shot back. "You don't understand. I'm not doing it for the money. And if you think I'm doing it for the money, you don't understand me at all."

When Doudna recounted the episode to me, it was the first time I had heard her so angry. Her steady tone had disappeared. "He claimed he didn't know what I was talking about, and it was ridiculous. It was bullshit. It was all a bunch of lies. And I could be wrong, Walter, but that was my feeling about it."

All of the founders of the company, including Zhang, sent her emails that day asking her to reconsider. They offered to make amends and do whatever was possible to heal the rift. But she refused.

"I'm done," she emailed back.

Immediately, she felt better. "It suddenly seemed like this big weight came off my shoulders."

When she explained the situation to Church, he suggested that, if she wanted, he would consider quitting as well. "I had had a phone call with George at his house on a Sunday," she says. "He vaguely offered to step down, but then he decided not to, and that was his decision."

I ask Church whether Doudna was right to mistrust the other founders. "They were conspiring behind her back, filing for patents without telling her," he agrees. But he says that Doudna should not have been surprised. Zhang was acting in his self-interest. "He probably had lawyers telling him what to do and say," says Church. "I try to understand why people do things." Everyone's actions, including those of Zhang and Lander, could have been predicted, he believes. "Everyone did what I would have expected them to do."

So why didn't he quit? I ask. He explains that it was not logical to be surprised by their behavior, so it was not logical to quit because of it. "I almost left with her, but then I thought, what would that accomplish? It would reward them by giving them the rest of the profits. I always advise people to stay calm. After I thought about it for a while, I decided it was better to be a little calm. I wanted to see a company succeed."

Shortly after she left Editas, Doudna was at a conference where she explained what happened to Charpentier. "Oh, that's interesting," Charpentier responded. "Would you like to get involved with CRISPR Therapeutics?" That was the company that she had founded with Novak.

"You know, it's like getting a divorce," Doudna replied. "I'm not sure I want to get involved again right away. I'm kind of done with companies now."

Within a few months, she decided that she would be most comfortable working with her trusted partner and former student Rachel Haurwitz, with whom she had started Caribou Biosciences in 2011. Caribou had created a spinoff called Intellia, with the mission of commercializing CRISPR-Cas9 tools. "I became very interested in Intellia, because the Caribou team was launching it with the academic scientists I most liked and trusted and respected," Doudna says. These included three great CRISPR pioneers, Rodolphe Barrangou, Erik Sontheimer, and Zhang's former collaborator Luciano Marraffini. They were all brilliant but had an even more important trait: "They were the people who do good science but are more importantly honorable straight-shooters."[10]

As a result, the pioneers of CRISPR-Cas9 ended up in three competing companies: CRISPR Therapeutics, founded by Charpentier and Novak; Editas Medicine, which included Zhang and Church and Doudna until she resigned; and Intellia Therapeutics, founded by Doudna, Barrangou, Sontheimer, Marraffini, and Haurwitz.

CHAPTER 29

Mon Amie

Drifting apart

Doudna's decision to go with a competing company reflected, and perhaps contributed to, the slight coolness that had developed between her and Charpentier. She had tried hard to maintain their relationship. For example, when they first started working together, one of their goals was to crystallize Cas9 and determine its exact structure. After Doudna and her lab succeeded in doing so in late 2013, she asked Charpentier if she wanted to be a coauthor on the resulting journal article. Charpentier, feeling it was a project she had brought to the Doudna Lab, responded that she would like that. This annoyed Jinek, but Doudna went along. "I was really trying to bend over backwards to be generous to her," she says, "and, frankly, I wanted to maintain our scientific and personal relationship."[1]

Partly as a way to keep their scientific partnership intact, Doudna suggested to Charpentier that they coauthor a review article for *Science* in 2014. Unlike a "research article," which is a featured paper on a new discovery, a "review article" is a survey of recent advances on a particular topic. Theirs was titled "The New Frontier of Genome Engineering with CRISPR-Cas9."[2] Doudna wrote a draft, and Charpentier made some edits. It helped to paper over, so to speak, any rift that might be developing between them.

Nevertheless, they began drifting apart. Rather than join Doudna in the quest to find ways to use CRISPR-Cas9 in humans, Charpentier told her that she planned to focus on fruit flies and bacteria. "I like basic research more than looking for tools," she says.[3] There was another underlying reason for the strain: from Doudna's perspective, she was an equal co-discoverer of the CRISPR-Cas9 system, but Charpentier viewed CRISPR-Cas9 as her own project, one that she had brought Doudna into late in the game. At times she spoke of it as "my work" and referred to Doudna as if she were a secondary collaborator. Now Doudna was basking in the limelight, giving interviews and making plans to pursue new CRISPR-Cas9 studies.

Doudna never quite understood Charpentier's proprietary feelings and couldn't figure out how to deal with the coolness that was evident beneath her warm and insouciant manner. She kept suggesting ways they could work together, and Charpentier would reply, "That sounds great." But then nothing would happen. "I wanted to continue collaborating, and Emmanuelle clearly didn't," Doudna says with sadness in her voice. "She never came out and said that to me. We just drifted apart." Eventually, Doudna became frustrated. "I came to feel that it was a passive-aggressive way of interacting," she says. "It was frustrating and it was hurtful."

Part of their problem was the different levels of comfort each had with publicity. When they met at awards ceremonies or conferences, the interactions could be awkward, especially at photo sessions where Charpentier exuded a subtly condescending and amused attitude when the limelight focused on Doudna. Eric Lander, Doudna's occasional antagonist at the Broad Institute, told me that when he talked to Charpentier she expressed resentment at the publicity Doudna got.

Rodger Novak saw Doudna as an American comfortable with acclaim, and his friend Charpentier, whose reputation he protected, as a more properly reticent Parisian. He pushed Charpentier to do more interviews and even get training in how to deal with the media. "It's just a different style of an individual not being on the West Coast but being European, a French person, who focuses more on science than on media hype," he later said.[4]

That is not fully accurate. Although she was comfortable with being a public figure and flattered by recognition, Doudna was not, in fact, someone who actively sought celebrity. She made a point of trying to share the limelight and prizes with Charpentier. Rodolphe Barrangou puts more of the blame on Charpentier. "Emmanuelle makes people feel uncomfortable, even when it comes time to pose for pictures or to be in a green room before a public appearance," he says. "It's baffling to me her lack of desire to share credit with others. I watch Jennifer try to share the light and even overcompensate, but Emmanuelle will seem slightly recalcitrant and resistant."[5]

Their difference in style was reflected in many ways, including their musical tastes. At one of the award ceremonies they attended together, they each got to choose the song that would play when they went onstage. Doudna chose Billie Holiday's bluesy rendition of "On the Sunny Side of the Street." Charpentier selected a technopunk piece from the French electronica duo Daft Punk.[6]

One substantive issue that came between them is one that historians know all too well. Almost every person in any saga tends to remember their own role as being a little more important than the other players see it. That's true in our own lives. We recall vividly the brilliance of our own contributions to a discussion; we're a bit hazier when recalling the contributions of others, or we tend to minimize their significance. As Charpentier views the CRISPR narrative, she was the one who first worked on Cas9, identified its components, and then brought Doudna into the project.

Take, for example, the pesky little issue that keeps cropping up in this tale of the ongoing role of tracrRNA, which not only helps to create the crRNA that guides to a targeted gene but then also, as Doudna and Charpentier revealed in their 2012 paper, sticks around to help the CRISPR-Cas9 complex cleave the targeted DNA. After they published the paper, Charpentier would occasionally suggest that she knew about the ongoing role of tracrRNA back in 2011, before she started collaborating with Doudna.

This began to annoy Doudna. "If you look at talks that she's given recently and the slides she has shown, my opinion is that she's been coached by lawyers and is trying to present the work as if they already knew that the tracrRNA was important for Cas9's function before we started our collaboration, and I think that's disingenuous, it's untrue," Doudna says. "I don't know whether that was her doing or coaching by lawyers, but I think she kind of tried to blur the line between what she did in her 2011 paper and what was figured out much later."[7]

When I ask Charpentier over dinner about the coolness that has developed between them, she is circumspect. She knows, after all, that I am writing a book with Doudna as the central character, and she has never tried to persuade me to shift my focus. With a dash of indifference, she admits that her March 2011 *Nature* paper did not, in fact, describe the full role of the tracrRNA, but she laughs and adds that Doudna should relax a bit and not be so competitive. "She doesn't need to be so stressed about getting proper credit for the tracrRNA and things," Charpentier says. "I find it unnecessary." She smiles as she describes Doudna's competitive streak, as if she finds that trait both admirable and amusing, but also faintly indecorous.

Their rift was exacerbated in 2017 when Doudna published a book on her CRISPR work, coauthored with Sam Sternberg, that was judicious but tended to use the first person more than Charpentier thought was seemly. "It's written in the first person even though her student did most of the writing," Charpentier says. "He should have been told to write in the third person. I know people who do the prizes and the Swedish mentality. They don't like people to write books too early on." By putting the words "prizes" and "Swedish" in the same sentence, she was referring to the most famous of them all.

Prizes

One force that kept Doudna and Charpentier bonded was scientific prizes. Their chances for winning them were best as a pair. Some carry awards of $1 million or more, but they have an even more important value than the money. They serve as a scorecard that the public, press, and future historians use to decide who deserves the most credit for important advances. Lawyers even cite them in arguments made in patent cases.

Each important science prize is given to a limited number of people (for the Nobel, the maximum is three in each field), so the awards do not reflect the full cast of players who contributed to a discovery. As a result, they can distort history and be a disincentive to collaboration, just like patents.

One of the largest and most glamorous of these awards, the Breakthrough Prize in Life Sciences, was given to Doudna and Charpentier as a pair in November 2014, a few months after Zhang beat them to the first patents. The citation heralded them "for harnessing an ancient mechanism of bacterial immunity into a powerful and general technology for editing genomes."

The prize, which carries a $3 million award for each recipient, had been established a year earlier by the Russian billionaire and early Facebook funder Yuri Milner, along with Sergey Brin of Google, Anne Wojcicki of 23andMe, and Mark Zuckerberg of Facebook. Milner, an ebullient fanboy of scientists, staged a glittering televised award ceremony that infused the glory of science with some of the glamor of Hollywood. The 2014 black-tie event, cohosted by *Vanity Fair*, was held in a spacecraft hangar at NASA's Ames Research Center in Mountain View, California, in the heart of Silicon Valley. The emcees included actors Seth MacFarlane, Kate Beckinsale, Cameron Diaz, and Benedict Cumberbatch. Christina Aguilera performed her hit "Beautiful."

Doudna and Charpentier, wearing elegant floor-length black gowns, were presented the prize by Cameron Diaz and Dick Costolo, then the CEO of Twitter. Doudna took the microphone first and paid tribute to the "puzzle-solving process that is science." Charpentier, with a puckish air, then turned to Diaz, who early in her career was a star in the television show *Charlie's Angels*. "We make three powerful women," Charpentier said, gesturing to Diaz and Doudna, and then turning to the bald and bespectacled Costolo added, "I was wondering if you were Charlie."

In the audience was Eric Lander, who had been a prizewinner the year before and had thus been given the duty to telephone Doudna and Charpentier to say they had won. As the director of the Broad Institute and Zhang's mentor, he was zealously engaged in the battle against them for CRISPR kudos. But he had formed a slight bond with Charpentier, or thought he had, by sharing what he believed was her resentment about the acclaim Doudna was garnering. At first, Doudna was nominated for the Breakthrough Prize on her own, Lander told me. But he was able to persuade the prize jury that Doudna's contributions were not as significant as those of Charpentier, Zhang, and the microbiologists who had originally discovered CRISPR in bacteria. "I got the people to understand that Jennifer is probably prize-worthy, but not for CRISPR but for her work on the structure of RNA," he says. "CRISPR was an ensemble act with a lot of people, and Jennifer's contribution was not the most important."

He was not able to prevail in having the prize go to Zhang, but he did help make sure that Charpentier was selected along with Doudna. He also thought that he had an understanding that Zhang would win the following year. When that didn't happen, he would blame Doudna for blocking it.[8]

The Breakthrough Prizes are limited to two winners in each field. The Gairdner Award in biomedical science, given by a Canadian foundation, is more expansive: it honors up to five researchers. That meant that when the foundation decided in 2016 to honor those who developed CRISPR, a broader array of scientists was represented: Doudna and Charpentier were joined by Zhang and the two Danisco yogurt researchers, Horvath and Barrangou. It also meant that some very important players were left out, including Francisco Mojica, Erik Sontheimer, Luciano Marraffini, Sylvain Moineau, Virginijus Šikšnys, and George Church.

Doudna was upset by the exclusion of her friend Church, so she did two things. She donated her prize money, about $100,000, to the Personal Genetics Education Project, which Church had set up with his wife, Ting Wu, a Harvard molecular biology professor. The project encourages people, especially young students, to understand their genes. She also invited them to the ceremony. She was doubtful Church would accept. After all, he had been left out of the honors and, perhaps more significant, was resistant to wearing a tuxedo. But in his gracious manner, Church did show up, impeccably dressed, with his wife. "I'd like to take this opportunity to celebrate the work of two people who have inspired me for a very long time, George Church and Ting Wu," Doudna said, and then she pointedly noted Church's "huge impact on the gene-editing field, including adapting the CRISPR-Cas system for gene editing in mammalian cells."[9]

Doudna and Charpentier completed a hat trick by winning a third major award in 2018, the Kavli Prize. Named after Fred Kavli, an American entrepreneur born in Norway, it carries many of the trappings of the Nobel Prize: there's a glamorous ceremony and for each recipient $1 million and a gold medal stamped with a bust of the prize's founder. The award can go to three scientists, and the committee chose to add Virginijus Šikšnys, a fitting recognition that had until then eluded the shy Lithuanian. "We dreamed of rewriting the language of life itself, and with the discovery of CRISPR we found a new powerful writing tool," said Norwegian actor Heidi Ruud Ellingsen, who cohosted the ceremony with the American actor and science geek Alan Alda. Doudna wore a short black dress, Charpentier a long one, and Šikšnys a sharp gray suit that looked as if it had been bought for the occasion. After being handed their medals by King Harald V of Norway, they bowed slightly amid a fanfare of trumpets.

Eric Lander

CHAPTER 30

The Heroes of CRISPR

Lander's tale

In the spring of 2015, while Emmanuelle Charpentier was visiting America, she had lunch with Eric Lander in his office at the Broad Institute. In his recollection, she was "in a funk" and resentful of some of the acclaim that Doudna was getting. "It became very clear to me that she is pissed at Jennifer," Lander recalls. "She believed that the credit was going to her more than to the microbiologists," such as Francisco Mojica, Rodolphe Barrangou, Philippe Horvath, and herself, who had originally figured out how CRISPR works in bacteria.

Perhaps Lander was right, or perhaps he was in part projecting his own resentments and stoking ones that Charpentier only vaguely felt. Lander is a persuasive personality who is good at getting people to agree with him. When I ask Charpentier about Lander's recollection, she gives a wry smile and suggests, with a tiny shrug, that the feelings were more Lander's than hers. Nevertheless, there was probably some truth to Lander's perceptions of her feelings. "She was subtle and French about it," he recalls.

His lunch conversation with Charpentier, Lander says, was the origin of what would become a detailed, vibrant, well-reported, and controversial journal article on the history of CRISPR. "After talking to

Emmanuelle, I decided to pull the thread and look back at the origins of CRISPR and give credit to the people who had done the original work but weren't getting the acclaim," he says. "I have this streak in me that I defend the underdog. I was brought up in Brooklyn."

I asked him whether he might have had other motivations as well, including a desire to downplay the role of Doudna and Charpentier, who were pitted against his protégé Feng Zhang for patents and prizes. For someone who is so feisty, Lander can also be laudably self-aware. He answers by referring to Michael Frayn's play *Copenhagen*, which applied the uncertainty principle to the motives of Werner Heisenberg when he visited Niels Bohr early in World War II and discussed the possibility of making an atom bomb. "Like the play *Copenhagen*, I cannot be sure of my own motivations," Lander says. "You don't know your own motives." Wow, I think.[1]

One of the appealing things about Lander is that he is merrily and jubilantly competitive, pushing Zhang to claim his credit, then driving the lawsuits to protect Zhang's patents. His bristly moustache and enthusiastic eyes are at all times expressive, conveying every changing emotion in a way that would delight a poker opponent. His relentless drive and passion to persuade—he reminds me of the late diplomat Richard Holbrooke—made him infuriating to rivals, but it also made him a hard-charging and effective team leader and institution builder. His paper on the history of CRISPR was an example of all of these instincts.

After months of reading all the scientific papers and interviewing many of the participants by phone, Lander published "The Heroes of CRISPR" in the journal *Cell* in January 2016.[2] At eight thousand words, it was vividly written and factually correct in its details. But it provoked a firestorm of responses from outraged critics who charged that it was skewed, in ways both subtle and heavy-handed, to tout the contributions of Zhang and minimize those of Doudna. It was history weaponized.

Lander's narrative began with Francisco Mojica and went through the other players I've discussed in this book, mixing personal color with scientific explanations of each step in the development of CRISPR. He described and praised the work of Charpentier in discovering the tracrRNA, but instead of then showing how she and Doudna went on to figure out the exact role of each component in 2012, he provided a long description of the work of the Lithuanian Šikšnys and his difficulty in getting published.

When Lander got to Doudna, he was pleasant enough. He called her "a world-renowned structural biologist and RNA expert," but he breezed by the work she did with Charpentier in just one paragraph out of the article's sixty-seven. Not surprisingly, Zhang was accorded a more lavish account. After stressing how difficult it was to move CRISPR-Cas9 from bacteria to human cells, Lander described in some detail, but without citing evidence, the work Zhang was doing in early 2012. As for Doudna's January 2013 paper showing the system working in human cells, published three weeks after Zhang's, Lander dismissed it in a sentence that included the dagger-like accusation "with assistance from Church."

The main theme of Lander's essay was important and correct. "Scientific breakthroughs are rarely eureka moments," he concluded. "They are typically ensemble acts, played out over a decade or more, in which the cast becomes part of something greater than what any one of them could do alone." Yet the article clearly had another thrust, one that was done with a velvet glove but was nonetheless an unmistakable diminishment of Doudna. Oddly for an academic journal, *Cell* did not disclose that Lander's Broad Institute was competing for patents with Doudna and her colleagues.

Doudna decided to be muted in her public reaction. She simply posted a comment online stating, "The description of my lab's research and our interactions with other investigators is factually incorrect, was not checked by the author, and was not agreed to by me prior to publication." Charpentier was similarly upset. "I regret that the description of me and my collaborators' contributions is incomplete and inaccurate," she posted.

Church was more specific in his criticisms. He pointed out that he, not Zhang, first demonstrated the use in human cells of an extended guide RNA that ended up working the best. He also disputed the assertions that Doudna had taken information from the preprint he had sent her.

Backlash

Doudna's friends rallied to her cause with a fury that would have impressed a Twitter mob. In fact, it included a Twitter mob.

The most vibrant and viral responses came from one of Doudna's high-octane colleagues at Berkeley, genetics professor Michael Eisen. "There is something mesmerizing about an evil genius at the height of their craft, and Eric Lander is an evil genius at the height of his craft," he wrote and posted publicly a few days after the article appeared. He called the piece "at once so evil and yet so brilliant that I find it hard not to stand in awe even as I picture him cackling loudly in his Kendall Square lair, giant laser weapon behind him poised to destroy Berkeley if we don't hand over our patents."

Eisen, who was upfront about the fact that he was Doudna's partisan friend, charged that Lander's piece was "an ingenious strategy" to promote the Broad and denigrate Doudna under the veneer of a historical perspective. "The piece is an elaborate lie that organizes and twists history with no other purpose than to achieve Lander's goals—to win Zhang a Nobel Prize and the Broad an insanely lucrative patent. It is, in its crucial moments, so disconnected from reality that it is hard to fathom how someone so brilliant could have written it."[3] I think that is not fair or true. My own view is that Lander may have been guilty of zeal as a mentor and a zest for spinning, but not dishonesty.

Other, more dispassionate scientists joined the criticism of Lander, with flames erupting on venues ranging from the scientific discussion board *PubPeer* to Twitter.[4] "'Shitstorm' would be one term of art for the reaction in the genome community to a commentary in *Cell* by Eric Lander," wrote Nathaniel Comfort, a professor of the history of medicine at Johns Hopkins. Comfort called Lander's piece "Whig history," suggesting that it was crafted in order "to use history as a political tool." He even created a Twitter hashtag, #Landergate, that became a rallying spot for those who thought Lander was insidiously slagging competitors of the Broad.[5]

In the influential *MIT Technology Review*, Antonio Regalado focused on Lander's assertions, not backed up with any citations, that Zhang had made great progress on developing CRISPR-Cas9 tools a year before the Doudna-Charpentier 2012 paper was published. "Zhang's discoveries weren't published at the time, and so they are not part of the official scientific record," Regalado wrote. "But they're very important if Broad wants to hold onto its patents. . . . No wonder, then, that Lander might like to see them described for the first time in an important journal such as *Cell*. I think that was a little Machiavellian on the part of Lander."[6]

Women scientists and writers, aware of the injustice done to Rosalind Franklin in some of the histories of DNA, were especially incensed at Lander, whose alpha-male style had never endeared him to feminists even though he has a laudable history of supporting women scientists. "His write-up serves as yet another instance of a woman being written out of scientific history," Ruth Reader, a science journalist, wrote in *Mic*. "This helps explain the urgency behind the backlash to Lander's report: Here again, a male leader appears to be usurping credit (and therefore financial gain) for a discovery that was the work of many." An article on *Jezebel*, which describes itself cheekily as "a supposedly feminist website," was headlined "How One Man Tried to Write Women Out of CRISPR, the Biggest Biotech Innovation in Decades." In it Joanna Rothkopf wrote, "The crediting issue evokes that of Rosalind Franklin."[7]

The flare-up against Lander, which occurred while he was on a trip to Antarctica and could not easily respond, became so newsworthy that mainstream publications covered it. Stephen Hall in *Scientific American* called it "the most entertaining food fight in science in years," and asked, "Why would such a shrewd and strategic thinker like Lander tempt such a public backlash by writing such a cleverly slanted history?" Hall quoted Church as saying of Lander, "The only person that could hurt him was himself," and then merrily declaring, "And you thought scientists couldn't talk smack."[8]

Lander responded by criticizing Doudna for not providing more input on the piece when he emailed her some passages right before it went to press. "I received input about the development of CRISPR from more than a dozen scientists around the world," Lander wrote in an email to Tracy Vence of *The Scientist*. "Dr. Doudna was the only one who declined, which is unfortunate. Nonetheless, I fully respect her decision not to share her perspective."[9] That final gauze-cloaked zinger was quintessential Lander.

The article helped to draw the battle lines in the CRISPR war. Doudna's admirers at Harvard, led by Church and her PhD advisor, Jack Szostak, were infuriated. "It's just an awful, awful, piece of writing," Szostak tells me. "Eric wants the credit for the genetic editing revolution to go to Feng Zhang and him, and not Jennifer. So he just totally belittled her contribution in a way that seems just pure animus."[10]

Even within his own institution, Lander's piece raised hackles. After several members of the staff questioned him about it, he sent them an email addressed "Dear Broadies." It was unapologetic. "The essay aims to describe the whole group of extraordinary scientists (many at the early stages of their careers) who took risks and made critical discoveries," he wrote. "I'm very proud of the essay and its messages about science."[11]

A couple of months after publication, as the controversy still simmered, I got enlisted as a peripheral player. Christine Heenan, who was then vice president of communications at Harvard, was asked by Lander to help smooth things over. I had known Eric for a long time, and I was (and am) one of his alloyed admirers. So Heenan asked me to host a discussion with him for the press and scientific community at the Washington headquarters of the Aspen Institute, where I worked. Her goal was to tamp down the controversy by getting Lander to say that he hadn't meant to minimize Doudna's contributions to the CRISPR field. Lander tried to do what Heenan urged, albeit not in a way that could be described as valiantly. "My intention is not to diminish anybody," he said, adding that Doudna was "a spectacular scientist." That was about it. When he was pressed by the *Washington Post*'s Joel Achenbach, he insisted that his article was factual and did not underplay Doudna's accomplishments. I caught Heenan's eye, and she shrugged.[12]

Eldora Ellison

CHAPTER 31

Patents

"Useful arts"

Ever since the Republic of Venice in 1474 passed a statute giving the inventors of "any new and ingenious device" the exclusive right to profit from it for ten years, people have been wrestling over patents. In the United States, they are enshrined in Article 1 of the Constitution: "The Congress shall have power to . . . promote the progress of science and useful arts by securing for limited times to authors and inventors the exclusive right to their respective writings and discoveries." A year after ratification, Congress passed an act that allowed patents on "any useful art, manufacture, engine, machine, or device, or any improvement thereon not before known."

As courts came to realize, it's complicated to apply such concepts, even to things as simple as a doorknob. In the 1850 case *Hotchkiss v. Greenwood*, which involved a patent application for the manufacture of doorknobs out of porcelain rather than wood, the U.S. Supreme Court began the process of defining what was "obvious" and "non-obvious" in assessing whether an invention was "not before known." Deciding on patents was particularly difficult when it involved biological processes. Nevertheless, biological patents have a long history. In 1873, for example, the French biologist Louis Pasteur was awarded the first known patent for a microorganism: a method for making "yeast free from organic germs of disease." Thus we have pasteurized milk, juice, and wine.

The modern biotechnology industry was born a century later, when a Stanford attorney approached Stanley Cohen and Herbert Boyer and convinced them to file for a patent on the method they had discovered for manufacturing new genes using recombinant DNA. Many scientists, including Paul Berg, the discoverer of recombinant DNA, were horrified at the idea of patenting a biological process, but the royalties that flowed to the inventors and their universities quickly made biotech patents popular. Stanford, for example, made $225 million in twenty-five years by granting hundreds of biotech companies non-exclusive licenses to the Cohen-Boyer patents.

Two major milestones occurred in 1980. The U.S. Supreme Court ruled in favor of a genetic engineer who had derived a strain of bacteria capable of eating crude oil, which made it useful in cleaning up oil spills. His application had been rejected by the Patent Office on the theory that you could not patent a living thing. But the Supreme Court ruled, in a 5–4 decision written by Chief Justice Warren Burger, that "a live, human-made micro-organism is patentable" if it is "a product of human ingenuity."[1]

Also that year, Congress passed the Bayh-Dole Act, which made it easier for universities to benefit from patents, even if the research was funded by the government. Until then, universities often were required to assign the rights to their inventions to the federal agencies that had funded them. Some academics feel that the Bayh-Dole Act cheats the public out of the proceeds from inventions funded with taxpayer money and distorts the way universities work. "Encouraged by a small number of patents that made huge sums, universities developed massive infrastructure to profit from their researchers," argues Michael Eisen, Doudna's colleague at Berkeley. He believes that the government should put all work funded by federal dollars into the public domain. "We all would benefit returning academic science to its roots in basic discovery oriented research. We see with CRISPR the toxic effects of turning academic institutions into money hungry hawkers of intellectual property."[2]

That's an appealing argument, but I believe that, on balance, American science has benefited from the current mix of federal funding and commercial incentives. To turn a basic scientific discovery into a tool or a drug can cost billions of dollars. Unless there is a way to recoup that, there won't be as much investment in research.[3] The development of CRISPR and the therapies it led to are a good example.

CRISPR patents

Doudna did not know much about patents. Little of her previous work had practical application. When she and Charpentier were finishing their June 2012 paper, she reached out to the woman at Berkeley in charge of intellectual property, who set her up with a lawyer.

For research professors in the U.S., the patents to their inventions are usually assigned to the academic institution, in Doudna's case Berkeley, with the inventors having a lot of say over how it will be licensed and taking a portion (in most universities about one-third) of the royalties. In Sweden, where Charpentier was then based, the patent goes directly to the inventor. So Doudna's application was filed jointly by Berkeley, Charpentier personally, and the University of Vienna, where Chylinski was based. Shortly after 7 p.m. on May 25, 2012, just as they were finishing their paper for *Science*, they filed their provisional patent application and used a credit card to pay the $155 fee for processing. It did not occur to them to spend a little extra to have the application expedited.[4]

The 168-page application, which included diagrams and experimental data, described CRISPR-Cas9 and made more than 124 claims for ways that the system could be used. All of the data in the application were from experiments done with bacteria. However, it mentioned delivery methods that could work in human cells, and it made the claim that the patent should cover the use of CRISPR as an editing tool in all forms of life.

As I noted earlier, Zhang and the Broad submitted their own patent application in December 2012, when his paper about editing in humans was accepted by *Science*.[5] It specifically described a process for using CRISPR in *human* cells. Unlike Berkeley, the Broad made use of a neat little provision in the patent process: it paid a small additional fee and agreed to a few conditions in order to expedite consideration under what was known as an Accelerated Examination Request or, more poetically, a Petition to Make Special.[6]

Initially, the Patent Office did not grant Zhang's application, asking for more information. Zhang responded by supplying a written declaration. In it, he made an allegation that infuriated Doudna. He pointed out that Church had sent her a preprint of his paper, and he implied that she used his data in her patent application. "I respectfully question the origin of the example," Zhang said. In one of their legal filings, Zhang and the Broad asserted, "It was only after the Church laboratory shared unpublished data that Dr. Doudna's laboratory reported they were able to adapt a CRISPR-Cas9 system" for use in human cells.

Doudna was outraged at Zhang's declaration because it implied that she had plagiarized Church's data. She called Church at his home on a Sunday afternoon, and he shared her anger at what his former student had alleged. "I'm happy to go public and say you didn't improperly use my data," Church told her. She had been polite to include a sentence about him in her acknowledgments, and it was "outrageous," he later told me, that Zhang would turn that small act of collegiality against her.[7]

Marraffini dropped

As Zhang was waiting for a ruling on his patent applications, he and the Broad did something unusual: they dropped the name of his collaborator Luciano Marraffini from the main application. The somewhat mystifying tale is a sad example of the distorting effects that patent law can have on scientific collaboration. It's also a tale of competitiveness, perhaps even greed, overwhelming kindness, and collegiality.

Marraffini is the soft-spoken Argentinian-born bacteriologist at Rockefeller University who collaborated with Zhang beginning in early 2012 and was a coauthor on his *Science* paper. When Zhang initially filed for his patents, Marraffini was listed as one of the co-inventors.[8]

A year later, Marraffini was called into the office of the president of Rockefeller and told, to his shock and profound sadness, that Zhang and the Broad had decided to narrow some of the patent applications and focus one of them only on the process of making CRISPR-Cas9 work in human cells. Marraffini did not contribute enough to that work to deserve being on the patent, the Broad unilaterally decided, so they were dropping him.

"Feng Zhang didn't even have the politeness to tell me directly," Marraffini says, shaking his head, still looking shocked and sad after six years. "I'm a reasonable guy. If they said my contribution was not worth an equal share, I would have accepted a smaller share. But they didn't even tell me." What particularly pains him is that he views the story of his work with Zhang as an inspiring American tale: two young rising stars who were immigrants, one from China and the other from Argentina, joining forces to show how CRISPR could be used in humans.[9]

When I ask Zhang about this, he likewise speaks quietly and sorrowfully, as if he's the one who is hurt. "I focused on Cas9 from the beginning," he insists when I ask if Marraffini should get some credit for getting him to concentrate on that enzyme. It may have been ungenerous to take Marraffini off the patent, but in Zhang's mind it was not unwarranted. Therein lies one of the problem with patents: they prod people to be less generous in sharing credit.[10]

Conflict

The Patent Office decided to grant Zhang's patent application on April 15, 2014, even though Doudna's application* was still being considered.[11] When she heard, she called Andy May, her business associate,

*I am using shorthand when I refer to the applications. When I talk about Doudna's, I am referring to the ones she did jointly with Charpentier, Berkeley, and the University of Vienna. Likewise, when I talk about Zhang's applications, I am referring to the ones he did with the Broad, MIT, and Harvard.

who was driving. "I remember pulling over in the car and taking the call and getting this blast," he says. "'How did this happen?' she asked. 'How did we get beaten?' She was livid, absolutely livid."[12]

Doudna's application was still languishing at the Patent Office. That raised a question: What happens if you apply for a patent and, before the decision gets made, another person is granted a similar patent? Under U.S. law, you have a year to request an "interference" hearing. So in April 2015, Doudna filed a claim that Zhang's patents should be disallowed because they interfered with the patent applications that she had previously submitted.[13]

Specifically, Doudna submitted a 114-page "Suggestion of Interference" detailing why some of Zhang's claims were "not patentably distinct" from her own pending claims. Even though her team's experiments had involved bacteria, she argued that their patent application "specifically states" that the system can be applied in "all organisms" and provides "detailed descriptions of numerous steps that could be taken to apply the system" to humans.[14] Zhang argued in his response declaration that Doudna's application "did NOT [emphasis in the original] have the features required for Cas9 binding and DNA target site recognition in a human cell."[15]

Thus the battle lines were drawn. Doudna and her colleagues had identified the essential components of CRISPR-Cas9 and engineered a technique to make it work using components from bacterial cells. Their contention was that it was then "obvious" how it would work in a human cell. Zhang and the Broad Institute countered that it was *not* obvious that the system would work in humans. It required another inventive step to make it work, and Zhang had beaten Doudna to it. In order to resolve this issue, the patent examiners in December 2015 launched an "interference proceeding" to be decided by a panel of three patent judges.

When Doudna's lawyers asserted it was "obvious" that a system that worked in bacteria would also work in humans, they were using a term of art. In patent law, the term "obvious" refers to a specific legal concept. Courts have declared that the "criterion for determination of obviousness is whether the prior art would have suggested to a person of ordinary skill in the art that this process would have a reasonable likelihood of success."[16] In other words, you don't deserve a new patent if you merely modified a prior invention in a way that was so obvious that a person with ordinary skill in the field could have done the same with a reasonable likelihood of success. Unfortunately, phrases such as "person of ordinary skill" and "reasonable likelihood of success" are fuzzy when applied to biology, where experiments are less predictable than in other forms of engineering. Unexpected things happen when you start fiddling with the innards of living cells.[17]

The trial

It took a full year for all the briefs, declarations, and motions to be filed, after which a hearing was held in December 2016 before a three-judge panel at the Patent and Trademark Office in Alexandria, Virginia. With its blond-wood dais and simple tables, the hearing room looks like a sleepy county traffic court. But on the day of the trial, a hundred journalists, lawyers, investors, and biotech fans, most of them bespectacled and looking a bit nerdy, began lining up at 5:45 a.m. to get seats.[18]

Zhang's lawyer opened the hearing by stating that the key issue was "whether the use of CRISPR in eukaryotic cells was obvious" after the Doudna-Charpentier 2012 article.[19] To make the case that it was not, he put up a series of posters with statements made earlier by Doudna and her team. The first was from an interview Doudna gave to a Berkeley Chemistry Department magazine: "Our 2012 paper was a big success, but there was a problem. We weren't sure if CRISPR-Cas9 would work in plant and animal cells."[20]

Zhang's lawyer then put up a quote that was not merely an offhand comment but a statement that Doudna and Martin Jinek made in the *eLife* paper that they had rushed into publication in January 2013. Their earlier paper had "suggested the exciting possibility" that the CRISPR system could be used for editing human genes, they wrote, but then they added, "However, it was not known whether such a bacterial system would function in eukaryotic cells." As Zhang's lawyer told the court, "These comments at the time belie this idea that this was all obvious."

Doudna's lawyers rebutted that her comments were simply the mark of a careful scientist. This did not impress the lead judge, Deborah Katz. "Are there any statements," she asked Doudna's lawyer, "in which anybody said they did believe it would work?" The best the lawyer could do was point to Doudna's statement that it was "a real possibility."

Fearing that he was playing a losing hand, Doudna's lawyer shifted the argument. Five labs had made the system work in eukaryotic cells within six months of the publication of the Doudna-Charpentier discovery, he said, which was an indication of how "obvious" such a step was. He displayed a chart showing that they all used well-known methods. "There's no special sauce here," he told the judge. "These labs would not have embarked on this quest unless they had a reasonable expectation of success."[21]

The three-judge panel ended up siding with Zhang and the Broad. "Broad has persuaded us that the parties claim patentably distinct subject matter," the judges declared in February 2017. "The evidence shows that the invention of such systems in eukaryotic cells would not have been obvious."[22]

Doudna's side appealed to the federal courts, beginning a process that took another nineteen months. In September 2018, the U.S. Court of Appeals for the Federal Circuit upheld the ruling of the patent board.[23] Zhang was entitled to his patent; it did not interfere with Doudna and Charpentier's application.

But as happens with many complex intellectual property cases, these rulings did not end the case or give Zhang a total victory. Because there was "no interference" between the two sets of applications, they could be considered separately, which meant that it was still possible that the Doudna-Charpentier application would be granted as well.

Patent priority dispute, 2020

That is what happened. In the final two sentences of its 2018 decision affirming Zhang's patent, the U.S. Court of Appeals had emphasized a significant point. "This case is about the scope of two sets of applied-for claims and whether those claims are patentably distinct," the judge wrote. "It is not a ruling on the validity of either set of claims." In other words, there was no "interference" between the patents granted to Zhang and the pending ones that had been applied for by Doudna and Charpentier. They could be considered as two distinct inventions, and it was possible that *both* could deserve patents or that the Doudna-Charpentier ones would take priority.

Of course such a result would be messy and somewhat paradoxical. If both sets of patents got granted and then seemed to overlap, that would fly in the face of the decision that there was no interference between them. But sometimes life, and in particular life inside of cells and courtrooms, can be paradoxical.

In early 2019, the U.S. Patent Office granted fifteen patents based on the applications that Doudna and Charpentier had filed in 2012. By then, Doudna had hired a new lead attorney, Eldora Ellison, who had blazed an educational path that was tailor-made for the age of biotech. She earned her undergraduate degree at Haverford in biology, then a doctorate at Cornell in biochemistry and cell biology, and finally a law degree at Georgetown. I often suggest to my students that they consider studying both biology and business, as Rachel Haurwitz did, or biology and law, as Ellison did.

When she analyzed the case for me over breakfast, Ellison was able to explain the nuances of both the biology and the law, and she readily cited from memory arcane footnotes in various scientific articles and court decisions. I came to the conclusion that Ellison would be great on the Supreme Court, which nowadays could use at least one justice who understands biology and technology.[24]

Ellison was able to prod the Patent Office in June 2019 to launch a new case.[25] Unlike the first case, which looked only at whether Zhang's patents interfered with the ones that Doudna had applied for, this new case would involve adjudicating the fundamental issue: which side had made the key discoveries first. This new "priority dispute" would attempt to pinpoint, using notebooks and other evidence, precisely when each applicant had invented CRISPR-Cas9 as an editing tool.

In a May 2020 hearing, done by phone because of the coronavirus closures, Zhang's lawyer argued that the issue had already been decided: it was not "obvious" that the CRISPR-Cas9 system discovered by Doudna and Charpentier in 2012 would work in human cells, and therefore Zhang was entitled to a patent for being the first to show how it would. Ellison responded that the legal issues in the new case were not the same. The patent that was granted to Doudna and Charpentier was for the use of CRISPR-Cas9 in all organisms, from bacteria to humans. The question, she said, was whether their patent application from 2012 contained enough evidence to show they had discovered this. She contended that even though their experimental data came from using bacterial components in a test tube, their patent application, when considered in its entirety, described how to use the system in any organism.[26] By late 2020, the case was still dragging along.

In Europe, there was initially a similar situation: Doudna and Charpentier were granted a patent, and then Zhang was also given one.[27] But at that point Zhang's dispute with Marraffini popped up again. After Zhang's applications were revised and Marraffini's name dropped, the European patent court ruled that Zhang could not use the date of his original application as his "priority date." As a result, other patent applications were deemed to have an earlier priority date, and the court revoked Zhang's patent. "Feng's European patent was nullified because of the way he took me off," Marraffini says.[28] By 2020, Doudna and Charpentier had been awarded the major patents also in Britain, China, Japan, Australia, New Zealand, and Mexico.

Were all of these patent battles worth it? Would Doudna and Zhang have been better off coming to a deal rather than battling in court? In retrospect, Doudna's business partner Andy May thinks so. "We would have saved a lot of time and money around all of the legal arguments if we had managed to come together," he says.[29]

To an unnecessary extent, the prolonged fight was driven by emotions and resentments. Instead, Doudna and Zhang could have followed the example of Jack Kilby of Texas Instruments and Robert

Noyce of Intel who, after five years of wrangling, agreed to share the patent rights for the microchip by cross-licensing their intellectual property to each other and splitting the royalties, which helped the microchip business grow exponentially and define a new age of technology. Unlike the CRISPR contestants, Noyce and Kilby obeyed an all-important business maxim: *Don't fight over divvying up the proceeds until you finish robbing the stagecoach.*

Part Four

CRISPR in Action

If ever man fell ill, there was no defense
—no healing food, no ointment, nor any drink—
but for lack of medicine they wasted away,
until I showed them how to mix soothing remedies.

—Prometheus, in Aeschylus's *Prometheus Bound*

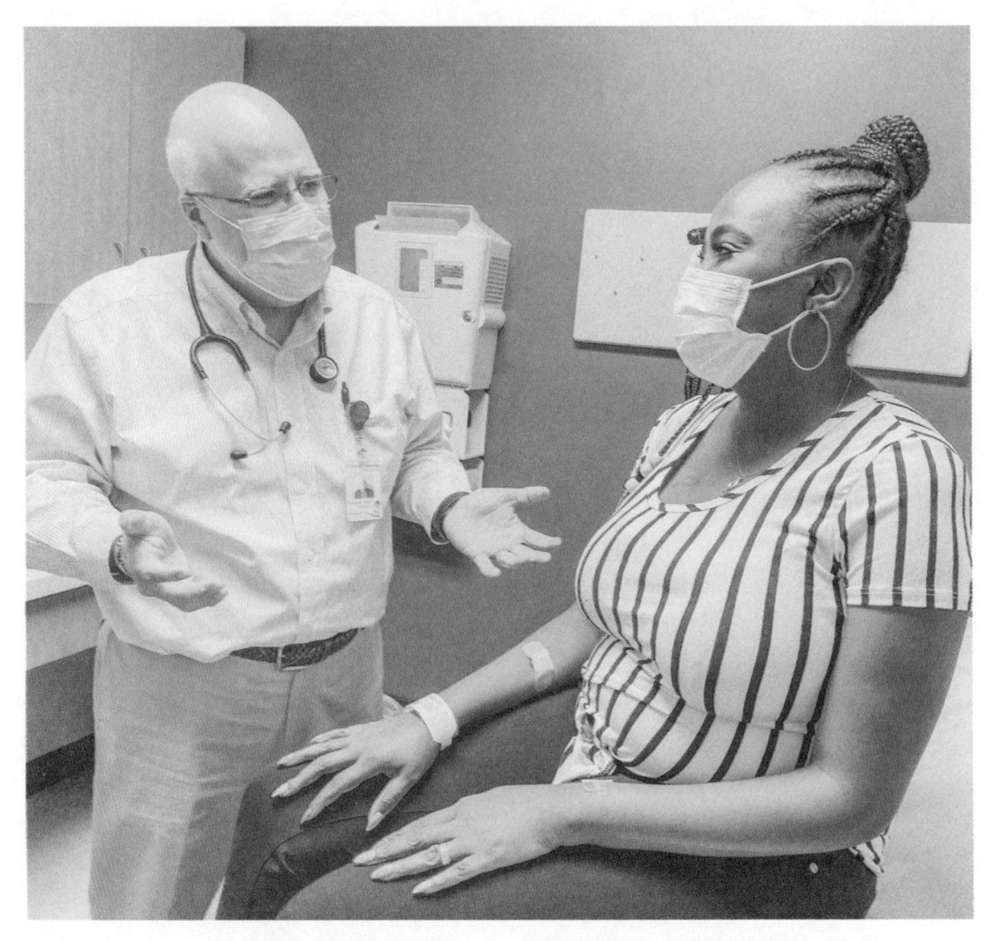

Dr. Haydar Frangoul of the Sarah Cannon Research Institute in Nashville with Victoria Gray

Chapter 32

Therapies

Sickle cell

In July 2019, a doctor at a Nashville hospital plunged the needle of a large syringe into the arm of a thirty-four-year-old African American woman from a small town in central Mississippi and infused her with stem cells that had been extracted from her blood and edited using CRISPR-Cas9. They were now being reinserted in an attempt to cure her of the sickle-cell disease that had plagued her with debilitating pain since she was a baby. Thus did Victoria Gray, a mother of four children, become the first person in the United States to be treated with a CRISPR gene-editing tool. The clinical trial was led by CRISPR Therapeutics, the company formed by Emmanuelle Charpentier. When Gray was injected, her heart rate shot up and for a while she had trouble breathing. "There was a little scary, tough moment for me," she told NPR reporter Rob Stein, who was allowed to follow her treatment. "After that, I cried. But it was happy tears."[1]

Much of the attention paid to CRISPR these days involves its potential to make inheritable (germline) edits in humans that will be passed along to all the cells of all of our future descendants and have the potential to alter our species. These edits are done in reproductive cells or early-stage embryos. This is what occurred with the CRISPR baby twins in China in 2018, and it is the controversial topic that I will discuss later in this book. But in this chapter I'm going to focus on what will be, at least for now, the most common and welcome uses of CRISPR: cases like that of Victoria Gray, in which CRISPR is used to edit some, but not all, of the body (somatic) cells of a patient and make changes that will *not* be inherited. This can be done by taking the cells out of the patient, editing them, and returning them (*ex vivo*) or by delivering the CRISPR editing tool into cells inside of the patient (*in vivo*).

Sickle-cell anemia is one of the best candidates for *ex vivo* gene editing because it involves blood cells that can be easily extracted and returned. The disease is caused by a mutation in a single letter out of more than three billion base pairs of a person's DNA, which causes a kink in the hemoglobin protein. A normal version of hemoglobin protein forms round and smooth blood cells, able to move easily through our vessels and carry oxygen from our lungs to the rest of our body. But the kinked hemoglobin protein forms long fibers that contort the red blood cells, which causes them to clump together and crumple into the shape of a sickle. Oxygen does not get to tissues and organs, causing severe pain and, in most cases, death by age fifty. Sickle-cell disease afflicts more than four million people worldwide, about 80 percent of them in sub-Saharan Africa, and about ninety thousand people in the U.S., mainly African Americans.

The simplicity of the genetic glitch and the severity of the syndrome make it a perfect candidate for gene editing. In the case of Victoria Gray, doctors extracted stem cells from her own blood and edited them, using CRISPR, to activate a gene that produces a type of blood cell that is normally made only during the fetal stage of life. That fetal-stage hemoglobin is healthy, so if the genetic modification works, patients can start producing their own good blood.

A few months after she was injected with her edited cells, Gray drove up to the Nashville hospital to see if the therapy was working. She was optimistic. Ever since she got the edited cells, she hadn't needed to get donor transfusions or had any attacks of pain. A nurse inserted a needle and drew multiple tubes of blood. After a nervous wait, her doctor came in to give her the news. "I am super-excited about your results today," he said. "There are signs that you are starting to make fetal hemoglobin, which is very exciting for us." About half of her blood was now fetal hemoglobin with healthy cells.

In June 2020, Gray got some even more exciting news: the treatment seemed to be lasting. After nine months, she still had not suffered any sickle-cell pain attacks, nor did she need any further blood transfusions. Tests showed that 81 percent of her bone marrow cells were producing the good fetal hemoglobin, meaning that the gene edits were sustained.[2] "High school graduations, college graduations, weddings, grandkids—I thought I wouldn't see none of that," she said after getting the news. "Now I'll be there to help my daughters pick out their wedding dresses."[3] It was an amazing milestone: CRISPR had apparently cured a genetic disease in humans. In Berlin, Charpentier listened to a recording of Gray's emotional NPR interview. "It was pretty amazing to realize as I heard her," she says, "that the little baby I helped to create, CRISPR editing, means that she will no longer suffer."[4]

Affordability

Applications of CRISPR such as this are likely to be lifesavers. They are also sure to be expensive. In fact, the treatment of a single patient could cost $1 million or more, at least initially. So the prospect of CRISPR doing great good is matched by its potential to bankrupt the healthcare system.

Doudna began to focus on this problem after a discussion that she had with a group of U.S. senators in December 2018. The meeting at the Capitol was held a few weeks after the announcement that twin "CRISPR babies" had been born in China with inheritable edits, and Doudna expected it to focus on that headline-making news. At first it did. But to her surprise, the discussion quickly shifted from the perils of inheritable gene-editing to the promise of using gene editing to treat diseases.

Doudna told the senators that CRISPR was on the verge of creating a cure for sickle-cell disease, which got them to perk up, but they immediately peppered her with questions about the cost. "We have 100,000 people in the U.S. affected by sickle cell," one senator pointed out. "How are we going to afford that if it's $1 million per patient? That just breaks the bank."

Doudna decided that making sickle-cell treatments affordable should become a mission of her Innovative Genomics Institute. "The Senate hearing was, for me, a watershed moment," she says. "I'd been thinking a lot about costs before that, but not in a focused way." When she arrived back at Berkeley, she convened a series of meetings of her team to discuss how to make wide access to sickle-cell treatments a new core part of their mission.[5]

The public-private partnership that led to the availability of the polio vaccine became an inspiration. She reached out to the Gates Foundation and the National Institutes of Health, which announced a partnership for a Cure Sickle Cell Initiative funded with $200 million.[6] The primary scientific goal of the initiative is to find a method to edit the sickle-cell mutation inside of a patient without needing to extract bone marrow. One possibility is to inject into the patient's blood a gene-editing molecule with an address label that directs it right to the cells in the bone marrow. The difficult part will be to find the right delivery mechanism, such as a virus-like particle, that won't trigger the patient's immune system.

If the initiative is successful, it will not only cure a lot of people of a dreadful disease; it will advance the cause of health justice. Most sickle-cell patients in the world are Africans or African Americans. These are populations that have been historically underserved by the medical community. Even though the genetic cause of sickle-cell disease has been understood for longer than any similar disorder, new treatments have lagged behind. For example, the fight against cystic fibrosis, which affects primarily white Americans and Europeans, has received eight times more funding from government, charities, and foundations. The great promise of gene editing is that it will transform medicine. The peril is that it will widen the healthcare divide between rich and poor. Doudna's sickle-cell initiative is designed to find ways to avoid that.

Cancer

In addition to treating blood disorders, such as sickle-cell anemia, CRISPR has been used to fight cancer. China has been the pioneer in this field, and it is two or three years ahead of the United States in devising treatments and getting them into clinical trials.[7]

The first person to be treated was a lung-cancer patient in Chengdu, a city of 14 million in the western Chinese province of Sichuan. In October 2016, a team removed from the patient's blood some of his T-cells, which are the white blood cells that help fight off diseases and confer immunity. The doctors then used CRISPR-Cas9 to disable a gene that produces a protein, known as PD-1, which stops the cell's immune response. Cancer cells sometimes trigger the PD-1 response, thus protecting themselves from the immune system. By using CRISPR to edit the gene, the patient's T-cells become more effective in killing the cancer cells. Within a year, China had seven clinical trials using this technique.[8]

"I think this is going to trigger 'Sputnik 2.0,' a biomedical duel on progress between China and the United States," said Carl June, a noted cancer researcher at the University of Pennsylvania who at the time was still struggling to get regulatory approval for a similar clinical trial. He and his colleagues were finally able to get their trial underway and reported preliminary results in 2020. Their method, used in three late-stage cancer patients, was more sophisticated than the one used in China. They knocked out the PD-1 gene and also inserted into the T cells a gene that targeted the patients' tumors.

Although the patients were not cured, the trials showed that the technique was safe. Doudna and one of her postdoctoral students published an article in *Science* explaining the Penn results. "Until now, it has been unknown whether CRISPR-Cas9–edited T cells would be tolerated and thrive once reinfused into a human," they wrote. "The findings represent an important advance in the therapeutic application of gene editing."[9]

CRISPR is also being used as a detection tool to identify precisely what type of cancer a patient has. Mammoth Biosciences, a company that Doudna founded with two of her graduate students, is designing diagnostic tools based on CRISPR that can be used on tumors to identify quickly and easily the DNA sequences associated with different types of cancers. Then precision treatments can be tailored for each patient.[10]

Blindness

The third use of CRISPR editing that was underway by 2020 was to cure a form of congenital blindness. In this case the procedure was performed *in vivo*—inside the patient's body—because eye cells cannot be extracted and returned the way blood and bone marrow cells can. The clinical trials were conducted in partnership with Editas Medicine, the company founded by Zhang and others.

The goal was to treat Leber congenital amaurosis, a common cause of childhood blindness. Those with the condition have a mutation in the gene that makes light-receptor cells in their eye. It causes a critical protein to be shortened, so that the light that hits the cells is not converted into nerve signals.[11]

The first use of the treatment occurred in March 2020, just before coronavirus shut down most clinics, at the Casey Eye Institute in Portland, Oregon. In the hour-long procedure, doctors used a tiny hair-width tube to inject three drops of fluid containing CRISPR-Cas9 into the lining that contains light-sensing cells directly beneath the retina of the patient's eyes. A tailored virus was used as the delivery vehicle to transport the CRISPR-Cas9 into the targeted cells. If the cells are edited as planned, the fix will be permanent, because unlike blood cells, the cells of the eye do not divide and replenish themselves.[12]

Coming soon

Work is also underway on some more ambitious uses of CRISPR gene editing that could make us less vulnerable to pandemics, cancers, Alzheimer's, and other diseases. For example, a gene known as *P53* encodes for a protein that suppresses the growth of cancerous tumors.

It helps the body respond to damaged DNA and prevents cancerous cells from dividing. Humans tend to have one copy of this gene, and cancers proliferate if something goes wrong with it. Elephants have twenty copies of this gene, and they almost never get cancer. Researchers are currently exploring ways to add an extra *P53* gene into humans. Likewise, the gene *APOE4* raises the risk of the devastating disease of Alzheimer's. Researchers are looking for ways to convert it into a benign version of the gene.

Another gene, *PCSK9*, encodes for an enzyme that facilitates the creation of LDL, the "bad" cholesterol. Some people have a mutated copy of the gene that leads to very low levels of this cholesterol, which results in an 88 percent reduction in risk for coronary heart disease. Before he decided to edit the gene for HIV receptors in the CRISPR babies he created, He Jiankui was studying ways to use CRISPR to make germline edits in the *PCSK9* gene of embryos to produce designer babies with far less risk of having heart disease.[13]

At the beginning of 2020, there were two dozen clinical trials for various uses of CRISPR-Cas9 in the pipeline. They included potential treatments for angioedema (a hereditary disease that causes severe swelling), acute myeloid leukemia, super-high cholesterol, and male pattern baldness.[14] In March of that year, however, most academic research labs were temporarily shut down because of the coronavirus pandemic. An exception was made for labs that were engaged in fighting the virus. Many CRISPR researchers, Doudna foremost among them, would shift their focus to creating detection tools and treatments for the disease, some of them making use of the tricks they had learned from studying how bacteria developed an immune response to ward off new viruses.

Josiah Zayner

Chapter 33

Biohacking

Wearing a black T-shirt and tight white jeans, Josiah Zayner stood in front of a roomful of biotechnologists at the Global Synthetic Biology Summit in San Francisco in 2017 and launched into a pitch about a do-it-yourself "frog genetic engineering kit" that he made in his garage. Available online for $299, it allowed users to cause a frog's muscles to double in size in a month by injecting CRISPR-edited DNA that turned off the gene that produces myostatin, a protein that inhibits muscle growth once an animal has reached its mature size.

It would also work on humans, Zayner said, flashing a conspiratorial smile. You could grow bigger muscles.

There was some nervous laughter and then a few shouts of encouragement. "What's holding you back?" someone hollered.

Zayner, a serious scientist wrapped in the persona of a rebel, took a swig of Scotch from a leather-covered hip flask. "Are you suggesting I should try it?" he responded.

There were more murmurs, a few gasps and laughs, then some more encouragement. Zayner reached into a medicine bag, pulled out a syringe, filled it from a vial of the edited DNA, and proclaimed, "All right, let's do it!" Sticking the needle into his left forearm, he winced a bit and then plunged the liquid into his veins. "This will modify my muscle genes to give me bigger muscles," he proclaimed.

There was scattered applause. He took another swig of Scotch from his hip flask. "I will let you know how it works out," he said.[1]

Zayner, with his bleached-blond forelock and ten piercings in each ear, thus became the poster boy for a new breed of biohacker, the spirited band of renegade researchers and merry hobbyists who want to democratize biology through citizen science and bring its power to the people. While conventional researchers worry about patents, biohackers want to keep the bio-frontier free of royalties, regulations, and restraints, similar to the way digital hackers felt about the cyber-frontier. In most cases, the biohackers are, like Zayner, accomplished scientists who forgo working at universities or corporations and instead become the rogue wizards of a rarefied part of the do-it-yourself maker's movement. In the drama of CRISPR, Zayner plays the role of one of Shakespeare's wise fools, such as Puck in *A Midsummer Night's Dream*, who speaks truth under the guise of showmanship, pokes fun at the pretensions of the high-minded, and pushes us forward by pointing out what fools these mortals be.

As a teenager, Zayner worked as a programmer for Motorola's cell phone network, but he got laid off when the tech bubble burst in 2000, so he decided to go to college. He earned his bachelor's degree in plant biology from Southern Illinois University and a doctorate in molecular biophysics at the University of Chicago, where he studied how light-activated proteins work. Instead of doing traditional postdoc studies, he wrote about using synthetic biology to help colonize Mars and found himself recruited to work for NASA. But he was not cut out for a hierarchical organization, so he quit to pursue the freedom of being a biohacker.

Before getting into CRISPR, Zayner tried a variety of synthetic biology experiments, including on himself. To treat his gastrointestinal problems, he performed a fecal transplant (don't ask) to transform his gut's microbiome. He did the procedure in a hotel room with two filmmakers documenting the scene, and (in case you really do want to know how it works) it became a short documentary called *Gut Hack* that can be found online.[2]

Zayner now runs from his garage an online biohacking supply store, The ODIN, which creates and sells "kits and tools that allow anyone to make unique and usable organisms at home or in a lab." Among its products, in addition to the frog-muscle kit, are a "DIY bacterial gene engineering CRISPR kit" ($169) and a "genetic engineering home lab kit" ($1,999).

Soon after Zayner started his business in 2016, he got an email from Harvard's George Church. "I like the stuff you're doing," Church wrote. They chatted, eventually met, and Church became the "business and scientific advisor" to The ODIN. "I think George is a collector of interesting people," Zayner says, correctly.[3]

Most of the biologists who work in academic labs are contemptuous of what they see as Zayner's shoddy methods. "Josiah's stunts demonstrate a reckless pursuit of publicity and a lack of scientific understanding," says Kevin Doxzen, who works in Doudna's lab. "Encouraging curiosity and inquiry within the public is a valuable pursuit, but selling kits that suggest you can engineer frogs in your kitchen, human cells in your living room, or bacteria in your garage attempts to simplify a technology that isn't simple. It saddens me to imagine high school teachers spending their shrinking budgets on kits that simply don't work." Zayner dismisses such criticism as coming from academic scientists trying to protect their priesthood. "We put the DNA sequences and all of our data and methods for our kits online for everyone to judge."[4]

The impromptu CRISPR procedure Zayner performed on himself at the San Francisco conference did not have a noticeable effect on the muscles of his somewhat scrawny body. That would have taken a prolonged series of treatments. But it did have an effect on the world of CRISPR regulation. By being the first person to try to edit his own DNA, he showed that the gene genie would someday be out of the bottle, which he insisted was a good thing.

Zayner wants to make the genetic engineering revolution as open and crowdsourced as the early digital revolution was, when coders like Linus Torvalds created the open-source operating system Linux and hackers like Steve Wozniak gathered at the Homebrew Computer Club and talked about liberating computers from the exclusive control of corporations and government institutions. Genetic engineering, he insists, is no harder than computer engineering. "I almost failed out of high school," he says, "but I was able to learn how to do this stuff." His dream is that millions of people around the world will take up amateur bioengineering. "We now all have this ability to program life," he says. "If millions of people took it up, that would immediately change medicine and agriculture, contributing so much to the world. By demonstrating how easy CRISPR is, I want to inspire people to do that."

Isn't it dangerous, I ask, for everyone to have access to this technology? "No, it's fucking exciting," he counters. "No great technology has flourished until people had complete access to it." He has a point. What truly caused the digital age to blossom was when computers became *personal*. It happened in the mid-1970s with the advent of the Altair and the Apple II, devices that democratized control of computing power. First hackers and then the rest of us got to play with our own computers and produce digital content. The digital revolution was kicked into an even higher orbit in the early 2000s with the birth of the smartphone. As Zayner says, "Once we have people doing biotechnology at home, like we did with computer programming, so many amazing things will be contributed."[5]

Zayner will probably have his way. CRISPR technology is on the verge of becoming easy enough that it will not be confined to well-regulated labs. It will also be advanced by rebels and rogues on the far edge of the frontier. In this way it may follow the path of the digital revolution, much of which, from Linux to Wikipedia, was driven by crowdsourcing. In the digital realm, there isn't a clear line separating amateur from professional coders. The same might soon be true of bioengineers.

Despite the dangers, there could be benefits if biotech followed this route. During a pandemic, it would be useful if societies could tap the biological wisdom and innovation of crowds. At the very least, it would be good to have citizens who could test themselves and their neighbors at home. Contact tracing and data collection could be crowdsourced. Today, there is a sharp line dividing officially sanctioned biologists from do-it-yourself hackers, but Josiah Zayner is dedicated to changing that. CRISPR and COVID could help him blur those lines.

Joseph Bondy-Denomy

Chapter 34

DARPA and Anti-CRISPR

Threat assessment

The possibility that CRISPR would be used by hackers or terrorists or foreign adversaries began to worry Doudna. She raised these concerns when she attended a 2014 conference where a researcher described how a virus could be engineered to carry CRISPR components into mice and edit a gene so that the mice would get lung cancer. A chill went through her. A tweak or a mistake in the guide could easily make it work in human lungs. At another conference a year later, she questioned a graduate student who had coauthored an article with Feng Zhang describing a similar CRISPR experiment that caused cancer in mice. These and other experiences led her to join an effort funded by the U.S. Defense Department to find ways to protect against the misuse of CRISPR.[1]

Ever since Cesare Borgia hired Leonardo da Vinci, military spending has driven innovation. This became true for CRISPR in 2016 when James Clapper, the U.S. Director of National Intelligence, issued the agency's annual "Worldwide Threat Assessment" and it included for the first time "genome editing" as a potential weapon of mass destruction. As a result, the Defense Advanced Research Projects Agency (DARPA), which is the Pentagon's well-funded research arm, launched a program called Safe Genes to support ways to defend against genetically engineered weapons. It dispensed $65 million worth of grants, making the military the largest single source of money for CRISPR research.[2]

The initial DARPA grants went to seven teams. George Church at Harvard received one to study the reversal of mutations that come from exposure to radiation. Kevin Esvelt at MIT was tapped to study gene drives, which can accelerate a genetic change through a population of organisms such as mosquitoes and mice. Amit Choudhary of Harvard Medical School got funding to develop ways to switch on and off genome editing.[3]

Doudna's grants, which would eventually total $3.3 million, covered a variety of projects, including looking for ways to block a CRISPR editing system. The goal was to create tools that, as the announcement put it, "might someday be capable of disabling weapons employing CRISPR." It sounded like the plot of a paperback thriller: terrorists or enemy states unleash a CRISPR system that can edit organisms, such as mosquitoes, to be super-destructive, and Dr. Doudna in a white lab coat has to rush in to save us.[4]

Doudna assigned the project to two young postdoctoral students who had just joined her lab, Kyle Watters and Gavin Knott. They focused on a method that some viruses use to disable the CRISPR systems of the bacteria they are attacking. In other words, bacteria developed CRISPR systems to ward off viruses, but then the viruses developed a way to shut down those defenses. It was an arms race the Pentagon could understand: missiles being countered by defense systems being countered by anti–defense systems. The newly discovered systems were dubbed "anti-CRISPRs."

Anti-CRISPR

Anti-CRISPRs were discovered in late 2012, just as Doudna and Zhang were racing to turn CRISPR-Cas9 into a human gene–editing tool, by a doctoral student at the University of Toronto, Joe Bondy-Denomy. He stumbled upon the discovery by trying something that should not have worked: he attempted to infect some bacteria with a virus that should have been defeated by the bacteria's CRISPR system. In a handful of cases, the attacking viruses survived.

At first he assumed he had botched the experiments. Then a thought occurred to him: perhaps the wily viruses had developed a way to disarm the bacteria's CRISPR defenses. That turned out to be right. The viruses had been able to infiltrate the bacteria's DNA with a little sequence that sabotaged their CRISPR system.[5]

His anti-CRISPRs didn't seem to work on CRISPR-Cas9, so the discovery got little attention at first. But in 2016, he and April Pawluk, who had worked with him on the original paper, identified anti-CRISPRs that disabled the Cas9 enzyme. That opened the floodgates for other researchers to join the hunt, and soon more than fifty anti-CRISPR proteins had been discovered. By then Bondy-Denomy had become a professor at the University of California, San Francisco, and he collaborated with Doudna's lab to show that the anti-CRISPRs could be delivered into human cells to modulate or stop CRISPR-Cas9 editing.[6]

It was a basic science discovery about the wonders of nature, showing how the amazing arms race between bacteria and viruses evolved. And once again, it became an example of basic science leading to useful tools. The anti-CRISPRs could be engineered to regulate gene-editing systems. That would be useful for medical applications that needed to time-limit a CRISPR edit, and they could be used as a defense against systems created by terrorists or malevolent enemies. Anti-CRISPRs could also be used to shut off gene drives, the CRISPR systems that are designed to make a genetic change that spreads rapidly through a fast-breeding population such as mosquitoes.[7]

Doudna was successful in delivering on the projects for DARPA, and her Innovative Genomics Institute at Berkeley was able over the next few years to receive grants for new research topics. Like Church's lab at Harvard, it was asked to study how to use CRISPR to protect against nuclear radiation. The leader of that $9.5 million project was Fyodor Urnov, who was an undergraduate at Moscow State University during the Chernobyl disaster. The mission was to save soldiers and civilians exposed to a nuclear attack or disaster.[8]

The labs that received Safe Genes grants gathered once a year with Renee Wegrzyn, the program manager of DARPA's Biological Technologies Office. Doudna went to one meeting in San Diego in 2018 and was impressed by how good Wegrzyn was at promoting collaboration among the labs that received military funding, just as DARPA had done in the 1960s when it was creating what became the internet. She was also struck by the incongruity of the conference. "We were eating outside in the beautiful weather under swaying palm trees," she says, "and we were talking about radiation sickness and genome editing being used to create weapons of mass destruction."[9]

Enlisting our hacker

On February 26, 2020, just as the COVID-19 plague was taking hold in America, a group of U.S. Army generals, Defense Department officials, and biotechnology executives walked past an imposing statue of a seated Albert Einstein and into a ground-floor room of the stately marble headquarters of the National Academy of Sciences in Washington, D.C. They were there to attend the conference, The Bio-Revolution and Its Implications for Army Combat Capabilities, sponsored by the army's Research and Technology Program. Among the fifty or so participants were some distinguished scientists, most notably George Church, as well as one outlier: Josiah Zayner, the biohacker with multiple ear piercings who had injected himself with a CRISPR-edited gene at a San Francisco synthetic biology conference.

"The building was nice, but the cafeteria was shit," Zayner says. And the conference? "It was really boring. A bunch of people who didn't really know what they were talking about." At one point, he scribbled in his notes, "Speaker sounds like she has taken Xanax."

Zayner likes being irreverent, and despite what he says, I got the feeling that he actually enjoyed the conference. He was not initially scheduled to give a talk, but he made such an impression that he was called on to speak impromptu. The military officials had been complaining that they had trouble recruiting quality scientists. "You need to open up your labs and maybe start a biohacker space to interact with the people more," Zayner told them. He pointed out that the military had done that with computer hackers. Government labs staffed by the do-it-yourself biology community, he said, could come up with solutions the military could use.

Some of the other speakers bought into the idea that the military should enlist help from, as they put it, "non-traditional communities." As one official said, "citizen science" can be tapped to improve the military's ability to identify threats. One of the industry scientists took note of the novel coronavirus spreading out of China, which was still a few days away from causing national alarm. They should imagine a world, he said, where such viral pandemics were common; in such situations, it could be useful to enlist citizen-scientists to figure out ways to deploy real-time detection methods and crowdsource the collection and analysis of data. It was an important point, one that Zayner and the biohacker community had been trying to make.

By the end of the meeting, Zayner was pleasantly surprised by the desire of officials to enlist the hacker community in the effort to deploy CRISPR to fight pandemics and to protect soldiers. "Everyone staring at me and surprised I came," he jotted in his notebook. Then, a little bit later: "People coming up to me thanking me for coming."[10]

Part Five

Public Scientist

This was a new room, rich with hope, terrible with strange danger. A dim folk memory had preserved the story of a greater advance: "the winged hound of Zeus" tearing from Prometheus' liver the price of fire. Was the world ready for the new step forward? Certainly, it will change the world. You have to make laws to fit it. And if plain people did not understand and control it, who would?

—Excerpted from James Agee's cover story, "Atomic Age," on the dropping of the atom bomb, *Time*, August 20, 1945

James Watson and Sydney Brenner at Asilomar

Herbert Boyer and Paul Berg at Asilomar

Chapter 35

Rules of the Road

Utopians vs. bioconservatives

For decades the idea of creating engineered humans belonged to the realm of science fiction. Three classic works warned of what might happen if we snatched this fire from the gods. Mary Shelley's 1818 novel, *Frankenstein; or, The Modern Prometheus*, was a cautionary tale about a scientist who engineers a humanlike creation. In H. G. Wells's *The Time Machine*, published in 1895, a traveler to the future discovers that humans have evolved into two species, a leisure class of Eloi and a working class of Morlocks. Aldous Huxley's *Brave New World*, published in 1932, describes a similarly dystopian future in which genetic modification produces an elite class of leaders with enhanced intellectual and physical traits. In the first chapter, a worker gives a tour of a baby hatchery:

> "We decant our babies as socialized human beings, as Alphas or Epsilons, as future sewage workers or future . . ." He was going to say "future World controllers," but correcting himself, said "future Directors of Hatcheries."

The idea of engineering humans moved from the realm of science fiction to the realm of science in the 1960s. Researchers began to crack the genetic code by figuring out the role played by some of the sequences of our DNA. And the discovery of how to cut and paste DNA from different organisms launched the field of genetic engineering.

The first reaction to these breakthroughs, especially among scientists, was an optimism that bordered on hubris. "We have become the latter-day Prometheus," biologist Robert Sinsheimer declared, with no sign that he understood the Greek myth. "Soon we shall have the power consciously to alter our inheritance, our very nature." He dismissed those who found this prospect troubling. Because the decisions about our genetic future would be guided by individual choice, he argued, this new eugenics would be morally different from the discredited eugenics of the first half of the twentieth century. "We should have the potential to create new genes and new qualities yet undreamed," he exulted. "This is a cosmic event."[1]

The geneticist Bentley Glass, in his address on becoming president of the American Association for the Advancement of Science in 1970, argued that the ethical problem was not that people would embrace these new genetic technologies but that they might reject them. "The right that must become paramount is the right of every child to be born with a sound physical and mental constitution," he said. "No parents will have a right to burden society with a malformed or a mentally incompetent child."[2]

Joseph Fletcher, a professor of medical ethics at the University of Virginia and lapsed Episcopal minister, agreed that genetic engineering could be considered a duty rather than ethically problematic. "Producing our children by 'sexual roulette' without pre-conceptive and uterine control, simply taking pot luck, is irresponsible, now that we can be genetically selective," he wrote in a 1974 book, *The Ethics of Genetic Control*. "As we learn to direct mutations medically, we should do so. Not to control when we can is immoral."[3]

Opposing this biotech utopianism was a group of theologians, technoskeptics, and bioconservatives who became influential in the 1970s. Princeton professor of Christian ethics Paul Ramsey, a prominent Protestant theologian, published *Fabricated Man: The Ethics of Genetic Control*. It is a turgid book with one vivid sentence: "Men ought not to play God before they learn to be men."[4] The social theorist Jeremy Rifkin, dubbed by *Time* America's "foremost opponent of genetic engineering," coauthored a book titled *Who Should Play God?* "Once, all of this could be dismissed as science fiction, the mad ravings of a Dr. Frankenstein," he wrote. "No more. We are not in the Brave New World yet, but we are well along the road."[5]

Even though human gene–editing technologies had not yet been devised, the battle lines had thus been defined. It became the mission of many of the scientists to find a middle ground rather than let the issue become politically polarized.

Asilomar

In the summer of 1972, Paul Berg, who had just published his seminal paper on how to make recombinant DNA, went to the ancient clifftop village of Erice on the coast of Sicily to lead a seminar on the new biotechnologies. The graduate students who attended were shocked by what he described, and they peppered him with questions about the ethical dangers of genetic engineering, especially the modification of humans. Berg had not focused on such questions, and he agreed to hold an informal discussion one evening on the ramparts of the old Norman-era castle overlooking the Straits of Sicily. Under a full moon, eighty students and researchers drank beer and wrestled with the ethical issues. The questions they asked were basic but hard for Berg to answer: What if we could genetically engineer height or eye color? What about intelligence? Would we do that? Should we? Francis Crick, the co-discoverer of DNA's double-helix structure, was there, but he stayed silent as he sipped his beer.[6]

The discussions led Berg to convene a group of biologists in January 1973 at the Asilomar conference center on the California coast near Monterey. Known as "Asilomar I" because it launched a process that would culminate two years later at the same conference site, the meeting focused mainly on lab safety issues. It was followed in April by a conference organized by the National Academies of Science at MIT, which discussed how to prevent the creation of recombinant

DNA organisms that could be dangerous. The more the participants discussed it, the less sure they became that any method would be foolproof. So they issued a letter—which was signed by Berg, James Watson, Herbert Boyer, and others—calling for a "moratorium" on the creation of recombinant DNA until safety guidelines could be formulated.[7]

This led to a memorable gathering that would become famous in the annals of scientists attempting to regulate their own field: the four-day Asilomar conference of February 1975. As the migration of monarch butterflies dappled the sky, 150 biologists and doctors and lawyers from around the world, plus a few journalists who agreed to turn off their tape recorders if the discussion got too heated, gathered to walk the dunes, sit at conference tables, and debate what restraints should be put on new genetic engineering technologies. "Their discussions suggest both the vitality of small boys with new chemistry sets and the electricity of back yard gossip," Michael Rogers of *Rolling Stone* wrote in a piece aptly titled "The Pandora's Box Conference."[8]

One of the primary organizers was a soft-spoken but gently commanding MIT biology professor named David Baltimore, who that year would win the Nobel Prize for his work showing that viruses containing RNA, such as coronaviruses, can insert their genetic material into the DNA of a host cell through a process known as "reverse transcription." In other words, the RNA can be transcribed into DNA, thus modifying the central dogma of biology, which states that genetic information travels in only one direction, from DNA to RNA. Baltimore would go on to become president of Rockefeller University and then Caltech, and his half-century career as a respected leader of policy councils would become a model for Doudna's own public involvement.

After Baltimore set the stage by explaining why the meeting had been convened, Berg described the science that was at issue: recombinant DNA technology made it "ridiculously simple" to combine DNA from different organisms and create new genes. Soon after he had published his discovery, Berg told the group, he started to get calls from researchers asking him to send them material so they could do their own experiments. When he asked the callers what they wanted to do, Berg recalls, "we'd get a description of some kind of horror experiment." He began to fear that some mad scientist would create a new microbe that could threaten the planet, like what Michael Crichton described in his 1969 bio-thriller, *The Andromeda Strain*.

During the policy debates, Berg insisted that the risks of using recombinant DNA to create new organisms were so hard to calculate that such research should be banned. Others found that position absurd. And Baltimore, as he would generally do throughout his career, sought to find a middle ground. He argued for restricting the use of recombinant DNA to viruses that had been "crippled" so that they could not spread.[9]

James Watson, true to form, played the cranky contrarian throughout. "They had worked themselves into a level of hysteria," he later told me. "I was for researchers doing whatever they wanted." At one point he got into a nasty clash with Berg, whose disciplined demeanor was a stark contrast to Watson's impetuousness. The argument got so heated that Berg threatened to sue Watson. "You have signed a letter saying there is a potential risk to this line of work," Berg reminded him, referring to their letter from the year before. "For you to now say you're not willing to institute any procedures which would protect the staff of Cold Spring Harbor, where you're the director, I could bring a suit against you for being irresponsible, and I will."

As the bickering among the elders intensified, some of the younger attendees sneaked out to the beach to smoke dope. By the evening before the conference was scheduled to end, no consensus had been reached. But a panel of lawyers helped spur the scientists by warning that their institutions would likely be held liable if anyone in any lab ever got infected with recombinant DNA. The university responsible might then have to shut down.

Later that night, Berg and Baltimore stayed up with a few colleagues eating takeout Chinese food in a beachside cabana. Using a blackboard they had commandeered, they spent hours trying to write a statement. Around 5 a.m., just before the sun rose, they emerged with a draft.

"The new techniques, which permit combination of genetic information from very different organisms, place us in an arena of biology with many unknowns," they wrote. "It is this ignorance that has compelled us to conclude that it would be wise to exercise considerable caution in performing this research." Then they described in detail the type of safeguards and restrictions that would be put on experiments.

Baltimore made copies of their provisional statement in time to have it distributed at the 8:30 a.m. session, at which point Berg took on the task of herding the scientists to support it. Someone insisted they vote on each paragraph. Berg knew that would be a disaster, and he vetoed the idea. But he did yield to the eminent molecular biologist Sydney Brenner, who asked for an up-or-down vote on the central recommendation being proposed: that the moratorium on genetic engineering research be lifted and that it should proceed with certain safeguards. "The pause is over," Brenner said. The room agreed. A few hours later, just as the bell rang for the final lunch, Berg asked for a vote on the document as a whole, which included detailed safety provisions that labs would have to follow. Most hands went up in favor. Ignoring those who still clamored to speak, he then asked if there were any opposed. Only four or five hands were raised, including that of Watson, who thought all the safeguards were silly.[10]

The conference had two goals: guarding against the hazards that could come from creating new forms of genes and guarding against the threat that politicians would ban genetic engineering altogether. On both fronts, the Asilomar process was successful. They were able to chart "a prudent path forward," an approach that Baltimore and Doudna would later replicate in the debates over CRISPR gene editing.

The restrictions agreed to at Asilomar were accepted by universities and funding agencies worldwide. "This unique conference marked the beginning of an exceptional era for science and for the public discussion of science policy," Berg wrote thirty years later. "We gained the public's trust, for it was the very scientists who were most involved in the work and had every incentive to be left free to pursue their dream that called attention to the risks inherent in the experiments they were doing. Restrictive national legislation was avoided."[11]

Others were less willing to join the mutual back-patting. Erwin Chargaff, a brilliant biochemist who had made key discoveries about the structure of DNA, looked back on the event as a charade. "At this Council of Asilomar there congregated the molecular bishops and church fathers from all over the world, in order to condemn the heresies of which they themselves had been the first and the principal perpetrators," he said. "This was probably the first time in history that the incendiaries formed their own fire brigade."[12]

Berg was right that Asilomar was a great success. It paved the way for genetic engineering to become a booming field. But Chargaff's mocking assessment pointed to another lasting legacy. Asilomar became notable for what the scientists did *not* discuss there. Their focus was on safety. None of them addressed the big ethical question, the one that Berg had stayed up late discussing in Sicily: How far should we go if and when methods of engineering our genes turned out to be safe?

Splicing Life, *1982*

Asilomar's lack of focus on ethical issues bothered many religious leaders. That prompted a letter to President Jimmy Carter signed by the heads of three major religious organizations: the National Council of Churches, the Synagogue Council of America, and the U.S. Catholic Conference. "We are rapidly moving into a new era of fundamental danger triggered by the rapid growth of genetic engineering," they wrote. "Who shall determine how human good is best served when new life forms are being engineered?"[13]

These decisions should not be left to scientists, the trio argued. "There will always be those who believe it appropriate to 'correct' our mental and social structures by genetic means. This becomes more dangerous when the basic tools to do so are finally at hand. Those who would play God will be tempted as never before."

Carter responded by appointing a presidential commission to study the issue. It came back in late 1982 with a 106-page report titled

Splicing Life that ended up being inconclusive mush. It merely called for further dialogue to reach societal consensus. "A goal of this Report is to stimulate thoughtful, long-term discussion—not preempt it with conclusions that would, of necessity, be premature."[14]

The commission's report did raise two concerns that were prescient. The first was a fear that genetic engineering was leading to increased corporate involvement in university research. Universities had historically focused on basic research and the open exchange of ideas, and the report warned, "These goals may run headlong into those of industry—the development of marketable products and techniques through applied research by maintaining a competitive posture, protecting trade secrets, and seeking patent protection."

The second concern was that genetic engineering would increase inequality. New biotech procedures would be expensive, so people who were born into privilege would likely get the most benefits. That could widen, and genetically encode, existing inequalities. "The possibilities presented by gene therapy and gene surgery may in fact call into question a central element of democratic political theory and practice: the commitment to equality of opportunity."

Preimplantation genetic diagnosis and Gattaca

After the development of recombinant DNA in the 1970s, the next big bioengineering advance—and set of ethical issues—came in the 1990s. It resulted from the confluence of two innovations: *in vitro* fertilization (the first test-tube baby, Louise Brown, was born in 1978) combined with genetic sequencing technology. This led, in 1990, to the first use of what became known as preimplantation genetic diagnosis.[15]

Preimplantation diagnosis involves fertilizing an egg with sperm in a Petri dish, doing tests on the resulting embryos* to determine their

*I use the word "embryo" in the broad sense. The single-cell organism resulting from a fertilized egg is a zygote. When the zygote divides to become a collection of cells that can implant in the wall of the uterus, it is called a blastocyst. About four weeks later, after the development of an amniotic sac, it becomes an embryo. After eleven weeks, it is usually referred to as a fetus.

genetic characteristics, and then implanting into a woman's womb the embryo with the most desired traits. It allows parents to choose the gender of their child and avoid having a child who carries a genetic disease or some other attribute the parents find undesirable.

The potential of such genetic screening and selection entered the popular imagination through the 1997 film *Gattaca* (the title is made up of the letters of the four DNA bases), starring Ethan Hawke and Uma Thurman. It tells of a future in which genetic selection is regularly used to ensure that children are enhanced with the best hereditary traits.

To promote the movie, the studio took out advertisements in newspapers that appeared as if they were for a real gene-editing clinic. Headlined "Children Made to Order," the ad read, "At Gattaca, it is now possible to engineer your offspring. Here's a checklist to help you decide what traits to pass on to your newborn." The list included gender, stature, eye color, skin color, weight, addictive susceptibility, criminal aggressive tendencies, musical ability, athletic prowess, and intellect. The final choice was "None of the above." The ad advises of that option, "For religious or other reasons, you may have reservations about genetically engineering your child. We respectfully invite you to reconsider. From where we sit, the human race could use a little improving."

At the bottom of the ad was a toll-free telephone number, which led to a recording offering callers three options: "Press one if you'd like to take the steps to ensure that your offspring is disease-free. Press two if you'd like to enhance intellectual and physical traits. Press three if you don't want to tamper with your kid's genetic makeup." Within two days, the toll-free number had received fifty thousand calls, but the studio, alas, did not track how many chose each of the options.

The hero of the movie, played by Hawke, was conceived without the benefits or burdens of preimplantation engineering, and he must battle genetic discrimination in order to fulfill his dream of becoming an astronaut. He is, of course, triumphant, since this is a movie. A particularly interesting scene occurs when his parents decide to make use of gene editing in having their second child. The doctor describes all the traits and enhancements he can engineer: better eyesight, desired eye and skin color, no predisposition toward alcoholism or baldness, and more. "Is it good to leave a few things to chance?" the parents ask. No, the doctor assures them, they are merely giving their prospective child "the best possible start."

That led film critic Roger Ebert to write, "When parents can order 'perfect' babies, will they? Would you take your chances on a throw of the genetic dice, or order up the make and model you wanted? How many people are prepared to buy a car at random from the universe of all available cars? That's how many, I suspect, would opt to have natural children." But then Ebert smartly expressed the worries that were beginning to form at the time: "Everybody will live longer, look better and be healthier in the Gattacan world. But will it be as much fun? Will parents order children who are rebellious, ungainly, eccentric, creative, or a lot smarter than their parents are? Don't you sometimes have the feeling you were born just in time?"[16]

Watson and others at UCLA, 1998

Once again, the irascible old DNA pioneer James Watson sat in the audience loudly mumbling provocative thoughts that he seemed gleefully unable to suppress. This time it was at a gene-editing conference hosted by UCLA professor Gregory Stock in 1998. French Anderson, a leader in using genetic engineering to create drugs, gave a mini-sermon on the need to distinguish between treating diseases, which he proclaimed to be moral, and providing children with genetic enhancements, which he said wasn't. Watson began to snort and stir. "No one really has the guts to say it," he interrupted, "but if we could make better human beings by knowing how to add genes, why shouldn't we do it?"[17]

The title of the gathering was "Engineering the Human Germline," and it focused on the ethics of making genetic edits that would be inherited. These "germline" edits were fundamentally different, medically and morally, from somatic-cell edits that affect only certain cells in an individual patient. The germline was a red line that scientists had been reluctant to cross. "This is the first gathering where people have talked openly about *germline* engineering," Watson said approvingly. "It seems obvious that germline therapy will be much more successful than somatic-cell edits. If we wait for the success of somatic therapy, we'll wait until the sun burns out."

It was absurd, Watson said, to treat the germline as "some great Rubicon and crossing it involved going against natural law." When he was challenged about the need to respect "the sanctity of the human gene pool," he erupted. "Evolution can be just damn cruel, and to say that we've got a perfect genome and there's some sanctity to it is utter silliness." His schizophrenic son, Rufus, was a daily reminder that the genetic lottery could be, as he put it, damn cruel. "The biggest ethical problem we have is not using our knowledge and not having the guts to go ahead and try to help someone," he insisted.[18]

For the most part, Watson was preaching to the choir. The opinions at the UCLA conference ranged from enthusiasm to unbridled enthusiasm for gene editing. When someone suggested that going down that slope might lead to unintended consequences, Watson was unwavering. "I think the slippery slope argument is just crap. Societies thrive when they're optimistic, not pessimistic, and the slippery slope argument sounds like one from a worn-out person who's angry at himself."

Lee Silver, a Princeton biologist, had just published *Remaking Eden*, which became a manifesto for the conference. He had coined the word "reprogenetics" to describe the use of technology to determine which genes a child would inherit. "In a society that values individual freedom above all else, it is hard to find any legitimate basis for restricting the use of reprogenetics," he wrote.[19]

Silver's work was important because it framed the issue as being about individual freedom and liberty in a market-based consumer society. "If democratic societies allow parents to buy environmental advantages for their children, how can they prohibit them from buying genetic advantages?" he prodded. "Americans would respond to any attempt at a ban with the question, 'Why can't I give my child beneficial genes that other children get naturally?' "[20]

Silver's techno-enthusiasm set the tone for what participants viewed as a historic moment. "For the first time we as a species have the ability to self-evolve," Silver told the group. "I mean, this is an incredible concept." He meant the word "incredible" to be a compliment.

As with the Asilomar conference, one of the goals of the UCLA conference was to fend off government regulation. "The main message we need to draw is to keep the state out of any form of genetic decision," Watson argued. The attendees accepted that view. "No state or federal legislation to regulate germline gene therapy should be passed at this time," organizer Gregory Stock wrote in his summation.

Stock went on to write a pro-editing manifesto, *Redesigning Humans: Our Inevitable Genetic Future.* "A key aspect of human nature is our ability to manipulate the world," he argued. "To turn away from germline selection and modification without even exploring them would be to deny our essential nature and perhaps our destiny." He emphasized that politicians should not try to interfere. "Policymakers sometimes mistakenly think that they have a voice about whether germinal technologies will come into being," he wrote. "They do not."[21]

The American enthusiasm for genetic engineering was a sharp contrast to the attitude in Europe, where policymakers and various commissions had increasingly turned against it, both in agriculture and in humans. The most notable expression came from a meeting convened by the Council of Europe in Oviedo, Spain, in 1997. The resulting Oviedo Convention was intended to be a legally binding treaty designed to prohibit the use of biological advances in ways that threatened human dignity. It barred genetic engineering in humans except "for preventive, diagnostic or therapeutic reasons and only where it does not aim to change the genetic make-up of a person's descendants." In other words, no germline editing. Twenty-nine European countries incorporated the Oviedo Convention into their laws, with Britain and Germany being notable holdouts. Even where it was not ratified it helped shape what is still a general consensus in Europe against genetic engineering.[22]

Jesse Gelsinger

The optimism among American researchers about genetic engineering was deflated in September 1999 by a tragedy that happened in Philadelphia to a sweet, handsome, and slightly rebellious eighteen-year-old high school student. Jesse Gelsinger suffered from a mild form of a disease of the liver caused by a simple genetic mutation. It caused his liver to have problems ridding his body of ammonia, which is a by-product of the breakdown of proteins. It usually kills victims as babies, but Gelsinger's milder form meant that he could survive by eating a very low-protein diet and taking thirty-two pills a day.

A team at the University of Pennsylvania was testing a genetic therapy for the disease. Such therapies do not involve actually editing the DNA of the cells inside the body. Instead, genes without the mutation are created in a lab, and then doctors put these good genes into a virus that serves as a delivery mechanism. In Gelsinger's case, the viruses with the good genes were injected into an artery that led into the liver.

It was unlikely that the therapy would help Gelsinger right away, because it was a trial designed to see how the therapy could be used to save babies. But it offered him hope that someday he would be able to eat hot dogs, and in the meantime, some babies would be saved. "What's the worst that can happen to me?" he said to a friend as he was leaving for the Philadelphia hospital. "I die, and it's for the babies."[23]

Unlike the seventeen other humans in the trial, Gelsinger had a massive immune response caused by the virus transporting the therapeutic gene, which resulted in a high fever followed by the breakdown of his kidneys, lungs, and other organs. In four days he was dead. Work on gene therapy ground to a halt. "We were all very much aware of what happened," Doudna recalled. "That made the whole field of gene therapy go away, mostly, for at least a decade. Even the term *gene therapy* became kind of a black label. You didn't want that in your grants. You didn't want to say, 'I'm working on gene therapy.' It sounded terrible."[24]

The Kass Commission, 2003

The debate over genetic engineering at the turn of the century—after the completion of the Human Genome Project and the cloning of Dolly the sheep—led to another U.S. presidential commission, this one created by President George W. Bush in 2003. It was chaired by Leon Kass, a biologist and social philosopher who had first expressed wariness about biotechnology thirty years earlier.

Kass is the most influential of the country's bioconservatives, the ethical traditionalists with a knowledge of biology who urge restraint when dealing with new genetic technologies. The son of secular Jewish immigrants, he earned a biology degree at the University of Chicago, where he was deeply influenced by its "great books" core curriculum. He got a medical degree from Chicago and a PhD in biochemistry from Harvard. With his wife, Amy, he went to Mississippi in 1965 as part of the cadre of civil rights workers registering Blacks to vote, an experience that reinforced his faith in traditional values. "In Mississippi I saw people living in perilous and meager circumstances, many of them illiterate, but sustained by religion, extended family and community attachment," he recalled.[25]

Upon returning to the University of Chicago as a professor, his writings ranged from scientific papers on molecular biology ("The Antibacterial Activity of 3-Decynoyl-N-Acetylcysteamine") to a book on the Hebrew Bible. After reading Huxley's *Brave New World,* he became more interested in "how the scientific project to master nature could, if we are not careful, lead to our dehumanization." Combining his appreciation for both science and the humanities, he began to tackle the issues raised by reproductive technologies such as cloning and *in vitro* fertilization. "I soon shifted my career from doing science to thinking about its human meaning," he wrote, "worrying about upholding our humanity against possible technological degradation."

His first published warning about bioengineering was a letter in *Science* in 1971 criticizing Bentley Glass's contention that "every child has the inalienable right to a sound heritage." Kass declared, "To make good such an 'inalienable right' means converting human reproduction into manufacture." The following year he wrote an essay explaining his wariness about genetic-engineering technologies. "The road to *Brave New World* is paved with sentimentality—yes, even with love and charity," he wrote. "Have we enough sense to turn back?"[26]

In 2001, the Kass Commission included many distinguished conservative or neoconservative thinkers, including Robert George, Mary Ann Glendon, Charles Krauthammer, and James Q. Wilson. Two prominent philosophers proved to be especially influential members. The first was Michael Sandel, a Harvard professor who is the contemporary successor to John Rawls in defining the concept of justice. At the time, he was writing an essay titled "The Case Against Perfection: What's Wrong with Designer Children, Bionic Athletes, and Genetic Engineering," which he published in *The Atlantic* in 2004.[27] The other key thinker was Francis Fukuyama, who in 2000 published *Our Posthuman Future: Consequences of the Biotechnology Revolution*, which was a forceful call for governments to regulate biotechnology.[28]

Not surprisingly, their final 310-page report, *Beyond Therapy*, was thoughtful, vibrantly written, and filled with qualms about genetic engineering. It warned of the dangers of using technology to go beyond merely treating diseases to using it to enhance human capabilities. "There are reasons to wonder whether life will really be better if we turn to biotechnology to fulfill our deepest human desires," the report declared.[29]

Focusing mainly on philosophical rather than safety concerns, the authors discussed what it meant to be human, to pursue happiness, to respect nature's gifts, and to accept the given. It argued the case, or more accurately it preached the case, that going too far to alter what is "natural" was hubristic and endangered our individual essence. "We want better children—but not by turning procreation into manufacture or by altering their brains to gain them an edge over their peers," they wrote. "We want to perform better in the activities of life—but not by becoming mere creatures of our chemists or by turning ourselves into tools designed to win or achieve in inhuman ways." One can almost sense a congregation nodding "Amen" while a few people in the back mutter, "Speak for yourself."

George Daley, Doudna, and David Baltimore at the 2015 international summit

CHAPTER 36

Doudna Steps In

The Hitler nightmare

In the spring of 2014, when the battle to win CRISPR patents and launch gene-editing companies was heating up, Doudna had a dream. More precisely, she had a nightmare. In it, a prominent researcher asked her to meet someone who wanted to learn about gene editing. When she went into the room, she recoiled. Sitting in front of her, with pen and paper ready to take notes, was Adolf Hitler with the face of a pig. "I want to understand the uses and implications of this amazing technology you've developed," he said. Doudna was jolted awake by the nightmare, she recalls. "As I lay in the dark, my heart racing, I couldn't escape the awful premonition with which the dream had left me." She began to have trouble sleeping at night.

Gene-editing technology had enormous power to do good, but the thought of using it to make alterations in humans that would be inherited by all future generations was unnerving. "Have we created a toolbox for future Frankensteins?" she asked herself. Or perhaps even worse, would it be a tool for future Hitlers? "Emmanuelle and I, and our collaborators, had imagined that CRISPR technology could save lives by helping to cure genetic disease," she later wrote. "Yet as I thought about it now, I could scarcely begin to conceive of all of the ways in which our hard work might be perverted."[1]

Happy Healthy Baby

Around that time, Doudna was confronted with an example of how people with good intentions could pave the way for gene editing. Sam Sternberg, one of the researchers on her close-knit CRISPR team, received an email in March 2014 from an aspiring young entrepreneur in San Francisco named Lauren Buchman, who had gotten Sternberg's name from a friend. "Hi, Sam," she wrote. "Nice to meet you by email. I see that you're located just across the Bridge. Any chance I could buy you a coffee and chat a bit about what you're up to?"[2]

"I'd be happy to meet sometime though my schedule is busy," Sternberg replied. "Maybe in the meantime, you could fill me in a bit on what your company is doing."

"I've started a company called Happy Healthy Baby," she explained in her next email. "We've seen a potential of Cas9 to aid in preventing genetic diseases in children conceived through IVF in the future. Ensuring that this is done with the highest level of scientific and ethical standards is first and foremost to us."

Sternberg was surprised but not totally shocked. By that time CRISPR-Cas9 had already been used to edit embryos implanted in monkeys. He was interested in digging a bit deeper into what Buchman's motivations were and how she was thinking about developing this concept, so he agreed to meet her at a Mexican restaurant in Berkeley. There Buchman pitched him on the idea of offering people the chance to use CRISPR to edit their future babies.

She had already registered the domain name HealthyBabies.com. Might he want to be a cofounder? This surprised Sternberg, and not simply because he shared with his lab pal Blake Wiedenheft a good-humored humility. He had no experience editing human cells, much less knowing the first thing about how to implant embryos.

When I first heard about Buchman's concept, I found it disconcerting. But when I tracked her down, I was surprised to find that she was actually quite thoughtful about the moral issues. Her sister was a leukemia survivor and could not, as a result of her treatment, have children. Buchman herself was trying to launch a career and worried about her biological clock running down. "I was a woman in my thirties," she recalls. "And we're all facing the same issue. We want a career and not to be mommy-tracked, and we are starting to deal with fertility clinics."

She knew that *in vitro* fertility clinics could screen for harmful genes before choosing an embryo to implant, but as a thirtysomething woman she also knew that producing a bunch of fertilized embryos was easier said than done. "You may end up producing only one or two embryos," she points out, "so preimplantation genetic screening is not always easy."

That's when she heard about CRISPR and got excited. "The idea that we could treat something in cells seemed so promising and wonderful."

She was sensitive to the social issues. "All tech can be used for good or for bad, but the early movers in new technologies have the opportunity to promote positive and ethical usage," she says. "I wanted to do gene-editing right, and to do it in the open, so there would be an established pattern for ethical procedures for patients who wanted to use it."

Some of the venture capitalists and biotech entrepreneurs she consulted ended up pitching her on weird ideas that freaked her out, such as enlisting biohackers to crowdsource the editing of patients' genes. "The more I heard, the more I thought 'I have to do this,'" she says, "because if I don't, these fringe folks with no regard for the impact or the ethics will take over the field."

Sternberg left the dinner at the Mexican restaurant before dessert. He had no interest in being a cofounder, but he was intrigued enough to agree to visit the company's workspace. "There was never a chance in a million that I was going to get involved, but I was curious," he says. He knew that Doudna was beginning to worry about these sorts of things, so he decided to visit the lab so that he could talk to someone who wanted to be in the driver's seat on the type of CRISPR application that would stir up controversy.

During his visit, Sternberg watched a promotional video for Happy Healthy Baby, filled with animation and stock footage of lab experiments, in which Buchman, sitting in a sunny room with big glass windows, explains the idea of gene-editing babies. He told her that he didn't see any chance that CRISPR would be approved for use on human babies in the U.S. for at least ten years. She replied that the clinics did not have to be in the U.S. There would likely be other countries where the procedure would be allowed, and people who could afford gene-edited babies would be willing to travel.

Sternberg decided not to get involved, but for a while George Church agreed to serve as an unpaid science advisor. "George suggested that I work with sperm cells rather than embryos," Buchman recalls. "He said it might be less controversial or troubling."[3]

Buchman eventually abandoned the venture. "I dug into the use cases, market regulations, and ethics, and it became obvious that I was too early to be working on this," she says. "The science wasn't ready, and society wasn't ready."

When Sternberg described his meetings to Doudna, he told her that Buchman had "a Promethean glint in her eye." Later, he used that phrase in a book he wrote with Doudna, which infuriated Buchman. Had the Happy Healthy Baby pitch occurred a few years earlier, Doudna and Sternberg wrote, they would have dismissed the idea "as pure fantasy" because "there was little chance of anyone pursuing such Frankenstein schemes." But the invention of CRISPR-Cas9 technology had changed that. "Now, we could no longer laugh off this kind of speculation. Making the human genome as easily manipulable as that of a bacterium was, after all, precisely what CRISPR had accomplished."[4]

Napa, January 2015

As a result of her Hitler dream and Sternberg's Happy Healthy Baby story, Doudna decided in the spring of 2014 to become more engaged in the policy discussions about how CRISPR gene-editing tools should be used. At first she considered writing an op-ed for a newspaper, but that did not seem adequate to the challenge. So she harked back forty years earlier to the process that led to the February 1975

Asilomar conference, the one that had come up with the "prudent path forward" guidelines for work on recombinant DNA. She decided that the invention of CRISPR gene-editing tools warranted convening a similar group.

Her first step was to enlist the participation of two of the key organizers of the 1975 Asilomar conference: Paul Berg, who had invented recombinant DNA, and David Baltimore, who had been involved in most of the major policy gatherings, beginning with Asilomar. "I felt that if we could get them both we would have a direct link to Asilomar and a stamp of credibility," she recalls.

Both agreed to participate, and the meeting was set for January 2015 at a resort in Napa Valley about an hour north of San Francisco. Eighteen other top researchers were invited, including Martin Jinek and Sam Sternberg from Doudna's lab. The focus would be on the ethics of making inheritable genetic edits.

At Asilomar the discussions had been mostly about safety, but Doudna made sure that the Napa conference tackled the moral questions: Did the premium that America put on individual liberty require that decisions about gene-editing of babies be left mainly to parents? To what extent would creating gene-edited babies—and abandoning the idea that our genetic endowments came from a random natural lottery—undermine our sense of moral empathy? Was there a danger in decreasing the diversity of the human species? Or, to frame the question from a more bioliberal perspective: If the technology was available to make healthier and better babies, would it be ethically wrong *not* to use it?[5]

A consensus quickly developed that it would be bad to completely ban germline gene editing. The participants wanted to leave the door open. Their objective became similar to that of Asilomar: finding a path forward rather than putting on the brakes. That would become the theme of most subsequent commissions and conferences organized by scientists: it was too early to do germline editing safely, but someday it would happen, and the goal should be to provide prudent guidelines.

David Baltimore warned of a development that made this Napa

meeting different from the Asilomar one forty years earlier. "The big difference today is the creation of the biotechnology industry," he told the group. "In 1975, there were no big biotechnology companies. Today, the public is concerned about commercial development, because there's less oversight." If the participants wanted to prevent a popular backlash against gene editing, he said, they would have to convince people to trust not only white-coated scientists but also commercially driven corporations. That could be a tough sell. Alta Charo, a bioethicist at the University of Wisconsin Law School, pointed out that the close relationship between academic researchers and commercial companies could taint the credibility of the academics. "Financial interests undermine the 'white coat' image of scientists today," she said.

One of the participants brought up the social justice argument. Gene editing would be expensive. Would only the wealthy have access? Baltimore agreed that was a problem, but he argued it was not a cause for banning the technology. "That argument doesn't cut very deep," he said. "That's how everything is. Look at computers. Everything gets cheaper when it gets done wholesale. It's not an argument against moving forward."

During the conference, word began to circulate about some editing experiments on non-viable embryos that were already happening in China. The technology, unlike that of building nuclear weapons, could spread easily and be used not only by responsible researchers but also by rogue doctors and biohackers. "Can we really put the genie back in the bottle?" one participant asked.

The group agreed that the use of CRISPR tools for *non-inheritable* gene editing in somatic cells was a good thing. It could lead to beneficial drugs and treatments. So they decided that it would be useful to agree to some restraints on germline editing in order to prevent a backlash. "We need to create a political safe space by going slow on germline editing so that we can continue working on somatic cell edits," one participant said.

In the end, they decided to call for a temporary halt on germline editing in humans, at least until the safety and social issues could be further understood. "We wanted the scientific community to hit the

Pause button until the societal, ethical, and philosophical implications of germline editing could be properly and thoroughly discussed—ideally at a global level," Doudna says.

Doudna drafted an initial version of the conference report, which she circulated to the other participants. After incorporating their suggestions, she submitted it in March to *Science.* It was titled "A Prudent Path Forward for Genomic Engineering and Germline Gene Modification."[6] Although she was the lead writer, the names of Baltimore and Berg were listed first. The happenstance of alphabetical order caused the two Asilomar pioneers to be at the fore.

The report clearly defined what was meant by "germline editing" and why crossing that threshold would be a major ethical as well as scientific step. "It is now possible to carry out genome modification in fertilized animal eggs or embryos, thereby altering the genetic makeup of every differentiated cell in an organism and so ensuring that the changes will be passed on to the organism's progeny," they wrote. "The possibility of human germline engineering has long been a source of excitement and unease among the general public, especially in light of concerns about initiating a 'slippery slope' from disease-curing applications toward uses with less compelling or even troubling implications."

As Doudna hoped, the journal article got major national attention. The *New York Times* ran a story on page 1 by Nicholas Wade, with a picture of Doudna at her Berkeley desk and the headline "Scientists Seek Ban on Method of Editing the Human Genome."[7] But the headline was misleading. Indeed, in most of the publicity about the Napa report, a key point was missed. Unlike some other scientists at the time,[8] the participants had purposely decided against calling for a ban or moratorium, which can over time become hard to lift. Their goal was to keep open the possibility of germline editing if it was safe and medically necessary. That was why, in the title of the piece, they called for "a prudent path forward," which had become the watchword of many of the scientific conferences on human germline gene editing.

Chinese embryo work, April 2015

During the Napa conference, Doudna heard an unnerving rumor: a group of Chinese scientists had used CRISPR-Cas9 to edit, for the first time, the genes in an early-stage human embryo, which in theory could create inheritable changes. The mitigating factor was that the embryos were not viable. They would not be implanted in a mother's uterus. Nevertheless, if true, the plans of well-intentioned policymakers would once again be disrupted by the zeal of eager researchers.[9]

The Chinese paper had not been published, but its existence had leaked. It had been rejected by the prestigious journals *Science* and *Nature,* and it was being shopped around. It was finally accepted by the somewhat obscure Chinese journal *Protein & Cell,* which published it online on April 18, 2015.

In the article, researchers at a university in Guangzhou described how they used CRISPR-Cas9 in eighty-six non-viable zygotes (precursors to an embryo) to cut out a mutated gene that causes beta thalassemia, a deadly blood disorder like sickle-cell anemia.[10] Although the embryos were never intended to be grown into babies, a line had been toed, if not crossed. For the first time, CRISPR-Cas9 had been used to make potential edits in the human germline, ones that could be inherited by future generations.

After Doudna read the article in her Berkeley office, she stared out at San Francisco Bay feeling, she later recalled, "awestruck and a bit queasy." Other scientists around the world were probably conducting similar experiments with the technology that she and Charpentier had created. That could lead, she realized, to some very unintended consequences. It could also provoke a public backlash. "The technology is not ready for clinical application in the human germline," she replied when a reporter for NPR asked her about the Chinese experiments. "That application of the technology needs to be on hold pending a broader societal discussion of the scientific and ethical issues."[11]

The Napa conference and the Chinese embryo-editing experiments aroused the interest of Congress. Senator Elizabeth Warren hosted a congressional briefing, and Doudna went to Washington to testify with her friend and fellow CRISPR pioneer George Church. The event was so popular that it was standing-room only. More than 150 senators, congressmen, staffers, and agency personnel crammed into the room. Doudna recounted the history of CRISPR, emphasizing that it had begun as pure "curiosity-driven" research about how bacteria fight off viruses. Using it in humans, she explained, required finding ways to get it to the right cells in the body, a task that was easier when the edits were made in early-stage embryos. "But using gene editing in such a way," she warned, "is also much more ethically controversial."[12]

Doudna and Church wrote back-to-back pieces in *Nature* presenting their perspectives on making inheritable gene edits. Although their positions conflicted to some extent, they reinforced the case that scientists were dealing with the issues seriously and did not require new government regulations. "Opinion on the use of human-germline engineering varies widely," Doudna wrote. "In my view, a complete ban might prevent research that could lead to future therapies, and it is also impractical given the widespread accessibility and ease of use of CRISPR-Cas9. Instead, solid agreement on an appropriate middle ground is desirable."[13] Church was more forceful in arguing that research, even in editing the human germline, should continue. "Rather than talk about the possibility of banning alteration of the human germline, we should instead be discussing how to stimulate ways to improve its safety and efficacy," he wrote. "Banning human-germline editing could put a damper on the best medical research and instead drive the practice underground to black markets and uncontrolled medical tourism."[14]

Church's bio-enthusiasm was given a boost in the popular press by one of his Harvard colleagues, the well-known psychology professor Steven Pinker. "The primary moral goal for today's bioethics can be summarized in a single sentence," he wrote in an op-ed for the *Boston Globe*. "Get out of the way." He took a brutal swipe at the entire profession of bioethicists. "A truly ethical bioethics should not bog down research in red tape, moratoria, or threats of prosecution based on nebulous but sweeping principles such as 'dignity,' 'sacredness,' or 'social justice,'" he argued. "The last thing we need is a lobby of so-called ethicists."[15]

The December 2015 International Summit

Following their Napa Valley meeting, Doudna and Baltimore urged the U.S. National Academy of Sciences and its sister organizations around the world to convene a globally representative group to discuss how to prudently regulate human germline editing. More than five hundred scientists, policymakers, and bioethicists—though very few patients or parents of afflicted children—gathered in Washington for three days at the beginning of December 2015 for the first International Summit on Human Gene Editing. In addition to Doudna and Baltimore, there were other CRISPR pioneers, including Feng Zhang, George Church, and Emmanuelle Charpentier. Cohosts included the Chinese Academy of Sciences and Britain's Royal Society.[16]

"We are here as part of a historical process that dates from Darwin and Mendel's work in the nineteenth century," Baltimore said in his opening remarks. "We could be on the cusp of a new era in human history."

A representative from Peking University assured the audience that China had in place safeguards to prevent germline gene editing: "The manipulation of the genes of human gametes, zygotes, or embryos for the purpose of reproduction is prohibited."

Because there were so many participants and journalists, the meeting consisted mainly of canned presentations rather than real debate. Even the conclusions had been pre-cooked. The most important was almost identical to what had been decided at the small Napa meeting at the beginning of the year. Human germline editing should be strongly discouraged until stringent conditions were met, but the words "moratorium" and "ban" were avoided.

Among the conditions the group adopted was that germline editing should not proceed until "there is broad societal consensus about the appropriateness of the proposed application." The need for a "broad societal consensus" was one that would be invoked often in discussions of the ethics of germline editing, as if a mantra. It was a laudable goal. But as the debate over abortion has shown, discussions do not always lead to broad societal consensuses. The organizers from the National Academy of Sciences realized that. Even as they called for public discussion of the issue, they created a twenty-two-person committee of experts to undertake a yearlong study on whether there should be a moratorium on germline DNA edits.

In their final report, issued in February 2017, the group did not call for a ban or a moratorium. Instead, it provided a list of criteria that should be met before germline editing should be allowed, among them: "absence of reasonable alternatives, restriction to preventing a serious disease or condition," and a few others that were not insurmountable in the foreseeable future.[17] Notably, it omitted one key restriction that was in the 2015 international summit report. There was no longer any mention of the need for a "broad societal consensus" before inheritable gene-editing would be permitted. Instead, the 2017 report called only for "broad ongoing participation and input by the public."

Many bioethicists were dismayed, but most scientists, including Baltimore and Doudna, felt that the report had found a sensible middle ground. Those engaged in medical research saw it as providing a yellow light, allowing them to proceed with caution.[18]

In Britain, the Nuffield Council, the nation's most prestigious independent bioethics organization, produced a report in July 2018 that was even more liberal. "Genome editing has the potential to give rise to transformative technologies in the field of human reproduction," it concluded. "So long as heritable genome editing interventions are consistent with the welfare of the future person and with social justice and solidarity, they do not contravene any categorical moral prohibition." The Council even went so far as to diminish the distinction between using gene editing to cure diseases and using it to provide genetic enhancements. "It is possible that genome editing could be used in the future for . . . enhancing senses or abilities," the guide to the report read. The report was seen, correctly, as paving the way for human germline gene editing. The headline in the *Guardian* was "Genetically Modified Babies Given Go Ahead by UK Ethics Body."[19]

Global regulations

Even though the U.S. National Academy of Sciences and Britain's Nuffield Council espoused a liberal approach to germline editing, some restrictions were imposed in both countries. Congress passed a provision barring the Food and Drug Administration from reviewing any treatment "in which a human embryo is intentionally created or modified to include a heritable genetic modification." President Barack Obama's science advisor, John Holdren, declared, "The Administration believes that altering the human germline for clinical purposes is a line that should not be crossed at this time," and the director of the National Institutes of Health, Francis Collins, announced, "The NIH will not fund any use of gene-editing technologies in human embryos."[20] In Britain, likewise, the editing of human embryos was restricted by various regulations. But in neither Britain nor the U.S. was there an absolute and clear law against germline gene editing.

In Russia, there were no laws to prevent the use of gene editing in humans, and President Vladimir Putin in 2017 touted the potential of CRISPR. At a youth festival that year, he spoke of the benefits and dangers of creating genetically engineered humans, such as super-soldiers. "Man has the opportunity to get into the genetic code created by either nature, or as religious people would say, by God," he said. "One may imagine that scientists could create a person with desired features. This may be a mathematical genius, an outstanding musician, but this can also be a soldier, a person who can fight without fear or compassion, mercy or pain."[21]

In China, the policies were more restrictive, or at least so it seemed. Although there were no clear laws explicitly outlawing inheritable genetic editing of human embryos, there were multiple regulations and guidelines that prevented—or were believed to prevent—it. For example, in 2003 the Ministry of Health issued "Technical Norms on Human Assisted Reproduction" that specified, "Genetic manipulation of human gametes, zygotes and embryos for reproductive purposes is prohibited."[22]

China has one of the world's most controlled societies, and few things happen in clinics without the government's knowledge. Duanqing Pei, a respected young stem-cell researcher who is the director general of Guangzhou Institutes of Biomedicine and Health, assured his fellow steering committee members at the international summit in Washington that germline gene editing of embryos would not happen in China.

That is why Pei and his like-minded friends from around the world were so shocked when they arrived in Hong Kong in November 2018 for the Second International Summit on Human Genome Editing and discovered that, despite all of their high-minded deliberations and carefully crafted reports, the human species had suddenly and unexpectedly been thrust into a new era.

Part Six

CRISPR Babies

A new species would bless me as its creator and source; many happy and excellent natures would owe their being to me.

—Mary Shelley, *Frankenstein; or, The Modern Prometheus*, 1818

He Jiankui taking a selfie with Doudna at Cold Spring Harbor Laboratory

Michael Deem

Chapter 37

He Jiankui

The eager entrepreneur

He Jiankui, the son of struggling rice farmers, was born in the Orwellian year 1984 and grew up in Xinhua, one of the poorest villages in a rural part of Hunan province in east-central China. The average family income there when he was a boy was $100 a year. His parents were so poor that they could not afford to buy him textbooks, so Jiankui* walked to a village bookstore to read them there. "I grew up in a small farming family," he recalled. "I picked leeches from my legs every day in the summer. I will never forget my roots."[1]

Jiankui's childhood instilled in him a hunger for success and fame, so he heeded the exhortations on the posters and banners at his school that he should dedicate himself to pushing forward the frontier of science. He would indeed end up pushing that frontier, though less by great science than great eagerness.

Spurred by his belief that science was a patriotic pursuit, young Jiankui built a rudimentary physics laboratory at home, where he

*His name, 贺建奎, is transliterated as He Jiankui and pronounced HUH JEE'-an-kway. His family or "last" name is He. Because it is confusing to refer to him as He, I refer to him by his given name, Jiankui.

relentlessly conducted experiments. After doing well in school, he was tapped to go to the University of Science and Technology in Hefei, 575 miles to the east, where he majored in physics.

He applied to four graduate schools in the United States and was accepted by only one of them: Rice University in Houston. Studying under Professor Michael Deem, a genetic engineer who would later become the subject of an ethics investigation, Jiankui became a star at creating computer simulations of biological systems. "Jiankui is a very high-impact student," Deem said. "He has done a fantastic job here at Rice, and I am sure he will be highly successful in his career."

Jiankui and Deem devised a mathematical model for predicting what strains of flu would emerge each year and, in September 2010, coauthored an undistinguished paper on CRISPR that showed how the spacer sequences matching viral DNA are formed.[2] Popular, gregarious, and an eager networker, Jiankui became president of Rice's Chinese Students and Scholars Association and an avid soccer player. "Rice is a place where you can really enjoy graduate school," he told the university magazine. "Outside of the lab, there's a lot to do. Oh, my God, Rice has six soccer fields! That's awesome."[3]

He got his PhD in physics but then decided that the future was in biology. Deem allowed him to go to conferences around the country and provided an introduction to the Stanford bioengineer Stephen Quake, who invited Jiankui to become a postdoc in his lab. Colleagues there remember him as funny and energetic, with a Texas-size passion for entrepreneurship.

Quake had founded a company to commercialize a gene-sequencing technology that he had developed, but it began sliding into bankruptcy. Believing that he could make the process commercially successful in China, Jiankui decided to start a company there. Quake was enthusiastic. "This has a chance to bring the phoenix back from the ashes," he exulted to one of his partners.[4]

China was eager to nurture biotech entrepreneurs. In 2011, it launched an innovative new university, the South University of Science and Technology, in Shenzhen, a booming city of 20 million people bordering on Hong Kong. Responding to a job opening posted on the university's website, Jiankui ended up getting hired there as a biology professor and announced on his blog that he was forming the "He Jiankui and Michael Deem Joint Laboratory."[5]

Chinese officials had designated genetic engineering as critical to the country's economic future and its competition with the U.S., and to that end they launched a variety of initiatives to encourage entrepreneurs and lure back researchers who studied overseas. Jiankui benefited from two of them: the Thousand Talents Recruitment Program and the Shenzhen government's Peacock Initiative.

When he formed his new company to build gene-sequencing machines based on Quake's technology in July 2012, Shenzhen's Peacock Initiative provided an initial round of $156,000 in funding. "Shenzhen's generosity in encouraging startups, especially venture capitalists, which is comparable to Silicon Valley, attracted me," Jiankui later told the *Beijing Review*. "I am not a professor in the traditional sense. I prefer to be a research-type entrepreneur."

Over the next six years, Jiankui's company would receive about $5.7 million in funding from government sources. By 2017, its gene sequencer was on the market and the company, of which Jiankui had a one-third stake, was valued at $313 million. "The development of the device is a major technical breakthrough and will significantly improve cost-effectiveness, speed and quality of gene sequencing," Jiankui said.[6] In a scientific article describing the use of the machine for sequencing genomes, he claimed that the results "show comparable performance to the Illumina," referring to the American company that dominates the market for DNA sequencers.[7]

With his smooth personality and thirst for fame, Jiankui became a minor scientific celebrity in China, where the state-run media was eagerly looking for innovators to tout as role models. The broadcast network CTV ran a series in late 2017 featuring the country's young science entrepreneurs. As inspiring patriotic music played, Jiankui was shown talking about his company's gene sequencer, which the narrator said works better and faster than American versions. "Somebody said we shocked the world with our machine," a smiling Jiankui declared to the camera. "Yes, they're right! I did that—He Jiankui! That's me who did that!"[8]

Jiankui initially used his gene-sequencing technology to diagnose genetic conditions in early-stage human embryos. But in early 2018, he began to discuss the possibility of not only reading human genomes but also editing them. "For billions of years, life progressed according to Darwin's theory of evolution: random mutation in DNA, selection and reproduction," he wrote on his website. "Today, genome sequencing and genome editing provide powerful new tools to control evolution." His goal, he said, was to sequence a human genome for $100, then move on to fixing any problems. "Once the genetic sequence is known, we can use CRISPR-Cas9 to insert, edit or delete the associated gene for a particular trait. By correcting the disease genes, we humans can better live in the fast changing environment."

He did, however, say that he was against using gene editing for some forms of enhancement. "I support gene editing for the treatment and prevention of disease," Jiankui wrote in a post on the social media site WeChat, "but not for enhancement or improving I.Q., which is not beneficial to society."[9]

Networker

He Jiankui's website and social media comments, which were in Chinese, did not garner much attention in the West. But as a promiscuous networker and conference hopper, he was beginning to develop a circle of acquaintances in the American scientific community.

In August 2016, he attended the annual CRISPR conference held at Cold Spring Harbor Laboratory. "The just-concluded Cold Spring Harbor Gene Editing Conference is the top event in this field," he bragged on his blog. "Feng Zhang and Jennifer Doudna and other leading figures attended the event!" Accompanying the post was a selfie Jiankui took with Doudna seated in the auditorium under the oil portrait of James Watson.[10]

A few months later, in January 2017, Jiankui sent Doudna an email.

As he had done with other top CRISPR researchers, he asked to meet with her when he next came to the United States. "I am working on the technology to improve the efficacy and safety of genome editing human embryos in China," he wrote. The email arrived when Doudna was helping to organize a small workshop on "the challenge and opportunity of gene editing." It had been two years since her Napa Valley conference, and the Templeton Foundation, which supports the study of big ethical questions, had provided funding for a series of discussions on CRISPR. Doudna invited twenty scientists and ethicists to a kickoff workshop in Berkeley, but few were from overseas. "We would be delighted to have your participation," she wrote back to Jiankui, who, not surprisingly, was equally delighted to accept.[11]

The meeting opened with a public lecture by George Church in which he spoke of the possible benefits of germline editing, including ones that would augment human capacities. Church showed a slide listing simple gene variations that offer beneficial effects. Among them was a variant of the *CCR5* gene that would make a person less receptive to the HIV virus that causes AIDS.[12]

On his blog, Jiankui wrote about the off-the-record meeting: "A lot of sharp issues caused fierce debates there, and the smell of gunpowder filled the air." Particularly interesting was his interpretation of the report from the international summit on gene editing, which had just come out. He called it a "yellow light for human genetic editing." In other words, instead of reading the report as a call to not proceed with heritable human embryo editing for the time being, he interpreted it as a signal that he could proceed cautiously.[13]

Jiankui's turn to present came on the second day of the meeting. His talk, titled "Safety of Human Gene Embryo Editing," was unimpressive. There was only one interesting part: his description of his work editing the *CCR5* gene, the one that Church had mentioned in his lecture as a potential candidate for future germline editing. Jiankui described how he had edited the gene, which produces a protein that can serve as a receptor for the HIV virus, in mice, monkeys, and nonviable human embryos discarded from fertility clinics.

Other Chinese researchers had already prompted international ethics discussions by using CRISPR to edit *CCR5* genes in non-viable human embryos, so nobody at the conference took much notice. "His talk made no impression on me," Doudna says. "I found him very eager to meet people and be accepted, but he hadn't published anything important, and he didn't seem to be doing any important science." When Jiankui asked Doudna if he could come to her lab as a visiting fellow, she was surprised at his audacity. "I deflected his request," she says. "I had absolutely no interest." What struck Doudna and others at the meeting was that Jiankui did not seem interested in the moral issues involved with making inheritable gene edits to embryos.[14]

Continuing to network and conference-hop, Jiankui returned to Cold Spring Harbor in July 2017 for its annual CRISPR conference. Wearing a striped shirt and with his dark hair youthfully mussed, he gave pretty much the same talk that he had given in Berkeley earlier that year, again eliciting yawns and shrugs. He ended on a cautionary note, with a slide showing a *New York Times* story on Jesse Gelsinger, the young man who died after receiving gene therapy treatments. "A single case of failure may kill the entire field," he concluded. There were three perfunctory questions. No one thought that his experiments had produced any scientific breakthroughs.[15]

Editing babies

In this July 2017 talk at Cold Spring Harbor, Jiankui described editing the *CCR5* gene in discarded, non-viable human embryos. What he did not say was that he had already made plans to edit the gene in viable human embryos with the intent of giving birth to genetically altered babies—in other words, making inheritable germline edits. Four months earlier, he had submitted a medical ethics application to Shenzhen's Harmonicare Women and Children's Hospital. "We plan to use CRISPR-Cas9 to edit the embryo," he wrote. "The edited embryos will be transferred to women and pregnancy will follow." His goal was to allow couples who suffered from AIDS to have babies who would be protected from the HIV virus, as would all of their descendants.

Because there were simpler ways to prevent AIDS infection, such as sperm-washing and screening for healthy embryos before implantation, the procedure was not medically necessary. Nor would it correct a clear genetic disorder; the *CCR5* gene is common and probably has multiple purposes, including helping to protect against West Nile virus. So Jiankui's plan did not meet the guidelines that had been agreed to at multiple international meetings.

But it did offer Jiankui the possibility, or at least he thought so, of achieving a major historical breakthrough and enhancing the glory of Chinese science. "This is going to be a great science and medicine achievement," he wrote in his application, comparing it to "the IVF technology which was awarded the Nobel Prize in 2010." The hospital ethics committee gave its consent unanimously.[16]

There are approximately 1.25 million HIV-positive people in China, a number that is still growing rapidly, and ostracism of victims is widespread. Working with a Beijing-based AIDS advocacy group, Jiankui sought to recruit twenty volunteer couples in which the husband was HIV positive and the wife was HIV negative. More than two hundred couples showed interest.

Two of the selected couples came to Jiankui's lab in Shenzhen one Saturday in June 2017 and, in a meeting that was videotaped, were informed about the proposed clinical trial and asked if they wished to participate. He walked them through the consent form. "As the volunteer, your partner is diagnosed to have AIDS or has been infected with HIV," it said. "This research project will likely help you produce HIV-resistant infants." The two couples agreed to participate, as did five more recruited at other sessions. They produced thirty-one embryos, sixteen of which Jiankui was able to edit. Eleven were implanted into the volunteers unsuccessfully, but by the late spring of 2018 he was able to implant twin embryos into one mother and one embryo into another.[17]

Jiankui's process involved taking sperm from the father, washing the cells to rid them of the HIV virus, and then injecting the sperm into the mother's eggs. This was probably enough to ensure that the resulting fertilized eggs were free of HIV. But his goal was to guarantee that the children would never later be infected. So he injected the fertilized eggs with CRISPR-Cas9 that targeted the *CCR5* gene. They were allowed to grow for five or so days in a Petri dish until they were an early-stage embryo more than two hundred cells large, and then their DNA was sequenced to see if the edits had worked.[18]

His American confidants

During his visits to the U.S. in 2017, Jiankui began hinting at his plans to a few of the American researchers he met, many of whom later expressed regret that they did not try harder to stop him or blow the whistle. Most notably, he confided in William Hurlbut, a neurobiologist and bioethicist at Stanford, who had co-organized the January 2017 Berkeley gathering with Doudna. They had, Hurlbut later told the journal *Stat*, "several long conversations, like four or five hours long, about science and ethics." Hurlbut realized that Jiankui was intent on making embryo edits leading to live births. "I tried to give him a sense of the practical and moral implications," he says, but Jiankui insisted that only "a fringe group" opposed making germline edits. If such edits could be used to avoid a dread disease, Jiankui asked, why would people be against it? Hurlbut viewed Jiankui as "a well-meaning person who wants his efforts to count for good" but who was spurred by a scientific culture "that puts a premium on provocative research, celebrity, national scientific competitiveness, and firsts."[19]

Jiankui also confided his plans to Matthew Porteus, an accomplished and respected stem-cell researcher at Stanford Medical School. "I was stunned and my jaw dropped," Porteus recalls. It turned from a polite conversation about scientific data into a half-hour lecture by Porteus about all the reasons he thought Jiankui's idea was terrible.[20]

"There's no medical need," Porteus said. "It violates all the guidelines. You're jeopardizing the entire field of genetic engineering." He demanded to know if Jiankui had run it by his senior people.

No, Jiankui said.

"You need to talk to these people, the officials in China, before you proceed any further," Porteus warned with rising anger.

At that point Jiankui became very quiet, his face flushed, and then he walked out of the office. "I don't think he was expecting such a negative reaction," Porteus says.

In hindsight, Porteus blames himself for not doing more. "I fear some people think I made an ass out of myself," he says. "I wish that, while he was in my office, I had insisted that we jointly send emails to various senior people in China." But it's unlikely that Jiankui would have permitted Porteus to tell other people. "He thought that if he told people ahead of time, they would try to stop him," Porteus says, "but once he succeeded in producing the first CRISPR babies everyone would recognize it as a great achievement."[21]

Jiankui also confided in Stephen Quake, the Stanford gene-sequencing entrepreneur who had supervised his postdoctoral work and helped him launch the Shenzhen-based company that used Quake's technology. As early as 2016, Jiankui told Quake that he wanted to be the first person to create gene-edited babies. Quake told him it was "a terrible idea," but when Jiankui persisted, Quake suggested that he do it with the proper approvals. "I will take your suggestion that we will get a local ethic approval before we move on to the first genetic edited human baby," Jiankui told Quake in an email, which was later reported by *New York Times* health writer Pam Belluck. "Please keep it in confidential."

"Good News!" Jiankui wrote Quake in April 2018. "The embryo with CCR5 gene edited was transplanted to the women 10 days ago, and today the pregnancy is confirmed!"

"Wow, that's quite an achievement!" Quake replied. "Hopefully she will carry to term."

After an investigation, Stanford cleared Quake, as well as Hurlbut and Porteus, of any wrongdoing. "The review found that the Stanford researchers expressed serious concerns to Dr. He about his work," the university declared. "When Dr. He did not heed their recommendations and proceeded, Stanford researchers urged him to follow proper scientific practices."[22]

The most involved and tainted of Jiankui's American enablers was Michael Deem, his PhD advisor at Rice. In a scene that was captured on videotape, Deem can be seen sitting at the table during the first of Jiankui's sessions where prospective parents were advised about giving their consent to the gene editing of their embryos. "When this couple gave their informed consent," Jiankui later said publicly, "it was observed by this United States professor." Deem spoke to the volunteers through a translator, a member of the Chinese team told *Stat*.

In an interview with the Associated Press, Deem admitted being in China during the meeting. "I met the parents," he said. "I was there for the informed consent of the parents." Deem also defended Jiankui's actions. But he then hired two Houston lawyers who issued a statement claiming that Deem was not involved in the informed-consent process, even though a scene from the video shows him sitting there. The lawyers also claimed, "Michael does not do human research and he did not do human research on this project." That seemed to be contradicted when it was revealed that Deem was a coauthor of the paper Jiankui wrote on his human gene–editing experiments. Rice said that it would launch an investigation, but after two years had not issued a finding. By the end of 2020, Deem's faculty page had been removed from the Rice website, but the university continued to refuse to offer any explanation.[23]

Jiankui's PR campaign

As the Chinese pregnancies progressed in mid-2018, Jiankui knew that his announcement would be earth-shaking news, and he wanted to capitalize on it. The goal of his experiment, after all, was not merely to protect two kids from AIDS. The prospect of achieving fame was also a motivation. So he hired Ryan Ferrell, a respected American public relations executive he had worked with on another project, who found Jiankui's plans to be so exciting that he left his agency and relocated temporarily to Shenzhen.[24]

Ferrell planned a multimedia announcement campaign. It included having Jiankui write an article on the ethics of gene editing for a journal, cooperate with the Associated Press on an exclusive story on the making of the CRISPR babies, and tape five videos that would be released on his website and YouTube. In addition, he would write a scientific piece, coauthored with Rice's Michael Deem, that he would try to publish in a prestigious journal such as *Nature*.

The ethics story, which Jiankui and Ferrell titled "Draft Ethical Principles for Therapeutic Assisted Reproductive Technologies," was intended for a new publication called the *CRISPR Journal*, edited by the CRISPR pioneer Rodolphe Barrangou and the science journalist Kevin Davies. In his draft, Jiankui listed five principles that should be followed when deciding whether to edit human embryos:

> *Mercy for families in need*: For a few families, early gene surgery may be the only viable way to heal a heritable disease and save a child from a lifetime of suffering. . . .
>
> *Only for serious disease, never vanity*: Gene surgery is a serious medical procedure that should never be used for aesthetics, enhancement, or sex selection. . . .
>
> *Respect a child's autonomy*: A life is more than our physical body. . . .
>
> *Genes do not define you*: Our DNA does not predetermine our purpose or what we could achieve. We flourish from our own hard work, nutrition, and support from society and our loved ones. . . .
>
> *Everyone deserves freedom from genetic disease*: Wealth should not determine health.[25]

Instead of following guidelines such as those established by the National Academy of Sciences, Jiankui had crafted a framework that, at least by his thinking, would justify his use of CRISPR to take out the receptor gene for HIV. He was following moral principles that had been propounded, sometimes quite convincingly, by some prominent Western philosophers. For example, Duke professor Allen Buchanan was the staff philosopher for President Reagan's Commission on Medical Ethics, was on the Advisory Council for the National Human Genome Research Institute under President Clinton, and is a fellow of the prestigious Hastings Center. Seven years before Jiankui decided to edit the *CCR5* gene in human embryos, Buchanan had supported the concept in his influential book *Better Than Human*:

> Suppose we learn that some desirable gene or set of genes already exists, but only in a small number of humans. This is precisely the situation for genes that confer resistance to certain strains of HIV-AIDS. If we rely on the "wisdom of nature" or "let nature take its course," this beneficial genotype may or may not spread through the human population. . . . Suppose it were possible to ensure that such beneficial genes spread much more quickly by intentional genetic modification. This could occur by injecting genes into the testicles or, more radically, by inserting them into a large number of human embryos, utilizing in vitro fertilization. We would get the benefits . . . without the carnage.[26]

Buchanan was not alone. At the time of Jiankui's clinical trial, many serious ethical thinkers, and not just gung-ho scientific researchers, had publicly argued, using the *CCR5* gene as a specific example, that gene editing to cure or prevent diseases could be permissible and even desirable.

Ferrell gave an Associated Press team—Marilynn Marchione, Christina Larson, and Emily Wang—exclusive access to Jiankui. They were even allowed to videotape a non-viable human embryo being injected with CRISPR in Jiankui's lab.

With Ferrell's guidance, Jiankui also prepared videos that featured him in his lab speaking directly to the camera. In the first one, he outlined his five ethical principles. "If we can protect a little girl or boy from certain disease, if we can help more loving couples start families, gene surgery is a wholesome development," he said. He also made a distinction between curing disease and making enhancements. "Gene surgery should only be used for treating serious disease. We should not use it for increasing I.Q., improving sports performance or changing skin color. That's not love."[27]

In the second video, he explained why he felt it was "inhuman for parents *not* to protect their children if nature gives us the tools to do so." The third video explained why he had chosen HIV as his first target. The fourth, which was in Chinese and delivered by one of his postdoctoral students, explained the scientific details of how the CRISPR edits were made.[28] They held off making the fifth video until they could announce the live births of the two babies.

Birth

The public relations campaign and release of the YouTube videos was planned for January, when the babies were due to be born. But one evening in early November 2018, Jiankui got a call saying that the mother had gone into labor prematurely. He dashed to the Shenzhen airport to fly to the city where the mother lived, taking some of the students in his lab. She ended up giving birth, after a caesarean section, to two apparently healthy girls, who were named Nana and Lulu.

The births happened so early that Jiankui had not yet submitted the official description of his clinical trial to Chinese authorities. On November 8, after the twins were born, it was finally submitted. It was written in Chinese, and for two weeks it went unnoticed in the West.[29]

He also finished the academic article he had been working on. Titled "Birth of Twins after Genome Editing for HIV Resistance," it was submitted to the prestigious journal *Nature*. It was never published, but the manuscript, a copy of which was given to me by one of the American researchers he sent it to, offers details about his science and glimpses into his mindset.[30] "Genome editing at the embryonic stage has potential to permanently cure disease and confer resistance to pathogenic infections," he wrote. "Here, we report the first birth from human gene editing: twin girls who had undergone CCR5 gene editing as embryos were born normal and healthy in November 2018." In the article, Jiankui defended the ethical value of what he had done. "We anticipate that human embryo genome editing will bring new hope to millions of families seeking healthy babies free from inherited or acquired life-threatening diseases."

Buried in Jiankui's unpublished paper were some disturbing pieces of information. In Lulu, only one of the two relevant chromosomes had been properly modified. "We confirmed Nana's CCR5 gene was edited successfully with frameshift mutations on both alleles and Lulu's was heterozygous," he admitted. In other words, Lulu had different gene versions on her two chromosomes, which meant that her system would still produce some of the CCR5 protein.

In addition, there was evidence that some unwanted off-target edits had been made and also that both embryos had been mosaics, meaning there had been enough cell division before the CRISPR editing was done that some of the resulting cells in the babies were unedited. Despite all of this, Jiankui later said, the parents chose to have both embryos implanted. Kiran Musunuru of the University of Pennsylvania later commented, "The first attempt to hack the code of life and, ostensibly, improve the health of human babies had in fact been a hack job."[31]

The news breaks

In the first few days after the babies were born, Jiankui and his publicist, Ferrell, tried to keep it under wraps until January, when they hoped that *Nature* would publish their scholarly paper. But the news was too explosive to hold. Just before Jiankui was scheduled to arrive at the Second International Summit on Human Genome Editing, to be held in Hong Kong, news of his CRISPR babies leaked.

Antonio Regalado, a reporter at the *MIT Technology Review*, combined a knowledge of science with the news instincts of a scoop-magnet journalist. He was in China in October, and he happened to be invited to a meeting with Jiankui and Ferrell just as they were making plans for the announcement. Although Jiankui did not reveal his secret, he did discuss the *CCR5* gene, and Regalado was a good enough reporter to suspect that something was afoot. By searching on the internet, he discovered the application Jiankui had submitted to the Chinese Clinical Trial Registry. "Exclusive: Chinese Scientists Are Creating CRISPR Babies" was the headline on his story, which went online November 25.[32]

With Regalado's story online, Marchione and her colleagues unleashed a well-balanced story brimming with details. Their lead sentence captured the drama of the moment: "A Chinese researcher claims that he helped make the world's first genetically edited babies—twin girls born this month whose DNA he said he altered with a powerful new tool capable of rewriting the very blueprint of life."[33]

All of the high-minded discussions that ethicists had been having about germline gene editing were suddenly preempted by an ambitious young Chinese scientist who wanted to make history. As with the birth of the first test-tube baby, Louise Brown, and the cloning of Dolly the sheep, the world had entered a new era.

That evening, Jiankui released the videos he had previously made along with a final one in which he made his momentous announcement on YouTube. Speaking calmly but proudly to the camera, he declared:

> Two beautiful little Chinese girls named Lulu and Nana came crying into the world as healthy as any other babies a few weeks ago. The girls are home now with their mom, Grace, and their dad, Mark. Grace started her pregnancy by regular IVF with one difference. Right after we sent her husband's sperm into her egg, we also sent in a little bit of protein and instructions for it to do gene surgery. When Lulu and Nana were just a single cell, this surgery removed the doorway through which HIV enters to infect people. . . . When Mark saw his daughters, the first thing he said was that he never thought he could be a father. Now he has found a reason to live, a reason to walk, a purpose. You see, Mark has HIV. . . . As a father of two girls, I can't think of a gift more beautiful and wholesome for the society than giving another couple a chance to start a loving family.[34]

He Jiankui coming onstage

With Robin Lovell-Badge and Matthew Porteus

Chapter 38

The Hong Kong Summit

On November 23, two days before He Jiankui's news broke, Doudna received an email from him. The subject line was dramatic: "Babies Born."

She was puzzled, then shocked, and then alarmed. "At first I thought it was fake, or maybe he was crazy," she says. "The idea that you would use 'Babies Born' as a subject line for something like this didn't seem real."[1]

He had included the draft of the manuscript he had submitted to *Nature*. When Doudna opened the attachment, she knew the situation was all too real. "It was a Friday, the day after Thanksgiving," she recalls. "I was over in our condo in San Francisco with family members and longtime friends when this email came in like a bolt out of the blue."

Doudna realized that the news would become even more dramatic because of its timing. In three days, five hundred scientists and policymakers were due to converge in Hong Kong for the Second International Summit on Human Genome Editing, the successor to the December 2015 summit in Washington. Doudna was one of the core organizers, along with David Baltimore, and He Jiankui was scheduled to be a speaker.

Doudna and the other organizers had not originally put Jiankui on the list of invited speakers. But they had changed their minds a few weeks earlier when they heard rumors that he had dreams or delusions of editing human embryos. Some on the planning committee felt that involving him in the summit might help dissuade him from crossing the germline.[2]

Upon getting Jiankui's shocking "Babies Born" email, Doudna tracked down Baltimore's cell phone number and reached him as he was leaving to fly to Hong Kong. They agreed that she would change her flight and arrive a day earlier than planned so that they could gather with some of the other organizers and decide what to do.

When she landed at dawn on the morning of Monday, November 26, and turned her phone back on, Doudna saw that Jiankui had been desperately trying to reach her by email. "The nanosecond I landed at the airport, I had just a ton of emails from Jiankui," Doudna told Jon Cohen of *Science*. He was driving to Hong Kong from Shenzhen, and he wanted to meet as soon as possible. "I have to talk to you right now," he emailed. "Things have really gotten out of control."[3]

She did not reply because she wanted to meet first with Baltimore and the other organizers. Soon after she checked into Le Méridien Cyberport hotel, where the conferees were staying, a bellman knocked at her door with a message from Jiankui, saying to call him right away.

She agreed to meet with Jiankui in the hotel lobby, but first she hastily convened some of the organizers in a fourth-floor conference room. Baltimore was already there, sitting with George Daley of Harvard Medical School, Robin Lovell-Badge of London's Francis Crick Institute, Victor Dzau of the U.S. National Academy of Medicine, and bioethicist Alta Charo from the University of Wisconsin. None of them had seen the scientific paper that Jiankui had submitted to *Nature*, so Doudna showed them the copy he had emailed to her. "Our group all scrambled to decide whether Jiankui should be allowed to remain on the conference program," Dzau recalls.

They quickly decided that he should. In fact, they decided it was important not to let him withdraw. They would give him a solo spot on the program and ask him to address the science and methods he used to make the CRISPR babies.

After fifteen minutes, Doudna went down to the lobby to meet Jiankui. She took with her Robin Lovell-Badge, who would be chairing Jiankui's session. The three of them sat on a couch, and Doudna and Lovell-Badge told Jiankui that they wanted his presentation to explain exactly how and why he had proceeded with his experiment.

Jiankui flummoxed them by insisting that he wanted to stick with his original slide presentation and not discuss the CRISPR babies. Lovell-Badge, whose usual hue is an English pale, turned almost white as he listened. Doudna politely pointed out that Jiankui was being ludicrous. He had triggered the most explosive scientific controversy in years, and there was no way he could avoid discussing it. That seemed to surprise Jiankui. "I think that he was oddly naïve as well as glory-seeking," she recalls. "He had intentionally caused an explosion and yet wanted to act like it hadn't happened." They convinced him to have an early dinner with some members of the organizing committee to discuss the issue.[4]

On her way out of the lobby, shaking her head in amazement, Doudna ran into Duanqing Pei, the American-educated stem-cell biologist from China who heads the Guangzhou Institute of Biomedicine and Health. "Have you heard?" Doudna asked him. When she told him the details, he had trouble believing it. Pei and Doudna had become friendly after many conferences, including the 2015 first international summit in Washington, and he had repeatedly told his American colleagues that there were regulations in China against germline editing in humans. "I assured people that, in our system, everything is carefully controlled and licensed, so this type of thing couldn't occur," Pei later told me. He agreed to come to the dinner with Jiankui that evening.[5]

Showdown over dinner

The dinner, a Cantonese buffet in the hotel's fourth-floor restaurant, was tense. When Jiankui arrived, he was defensive, even a bit defiant, about what he had done. He pulled out his laptop to show his data and the DNA sequencing he had performed on the embryos. "We were increasingly horrified," Lovell-Badge recalled. They peppered him with questions: Had there been oversight on his consent process? Why did he believe germline embryo editing was medically necessary? Had he read the guidelines that the international academies of medicine had adopted? "I feel I complied with all those criteria," Jiankui answered. His university and hospital knew all about what he was doing and had approved, he insisted, "and now that they're seeing the negative reaction, they're denying it and hanging me out to dry." When Doudna walked through the reasons why germline editing was not "medically necessary" to prevent HIV infection, Jiankui got very emotional. "Jennifer, you don't understand China," he said. "There's an incredible stigma about being HIV positive, and I wanted to give these people a chance at a normal life and help them have kids when they otherwise might not have."[6]

The dinner became increasingly fraught. After an hour, Jiankui shifted from being plaintive to being angry. He stood up abruptly and tossed some bills on the table. He had been receiving death threats, he said, and now he was going to move to an undisclosed hotel where the press couldn't find him. Doudna chased after him. "I think it's very important that you appear on Wednesday and present your work," she said. "Will you come?" He paused and then agreed to, but he wanted security. He was afraid. Lovell-Badge promised to have Hong Kong University provide police protection.

One reason Jiankui was defiant was that he had thought he would be hailed as a Chinese hero, perhaps even a global one. Indeed, the first Chinese news reports did so. The *People's Daily*, a government organ, ran a story that morning with the headline "The World's First Gene-Edited Babies Genetically Resistant to AIDS Were Born in China," and it called Jiankui's work "a milestone accomplishment China has achieved in the area of gene-editing technologies." But the tide quickly turned as scientists, even in China, began to criticize his actions. Later that evening, the *People's Daily* deleted the story from its website.[7]

After Jiankui left the hotel restaurant, the organizers stayed at the table discussing how to handle the situation. Pei looked at his smartphone and reported that a group of Chinese scientists had put out a statement condemning Jiankui. Pei began translating it for the others at the table. "Direct human experimentation can only be described as crazy," they declared. "This is a huge blow to the global reputation and development of Chinese science, especially in the field of biomedical research." Doudna asked Pei whether the statement had come from the Chinese Academy of Sciences. No, Pei replied, but a group of more than a hundred prestigious Chinese scientists had signed it, which meant that the statement had official blessing.[8]

Doudna and her dinner partners realized that they, as the conference organizers, should put out a statement as well. But they did not want to make it too strong for fear that it would provoke Jiankui to cancel his talk. Truth be told, Doudna admitted, their motives were not merely scientific. The global buzz was huge, eyes were on Hong Kong, and it would be quite a letdown if Jiankui drove back to Shenzhen and they all missed the chance to be part of a historic moment. "We put out a very short statement that was quite bland and got criticism for that," she says, "but we wanted to ensure that he would show up."

As Doudna and her colleagues were having dinner, Jiankui's extensive publicity plan was unfolding: the YouTube videos were released, the AP story he had cooperated with went viral, and the high-minded ethics piece that he wrote was finally published online by the *CRISPR Journal* editors (though they later retracted it). "We were all struck by the fact that he was pretty young and came across as an interesting combination of hubris and remarkable naïveté," Doudna says.[9]

Jiankui's presentation

At high noon on Wednesday, November 28, 2018, it was finally time for He Jiankui to present.[10] Robin Lovell-Badge, the moderator, came to the podium looking nervous. With his sandy blond hair, which his hands kept mussing nervously, and his horn-rimmed glasses, he looked like an even nerdier version of Woody Allen. He also looked haggard. He later told Doudna that he had not slept at all the night before. Reading from his notes, he instructed the audience to be polite, as if he was afraid that the conferees might rush the stage. "Can you please allow him to speak without interruptions," he said, then waved his hand as if wiping a machine and added, "I have the right to cancel the session if there is too much noise or interruption." But the only sounds were from the clicking cameras of the dozens of photographers standing in the back.

Lovell-Badge explained that Jiankui had been scheduled to speak before the news of his CRISPR babies was known. "We didn't know the story that was going to break over the last couple of days," he said. "In fact he sent me the slides he was going to show in this session, and they did not include any of the work he is going to talk about." Then, looking around nervously, he announced, "I would like, if he can hear me, to invite Jiankui He to the stage to present his work."[11]

At first no one appeared. The audience seemed to be holding its breath. "I'm sure people were wondering if he was actually going to appear," Lovell-Badge later recalled. Then, from directly behind Lovell-Badge, who was standing on the right side of the stage, a young Asian man appeared in a dark suit. There was scattered, tentative applause and a bit of confusion. The man fiddled with a laptop to get the right slide up, then adjusted the microphone. The audience members began to laugh nervously as they realized that it was the audiovisual technician. "Look, I don't know where he is," Lovell-Badge said, waving his notebook.

For an eerie thirty-five seconds, which in cases like this is a *very* long time, there was a charged silence in the room, but no movement. Finally, somewhat tentatively, a slight man wearing a striped white shirt and carrying a bulging tan briefcase stepped out from the far side of the stage. In the somewhat formal atmosphere of Hong Kong (Lovell-Badge was wearing a suit), he looked incongruous with his wide-open collar and no jacket or tie. "He looked more like a commuter hurrying to catch the Star Ferry in the Hong Kong humidity than a scientist at the center of a massive international storm," the science editor Kevin Davies later reported.[12] Lovell-Badge, relieved, waved him over and, when Jiankui got to the podium, whispered in his ear, "Please not too long, we need time to ask you questions."

As Jiankui started to speak, a barrage of camera clicks and flashes from the paparazzi drowned him out and seemed to startle him. David Baltimore stood up in the front row, turned to the press section, and berated them. "The clicking of the cameras was so loud that we couldn't hear what was happening onstage," he says. "So I took over the meeting for a moment and got them to stop."[13]

Jiankui glanced around sheepishly, his smooth face making him look even younger than his thirty-four years. "I must apologize that my results leaked unexpectedly, taking away the chance for peer review before being presented to this conference," he began, and then went on, without seeming to be aware of the contradiction, to "thank the Associated Press who we engaged months before the birth of the humans for accurately reporting the study's outcome." Reading slowly from his speech, with little emotion, he described the scourge of HIV infection, the deaths and discrimination that resulted, and how a *CCR5* gene mutation could prevent the infection of babies born to HIV-positive parents.

After twenty minutes of showing slides and discussing his process, it was time for questions. Lovell-Badge invited Matthew Porteus, the Stanford stem-cell biologist who knew Jiankui, onstage to help with the questioning. Instead of asking Jiankui about the huge issue of why he would violate international norms by making germline edits in a human embryo, Lovell-Badge began with a long question, and then another, about the evolutionary history and possible roles of the *CCR5* gene. Porteus followed up with multiple detailed questions about how many couples, eggs, embryos, and researchers were involved in Jiankui's clinical trial. "I was disappointed that the discussion onstage didn't focus on the main issues," Doudna later said.

Finally, the audience was invited to comment and ask questions. Baltimore rose first and went right to the point. After describing the international guidelines that were supposed to be met before any germline editing of humans, he declared, "That has not happened." He called Jiankui's actions "irresponsible," secretive, and not "medically necessary." David Liu, the prominent biochemist at Harvard, spoke next and challenged Jiankui about why he felt embryo editing was warranted in this case. "You could do sperm washing and generate uninfected embryos," Liu said. "What is the unmet medical need for these patients?" Speaking softly, Jiankui responded that he was not just trying to help the twin girls but wanted to find a way "for millions of HIV children" who might someday need protection from being infected with HIV from their parents even after being born. "I have personal experience with people in an AIDS village where thirty percent of villagers were infected, and they have to give up their children to aunts and uncles for fear of infecting them."

"There's a consensus to not allow genome editing on germline cells," a professor at Peking University pointed out. "Why did you choose to cross this red line? And why did you conduct these [procedures] in secret?" When Lovell-Badge took it upon himself to rephrase the question, he asked only about the secrecy part, which Jiankui deflected by describing how he had consulted with a lot of researchers in the U.S., and so he never directly addressed the key historic issue involved. The final question was submitted by a journalist: "If this was going to be your baby, would you have gone ahead with this?" Jiankui's answer: "If it was my baby in this situation, I would have tried it." Then Jiankui picked up his briefcase, exited the stage, and was driven back to Shenzhen.[14]

Sitting in the audience, Doudna began to sweat. "I was feeling a combination of nervous energy and being sick to my stomach," she recalls. Here was the amazing gene-editing tool, CRISPR-Cas9, that she had co-invented, being used to produce, for the first time in history, a genetically designed human being. And it had been done before the safety issues had been clinically tested, the ethical issues resolved, or a social consensus had formed over whether this was the way for science—and for humans—to evolve. "It was quite emotional for me feeling the incredible disappointment and disgust at the way that it had been handled. I was concerned that the race to do this had been motivated not by medical need or by the desire to help people but by a desire for attention and to be first."[15]

The question that she and the other organizers faced was whether they were partly to blame. For years they had been crafting criteria that should be met before there was any editing of humans. But they had stopped short of calling clearly for a moratorium or prescribing a clear process for approval of a trial. Jiankui could claim, as he did, that in his own mind he had followed these criteria.[16]

"Irresponsible"

Later that evening, Doudna went to the bar of the hotel and huddled with a few of her exhausted fellow organizers. Baltimore showed up, and they ordered beer. He believed, more than the others, that there had been a failure by the scientific community to do enough self-regulation. "One thing is clear," he said. "If this guy really did what he claims to have done, this is actually not very hard to do. That's a sobering thought." They decided that they had to issue a statement.[17]

Doudna, Baltimore, Porteus, and five others commandeered a small meeting room and began hammering out a draft. "It was many hours going over line by line and discussing what the point of each sentence was," Porteus recalls. Like the others, he wanted to express strong disapproval of what Jiankui had done and yet avoid using the word "moratorium" or doing anything that might hamper the progress of gene-editing research. "I find the word 'moratorium' not very productive because it doesn't give you a sense of how you move past it," Porteus says. "I know it's a term that appeals to people because it puts a nice dark line that thou shall not cross. But to just say there should be a moratorium cuts off conversation and doesn't allow us to think through how one might get there in a responsible way."

Doudna was tugged in two directions. She was appalled at what Jiankui had done, because it was premature and unnecessary as a medical procedure and a grandstanding act that could spark a backlash against all gene-editing work. Yet she had come to believe, and hope, that CRISPR-Cas9 would prove to be a powerful tool for human well-being, including someday through making germline edits. During the discussion of the draft statement, that became the consensus at the table.[18]

So they decided, once again, to steer a middle course. There was a need for more specific guidelines on when germline gene editing should be done, but it was also important to avoid rhetoric that would lead to national bans and moratoria. "The sense at the meeting was that the technology had advanced to the stage where we need to have a clear pathway to clinical use of gene editing in embryos," Doudna says. In other words, instead of trying to stop any further uses of CRISPR to make gene-edited babies, she wanted to pave the way to making it safer to do so. "To put your head in the sand or say we need a moratorium is just not realistic," she argues. "Instead we should say, 'If you want to move into the clinic with gene editing, these are the specific steps that need to be taken.'"

Doudna was influenced by George Daley, dean of Harvard Medical School, a longtime friend who was part of these deliberations. He strongly believed that CRISPR could be used someday to make inheritable edits; research was then underway at Harvard to study germline edits in sperm that might prevent Alzheimer's. "George appreciates the potential value of human germline editing in embryos and has wanted to maintain the potential for this to be used in the future," Doudna says.[19]

So the statement that Doudna, Baltimore, and the other organizers crafted was very restrained. "At this summit we heard an unexpected and deeply disturbing claim that human embryos had been edited and implanted, resulting in a pregnancy and the birth of twins," they wrote. "The procedure was irresponsible and failed to conform with international norms." But there was no call for a ban or moratorium. Instead, the statement simply said that the safety risks were currently too great to permit germline editing "at this time." It then proceeded to stress, "Germline genome editing could become acceptable in the future if these risks are addressed and if a number of additional criteria are met." The germline was no longer a red line.[20]

Chapter 39

Acceptance

Francis Collins, Doudna, and Senator Richard Durbin at a congressional hearing

Josiah Zayner celebrates

Josiah Zayner, the biohacker who had injected himself with a CRISPR-edited gene a year earlier, was so excited that he stayed up all night watching a livestream of He Jiankui's announcement in Hong Kong. He watched on his laptop in bed with a blanket over his legs and the lights off, with just the glow of the laptop on his face, because his girlfriend was asleep next to him. "I'm just sitting there waiting for him to take the stage, and I got a tingle down my spine and goosebumps knowing that something exciting was about to happen," he says.[1]

When Jiankui described the CRISPR-edited twins he had wrought, Zayner said to himself, "Holy shit!" It was, he felt, not just a scientific achievement but a milestone for the human race. "We did it!" he exulted. "We genetically engineered an embryo! Our humanity has just been changed forever!"

There was no way to go back now, he realized. It was like when Roger Bannister broke the four-minute mile. Now that it had happened, it would happen again. "I view it as one of the most groundbreaking things that's been done in science. In all of human history, we didn't get to decide what genes we have, right? Now we do." And in a personal way, it validated what Zayner felt was his own mission. "For days I was so excited I couldn't sleep, because it affirmed to me why I do what I do, which is to try to make sure that people can push humanity forward."

Push humanity forward? Yes, sometimes it's the rebels who do so. As Zayner speaks, his flat tones and crazy excitement remind me of a day when Steve Jobs sat in his backyard and recited from memory the lines he had helped craft for Apple's "Think Different" commercial about the misfits and rebels and troublemakers who are not fond of rules and have no respect for the status quo. "They push the human race forward," Jobs said. "Because the people who are crazy enough to think they can change the world, are the ones who do."

One reason it would be hard to prevent future CRISPR babies, Zayner later explained in an essay for *Stat*, was that the technology would soon be within the reach of accomplished misfits. "People are already editing human cells using a $150 inverted microscope," he wrote, and online companies like his own sold Cas9 protein and guide RNAs. "The requirements of embryo injection are minimal: a microinjector, micropipette, and microscope. All of these can be purchased on eBay and assembled for a few thousand dollars." Human embryos can be bought from fertility clinics for about $1,000, he said. "You can probably have the embryo transferred to a human by a medical doctor in the U.S. if you don't tell him or her what you've done, or you can do it in another country. . . . So it won't be long until the next human embryo is edited and implanted."[2]

The great thing about germline gene editing, Zayner says, is that it can remove a disease or genetic abnormality permanently from the human race. "Not just cure it in a patient," he says, "but completely remove miserable death-sentence diseases like muscular dystrophy from the future of humanity, forever." He even supports using CRISPR to make enhancements in children. "If I could have my children be less prone to being obese or having genes that make them perform better athletically and stuff, why would I say no?"[3]

For Zayner, the issue is also personal. When I talked to him in mid-2020, he and his partner were trying to conceive a child through *in vitro* fertilization, and they took advantage of preimplantation genetic diagnosis to select the sex of their child. The doctors also screened for a few major genetic diseases, but they would not give Zayner full genome sequences and markers of the prospective embryos. "We don't get to choose the genes that go into our baby, which is crazy," he says. "Instead we let it be done by chance. I think it's okay to choose the genes you want for your children. It's scary and it's going to create *Homo sapiens* version 2.0. But I also think it's really, really, really exciting."

As I start to push back, Zayner stops me cold by citing a personal example of the type of genetic disposition he would like to edit. "I suffer from bipolar disorder," he says. "It's terrible. It inflicts serious issues on my life. I would love to get rid of it." Does he worry, I ask, that eliminating the disorder would change who he is? "People try to make up these lies that it helps you be more creative and all this other bullshit, but it's a disease. It's an illness that causes suffering, a shitload of suffering. And I think we could probably figure out ways to be creative without this disease."

Zayner knows that there are multiple genes that contribute in mysterious ways to psychological disorders, and we don't know enough now to fix them. But in theory, if it could be made to work, he feels that he would want to use germline gene editing to make sure his own children are less likely to suffer. "If I could edit the genes in a way that would reduce the probability that my child would be bipolar, if I could reduce that a little bit for my child, like, how could I not do that? Like, how can I want my child to just grow up and have to suffer like I have? I just don't think I could."

What about less medically needed edits? "Sure, I would make my kids six inches taller and more athletic if I could," he says. "And more attractive. People who are taller and more attractive are more successful, right? What would you want for your child? Obviously for my kids, I would want the world for them." He guesses, correctly, that I grew up in a household with parents who provided me with the best possible education. "Is that any different," he asks, "than wanting to provide a kid with the best genes?"

No backlash

When Doudna returned home from Hong Kong, she found that her teenage son could not understand why there was so much fuss about Jiankui's gene editing. "Andy was very cavalier, which makes me wonder whether future generations will see this as such a big deal," she says. "Maybe they'll see it like IVF, which was very controversial when it first arose." Her parents, she recalls, were shocked when the first test-tube baby was born in 1978. She was fourteen, had just read *The Double Helix*, and remembers discussing with them why they thought creating babies by *in vitro* fertilization was unnatural and felt wrong. "But then it came to be accepted and my parents accepted it—they had friends who could only have kids by IVF and were delighted that the technology existed."[4]

As it turned out, the political and public reaction to the CRISPR babies was in line with Andy's. Two weeks after returning from Hong Kong, Doudna attended that meeting on Capitol Hill with eight senators to discuss gene editing. Such meetings are usually a forum for politicians to express their shock and dismay about something they don't fully understand and then call for more laws and regulations. Quite the opposite occurred at the Senate briefing, which was hosted by Illinois Democrat Dick Durbin and included South Carolina Republican Lindsey Graham, Rhode Island Democrat Jack Reed, Tennessee Republican Lamar Alexander, and Louisiana Republican Bill Cassidy (a doctor). "I was pleased that all of those senators, all of them, were encouraging of the general idea of editing as an important technology," Doudna says. "I was surprised none of them were demanding more regulations. They just wanted to figure out, 'Where do we go from here?'"

Doudna and National Institutes of Health director Francis Collins, who accompanied her, explained that there were already regulations in place to restrict the use of gene editing in embryos. The senators were more interested in trying to understand the value CRISPR might have in medicine and agriculture. Rather than focus on the just-born Chinese CRISPR babies, they asked detailed questions about how CRISPR might work, both in somatic therapies and germline editing, to cure sickle-cell anemia. "They were electrified by the sickle-cell potential, and for other debilitating single-gene diseases such as Huntington's and Tay-Sachs," Doudna recalls. "They talked about what it meant for sustainable health care."[5]

Two international commissions were created to deal with the issue of germline editing. The first was organized by the national science academies that had been part of the process since 2015. The other was convened by the World Health Organization. Doudna feared that having two groups might lead to conflicting messages, thus allowing future He Jiankuis to make their own interpretations of the guidelines. So I met with Victor Dzau, president of the U.S. National Academy of Medicine, and Margaret Hamburg, co-chair of the WHO commission, to see how they would divide up responsibility. "The national academies group is focusing on science," Hamburg said. "The WHO is looking at how to create a global regulatory framework." Even though there will be two reports, Dzau said, it will be better than in the past, when the scientific academies in different countries were creating different guidelines.

Nevertheless, Hamburg conceded that this was unlikely to prevent countries from crafting their own rules. "They have different attitudes and regulatory standards, like they do on genetically modified foods, that reflect their different social values," she explained. That could, unfortunately, lead to genetic tourism. Privileged people who want enhancements will travel to the countries that offer them. She acknowledged that it would be hard for the WHO to police compliance: "This is not like nuclear weapons where you can have guards and padlocks to enforce a security regimen."[6]

The moratorium issue

As the two commissions were getting to work in mid-2019, a public dispute erupted in the scientific community that once again pitted Doudna against the Broad Institute's hard-charging Eric Lander. It was over the use of the word "moratorium," which most scientific committees over the years had avoided.

In some ways the dispute over whether to call for an official moratorium was semantic. The conditions that had been specified for permissible embryo gene editing—that it be safe and "medically necessary"—could not be met for the time being. But some argued that Jiankui's actions showed the need for a clearer and brighter stoplight. Among them were Lander, his protégé Feng Zhang, Paul Berg, Francis Collins, and Doudna's scientific collaborator Emmanuelle Charpentier. "If you use the m-word," Collins explained, "it has a little more clout."[7]

Lander liked being a public intellectual and policy advisor. Articulate, funny, gregarious, and magnetic—at least to those not turned off by his intensity—he was very good at advocating positions and convening groups of earnest chin-strokers. But Doudna suspected that he stirred up the moratorium issue, at least in small part, because she and David Baltimore, rather than the publicity-shy Zhang, had taken the limelight as the foremost public policy thinkers about CRISPR. "Eric and the Broad Institute have a very big bullhorn," she says. "Their call for a moratorium was a way for them to capture a lot of headlines about something they didn't step up to the plate about early on."

Whatever his motives (and I tend to think they were sincere), Lander set about rounding up support for an article to be published in *Nature* titled "Adopt a Moratorium on Heritable Genome Editing." Zhang of course signed up, as did Doudna's erstwhile collaborator Charpentier. So did Berg, whose recombinant DNA discoveries had prompted Asilomar forty-four years earlier. "We call for a global moratorium on all clinical uses of human germline editing—that is, changing heritable DNA (in sperm, eggs or embryos) to make genetically modified children," the article began.[8]

Lander coordinated the essay with his friend Collins, with whom he had worked on the Human Genome Project. "We have to make the clearest possible statement that this is a path we are not ready to go down, not now, and potentially not ever," Collins said in an interview the day the Lander article was released.

Lander emphasized that the issue should not be left to individual choice and the free market. "We're trying to plan the world we're going to leave for our children," he said. "Is it a world where we're deeply thoughtful about medical applications, and we're using it in serious cases, or is it a world where we just have rampant commercial competition?" Zhang made the point that the issues surrounding gene editing needed to be settled by society as a whole and not by individuals. "You can imagine a situation where parents will feel pressure to edit their children because other parents are," he said. "It could further exacerbate inequality. It could create a total mess in society."[9]

"Why is Eric so intent on publicly pushing for a moratorium?" Margaret Hamburg, the co-chair of the World Health Organization group, asked me. It was a sincere question. Lander's reputation was such that even when he did something that seemed straightforward, others suspected his motives. The call for a moratorium, she felt, seemed like showboating; it was unnecessary, since both the WHO and the national academies were already embarked on figuring out proper guidelines rather than calling a halt to germline editing.[10]

Baltimore too expressed puzzlement. Lander had tried to recruit him to sign the letter, but as with the discussion of recombinant DNA forty years earlier at Asilomar, Baltimore was more interested in finding "a prudent path forward" for what could be a lifesaving advance rather than declaring a moratorium that may be difficult to lift once in place. He suspected that Lander might be pushing the moratorium to curry favor with Collins, the director of the National Institutes of Health, which provides a lot of funding for academic labs.

As for Doudna, her opposition to a moratorium became stronger the more that Lander pushed it. "Since germline editing has already been done with the Chinese babies, I think to put out a call for a moratorium at this stage is just unrealistic," she says. "If you call for a moratorium, you effectively take yourself out of the conversation."[11]

Doudna's view prevailed. In September 2020, a two-hundred-page report was issued by the international academies of science commission formed after Jiankui's shocking announcement. It did not call for a moratorium, nor mention that word, even though Lander was one of the eighteen commission members. Instead, it said that heritable human genome editing "might in the future provide a reproductive option" for couples who have genetic diseases. The report noted that making inheritable gene edits was not yet safe and usually not medically necessary, but it came down in favor of "defining a responsible pathway for clinical use of heritable human genome editing"—in other words, continuing to pursue the goal of "a prudent path forward" that was endorsed at the January 2015 Napa Valley conference that Doudna organized.[12]

He Jiankui convicted

Instead of being acclaimed a national hero, as he had fantasized, He Jiankui was put on trial at the end of 2019 in the People's Court of Shenzhen. The proceedings had many elements of a fair trial: he was permitted to have his own attorneys and to speak in his own defense. But the verdict was not in doubt since he had pleaded guilty to the charge of "illegal medical practice." He was sentenced to three years in prison, fined $430,000, and banned for life from working in reproductive science. "In order to pursue fame and profit, [he] deliberately violated the relevant national regulations and crossed the bottom lines of scientific and medical ethics," the court declared.[13]

The official Chinese news report on the trial also revealed that a third CRISPR baby engineered by Jiankui had been born to a second woman. There were no details about the baby nor about the current status of Lulu and Nana, the original CRISPR-edited twins.

When Doudna was asked by the *Wall Street Journal* to comment on the conviction, she was careful to criticize Jiankui's work but not to denounce germline gene editing. The scientific community would have to sort out the safety and ethical issues, she said. "To me, the big question is not will this ever be done again," she said. "I think the answer is yes. The question is when, and the question is how."[14]

Part Seven

The Moral Questions

If scientists don't play God, who will?

—James Watson, to Britain's Parliamentary and Scientific Committee, May 16, 2000

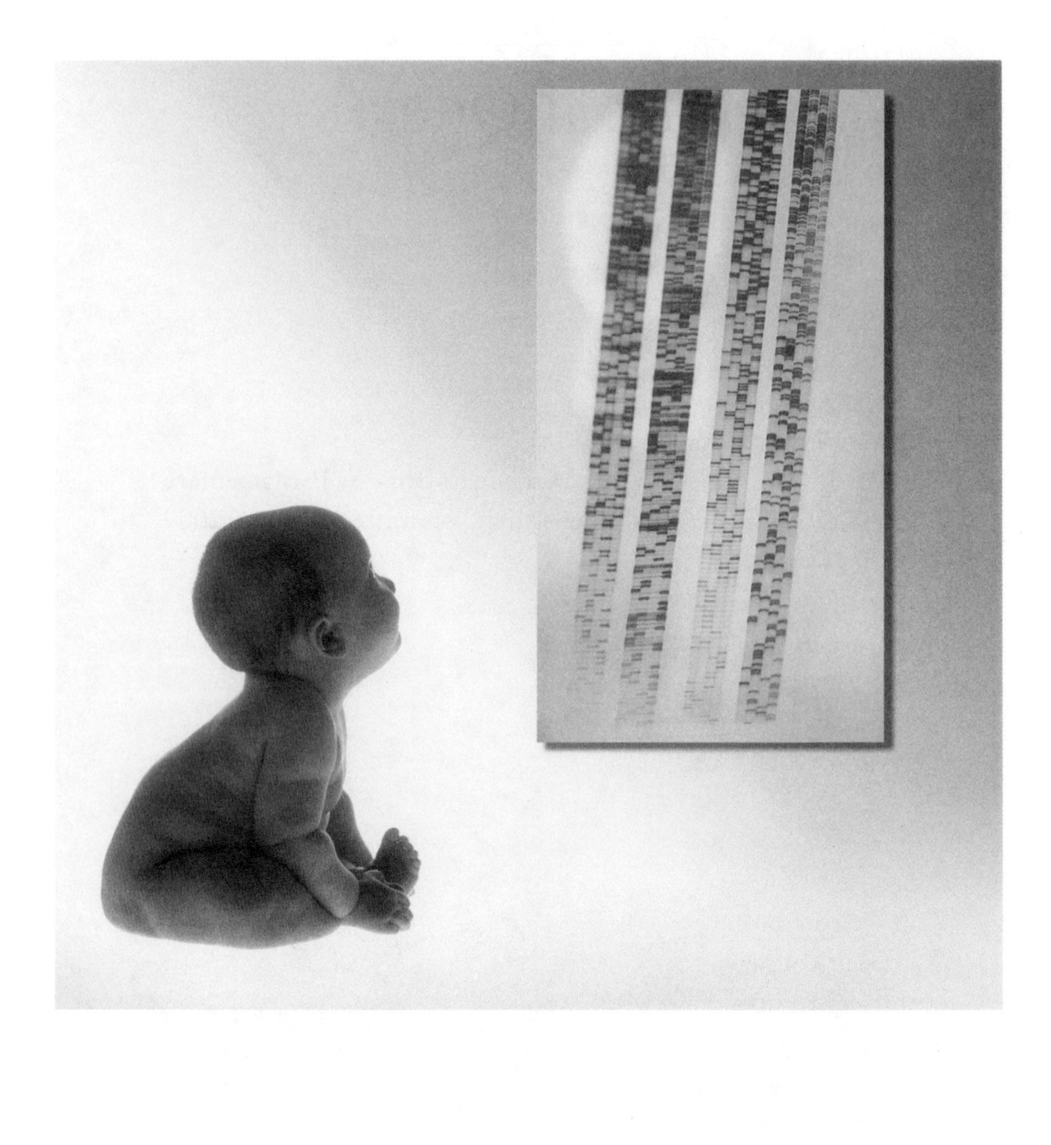

Chapter 40

Red Lines

The stakes

When He Jiankui produced the world's first CRISPR babies, with the goal of making them and their descendants immune to an attack by a deadly virus, most responsible scientists expressed outrage. His actions were deemed to be at best premature and at worst abhorrent. But in the wake of the 2020 coronavirus pandemic, the idea of editing our genes to make us immune to virus attacks began to seem a bit less appalling and a bit more appealing. The calls for a moratorium on germline gene editing receded. Just as bacteria have spent millennia evolving ways to develop immunity to viruses, perhaps we humans should use our ingenuity to do the same.

If we could safely edit genes to make our children less susceptible to HIV or coronaviruses, would it be wrong to do so? Or would it be wrong *not* to do so? And what about gene edits for other fixes and enhancements that might be possible in the next few decades? If they turn out to be safe, should governments prevent us from using them?[1]

The issue is one of the most profound we humans have ever faced. For the first time in the evolution of life on this planet, a species has developed the capacity to edit its own genetic makeup. That offers the potential of wondrous benefits, including the elimination of many deadly diseases and debilitating abnormalities. And it will someday offer both the promise and the peril of allowing us, or some of us, to boost our bodies and enhance our babies to have better muscles, minds, memory, and moods.

In the upcoming decades, as we gain more power to hack our own evolution, we will have to wrestle with deep moral and spiritual questions: Is there an inherent goodness to nature? Is there a virtue that arises from accepting what is gifted to us? Does empathy depend on believing that but for the grace of God, or the randomness of the natural lottery, we could have been born with a different set of endowments? Will an emphasis on personal liberty turn the most fundamental aspects of human nature into consumer choices made at a genetic supermarket? Should the rich be able to buy the best genes? Should we leave such decisions to individual choice, or should society come to some consensus about what it will allow?

Then again, are we getting a bit overdramatic with all of this hand-wringing? Why in the world would we not seize the benefits that will come from ridding our species of dangerous diseases and enhancing the capacities of our children?[2]

The germline as a red line

The primary concern is germline editing, those changes that are done in the DNA of human eggs or sperm or early-stage embryos so that every cell in the resulting children—and all of their descendants—will carry the edited trait. There has already been, and rightly so, general acceptance of what is known as somatic editing, the changes that are made in targeted cells of a living patient and do not affect reproductive cells. If something goes wrong in one of these therapies, it can be disastrous for the patient but not for the species.

Somatic editing can be used on certain types of cells, such as those of the blood, muscles, and eyes. But it is expensive, doesn't work on all cells, and may not be permanent. Germline edits could make a fix in all of the cells of the body. Thus it holds a lot more promise. And a lot more perceived peril.

Until the creation of the first CRISPR babies in 2018, there were two main medical methods for selecting the genetic traits of a child. The first was prenatal testing, which involves performing genetic tests on embryos as they are growing in the womb. Nowadays, such tests can detect Down's syndrome, sex, and dozens of congenital conditions. Parents can decide to abort the embryo if they don't like the traits. In the U.S., a prenatal diagnosis of Down's syndrome results in an abortion approximately two-thirds of the time.[3]

The development of *in vitro* fertilization led to another advance in genetic control: preimplantation genetic diagnosis. Couples can, if they are able, produce multiple fertilized eggs and have them tested in a lab dish, before they get implanted, for genetic characteristics. Do they have the mutations for Huntington's or sickle cell or Tay-Sachs? Or someday we can ask, as happens in the movie *Gattaca*, do they have the desired genes for height, memory, and muscle mass? With preimplantation diagnosis, those fertilized eggs with the parents' desired traits can be implanted and the rest discarded.

Both of these techniques raise some of the same moral issues as germline gene editing. For example, James Watson, the outspoken co-discoverer of DNA, once opined that a woman should have the right to abort a fetus based on any preference or prejudice, including not wanting a child that would be short or dyslexic or gay or female.[4] This caused a lot of people to recoil, understandably. Nevertheless, preimplantation genetic diagnosis is now considered morally acceptable, and parents are generally free to make their own choices about what criteria to use.

The question is whether germline gene editing will someday be considered just another in a long continuum of once controversial biological interventions, such as prenatal or preimplantation screening, that have gradually been accepted. If so, does it make sense to treat germline editing as something distinct, subject to a different set of moral standards?

Call this the continuum conundrum. There are ethicists who are good at making distinctions and those who are good at debunking distinctions. Or to put it another way, there are ethicists who discern lines and others who blur them. The ones who like to blur the lines often go on to pronounce that the lines are so blurry there is no rationale for treating the categories differently.

Take the atom bomb, as an analogy. When Secretary of War Henry Stimson was wrestling with whether to drop it on Japan, some argued that it was an entirely new category of weapon, a line that should not be crossed. Others said it was not fundamentally different, and indeed might be less brutal, than the massive firebombing campaigns that had been waged on Dresden and Tokyo. The latter side prevailed, and the bomb was dropped. Later, however, atomic weaponry came to be seen as being in a distinct category, and it hasn't been used since.

In the case of gene editing, I think the germline is indeed a real line. There may not be a razor-sharp line differentiating it from other biotechnologies, but as Leonardo da Vinci taught us with his *sfumato*, even slightly blurry lines can be definitive. Crossing the germline takes us to a distinct new realm. It involves engineering a genome rather than nurturing one that was produced naturally, and it introduces a change that will be inherited by all future descendants.

Nevertheless, this doesn't mean the germline should never be crossed. It simply means that we can view the germline as a firebreak that gives us a chance to pause, if we decide we ought to, the advance of genetic engineering techniques. The question becomes: Which cases, if any, should cause us to cross this germline?

Treatment vs. enhancement

Another line we might consider, in addition to that between somatic and germline editing, involves the distinction between "treatments" designed to fix dangerous genetic abnormalities and "enhancements" designed to improve human capacities or traits. At first glance, treatments seem easier to justify than enhancements.

But the treatment-vs.-enhancement distinction is a blurry one. Genes might predispose or predetermine certain kids to be short or obese or have attention deficits or be depressive. At what point do genetic modifications to fix such traits cross the line from health treatment to enhancement? What about genetic modifications that help prevent a person from getting HIV or coronavirus or cancer or Alzheimer's? Perhaps for these we need a third category called "preventions" in addition to the ill-defined "treatments" and "enhancements." And to those we might even add a fourth category, called "super-enhancements," which would include giving humans new capabilities that the species has not had before, such as the ability to see infrared light or hear super-high frequencies or avoid the bone, muscle, and memory loss that comes with age.

As you can see, the categories can get complex, and they don't necessarily correlate with what might be desirable and ethical. In order to chart our way through this moral minefield, it may be useful to do some thought experiments.

David Sanchez looks at a CRISPR cure for sickle cell

Chapter 41

Thought Experiments

Huntington's disease

Before our knees jerk and we stumble into hard-and-fast pronouncements—*Somatic editing is fine but inheritable germline edits are bad*; *Treatments are fine but enhancements are bad*—let's explore some specific cases and see what questions they raise.

If ever there was a case for editing a human gene, it would be for getting rid of the mutation that produces the cruel and painful killer known as Huntington's disease. Caused by an abnormal repetition of letters in a DNA sequence, it eventually leads to the death of brain cells. Beginning in middle age, victims start to twitch uncontrollably. They cannot focus. They lose their jobs. Eventually they are unable to walk, then talk, then swallow. Sometimes dementia sets in. It is an agonizing death in very slow motion. And it is devastating for the families—especially the kids, who watch their parent's gruesome decline, face the pity or ridicule of their schoolmates, and eventually learn that they have at least a 50 percent chance of suffering the same fate. One must be a fanatic believer in salvation through suffering to think that any good comes from its existence.[1]

Huntington's is a rare dominant disease; even one copy of the mutation spells doom. Symptoms usually arise only after a person's childbearing years, so its victims often have children before they know they have the genetic disease. Therefore, it's not weeded out by natural selection. The evolutionary process cares little about what happens to us after we have children and get them to a safe age, so there are a whole bunch of middle-aged maladies, including Huntington's and most forms of cancer, that we humans would want to eliminate, even though nature sees no need to.

Fixing Huntington's is not a complex edit. The wild sequence of excess DNA serves no good purpose. So why not edit it out in the germline of afflicted families—and out of our species once and for all?

One argument is that it would be better, where possible, to find an alternative approach to germline gene editing. In most cases—except when both parents have the disease—it might be possible to assure healthy children through preimplantation genetic diagnosis. If the parents can produce enough fertilized eggs, the ones with Huntington's can be weeded out. But producing a lot of viable eggs, as anyone who has been through fertility treatments knows, isn't always easy.

Another alternative is adoption. That, likewise, is not always easy these days. In addition, prospective parents often want to have a genetically related child. Is that a reasonable desire or just vanity?[2] Whatever some ethicists may say, most parents would feel it is reasonable. Millions of years of struggle by organisms, from bacteria to humans, to find ways to pass on their genes show that the impulse to produce genetically related offspring is among the most natural on this planet.

In making a gene edit to eliminate Huntington's, nothing has been altered except the elimination of the horrific mutation. So should it be permissible to do so, especially in cases where preimplantation screening is difficult? Even if we decide to set a high bar for the use of germline editing, it seems (at least to me) that Huntington's is a genetic malady we should try to eliminate from the human race.

If so, what other genetic problems should parents have the right to prevent from being passed along to their babies? Because this slope is slippery, let's take it step by step.

Sickle cell

Sickle-cell anemia is an interesting next case to consider because it raises two complexities, one medical and the other moral. Like Huntington's, sickle cell is caused by a simple mutation. In people who inherit a bad copy of the gene from both parents, the mutation distorts red blood cells, which deliver oxygen to the tissues of the body, into the shape of a sickle. Because these sickled cells die more quickly and have a harder time moving through the body, the disease can lead to fatigue, infections, spasms of pain, and early death. It tends to strike Africans and African Americans.

By 2020, trials were underway for somatic sickle-cell therapies, including the one described earlier involving the Mississippi woman Victoria Gray, who was part of a clinical trial in Nashville. Blood stem cells are removed from patients, edited, and then reinserted into the body. But this is an extraordinarily expensive procedure, not feasible for the more than four million afflicted globally. If the sickle-cell mutation could be fixed in the germline, by editing eggs or sperm or early-stage embryos, that would be a cheaper, one-time cure that would be inherited and could eventually eliminate the disease from our species.

So, does it fall into the same category as Huntington's? Is it a disease that should be eliminated using inheritable edits?

Well, as with many such genes, there's a complexity. People who get a copy of the gene from only one parent do not develop the disease, but they do develop immunity to most forms of malaria. In other words, the gene was (and in some places still is) useful, especially in sub-Saharan Africa. Now that there are treatments for malaria, it's less useful. But it is a reminder, when we think of messing with Mother Nature, that genes may play multiple roles and have evolutionary reasons for existing.

Let's suppose that researchers show that editing out the sickle-cell mutation is safe. Would there then be any reason to prohibit patients from having the gene edited out when they conceive children?

At this point in the discussion, a delightful kid named David Sanchez pops up to add another bit of complexity. He's a plucky, charming, reflective, African American teenager in California who loves to play basketball, except when his sickle-cell anemia causes him to double over in pain. At one point he developed a chest syndrome when the sickled cells blocked the blood to his lungs, and he had to drop out of high school. In a powerful 2019 documentary about CRISPR, *Human Nature*, he is an unlikely star. "My blood just does not like me very much, I guess," he says. "Sometimes you have a little sickle-cell crisis. Sometimes you have a really bad one. But I'm not just going to not play basketball."[3]

Every month, Sanchez's grandmother takes him to Stanford University Children's Hospital, where he gets an infusion of healthy cells from a blood donor. That gives him temporary relief. Matthew Porteus, the gene-editing pioneer at Stanford, has been helping to treat him. At one point he explained to Sanchez that, someday in the future, germline gene editing might eliminate the disease. "Maybe one day with CRISPR," Porteus told him, "they could go in and change the gene in the embryo so that the kid, when it's born, doesn't have sickle cell."

Sanchez's eyes lit up. "I guess that's kind of cool," he said. Then he paused. "But I think that should be up to the kid later." Asked why, he reflected for a moment and then continued slowly. "There's a lot of things that I learned having sickle cell. Because I had it, I learned patience with everyone. I learned how just to be positive."

But would he like to have been born without sickle cell? Again, he pauses. "No, I don't wish that I'd never had it," he says. "I don't think that I would be me if I didn't have sickle cell." Then he bursts into a big and lovely smile. He was born to be in such a documentary.

Not everyone with sickle cell is like David Sanchez. Even David Sanchez may not always be like the David Sanchez in the documentary. Despite what he said on camera, it is hard for me to imagine a kid choosing to have sickle cell rather than not having it. It's even more difficult to imagine parents, especially ones who have themselves endured a life with sickle cell, deciding that they want their kids to have it. After all, Sanchez is enrolled in a program to keep his sickle-cell anemia at bay.

The question gnaws at me, so I arrange to pose some questions to Sanchez.[4] This time his thinking is a bit different than when he was interviewed for the documentary. On complex personal issues like this, our thoughts understandably tend to fluctuate. Would you like to find a way, I ask him, to make sure your children are born without sickle cell? "Yes," he responds. "If that's an option, then of course."

What about the patience and the positive attitude that, as he told the documentary producers, he learned by having sickle cell? "Empathy is something that's really important to humans," he responds. "That is something I learned from sickle cell, and that is something I would really want to convey to my kids if they could be born without sickle cell. But I wouldn't want my kids or others to go through what I went through." The more he learns about CRISPR, the more excited he becomes about how it may cure him and protect his children. But it's complicated.

Character

David Sanchez's wise words bring up a larger question. Challenges and so-called disabilities often build character, teach acceptance, and instill resilience. They may even be correlated to creativity. Take Miles Davis. The pain of sickle cell drove him to drugs and drink. It may have even driven him to his death. It also, however, may have driven him to be the creative artist who could produce *Kind of Blue* and *Bitches Brew*. Would Miles Davis have been Miles Davis without sickle cell?

This is not a new question. Franklin Roosevelt was forged by polio. The challenge transformed his character. Likewise, I knew a guy who was one of the last kids to be touched by polio before Salk and Sabin came up with their vaccines in the late 1950s. He achieved success, I think, partly because of his great depth of character, and he taught all of us about grit and gratitude and humility. My favorite novel, Walker Percy's *The Moviegoer*, tells of the transformative effect the disabled boy Lonnie has on the other characters.

The bioethicist Rosemarie Garland-Thomson, who was born with distorted arms, tells of the friendship circle she has with three other women born with genetic conditions, one blind, one deaf, and one with muscular impairment. "Our genetic conditions gave us a head start in accessing multiple opportunities for expression, creativity, resourcefulness, and relationships—for human flourishing," she writes.[5] Similarly, Jory Fleming is an amazing young man who was born with severe autism as well as other challenging health conditions. He could not cope in class, so he was homeschooled. As he grew older, he taught himself how to deal with the fact that his internal world was different from those of other people. He ended up winning a Rhodes Scholarship to Oxford. In his 2021 memoir, *How to Be Human*, he reflects on whether gene editing should be used, if it becomes feasible, to eliminate some of the causes of autism. "You'd be removing an aspect of the human experience," he writes, "but for what benefit exactly?" Autism, he argues, is a difficult condition to have, but the challenges largely come because the world is not good at accommodating people whose emotional lives are different. Those differences can actually provide a useful perspective for the rest of us, including on how to make decisions that are not unduly influenced by emotion. "Should society change to recognize the benefits of autism instead of just the challenges?" he asks. "Certainly, my experience has been very challenging, and it has been also rewarding. And who knows, hopefully, I'll be able to do something with my life that benefits other people in some way."[6]

It's an interesting dilemma. Once a vaccine was discovered to stop polio, we humans quickly and easily decided to use it to eliminate that disease from our species, even at the risk of allowing future Franklin Roosevelts to remain unforged. Using gene editing to prevent disabilities may make society less diverse and creative. But does this give governments the right to tell parents they can't use such technologies?

Deafness

That raises the question of what attributes should be labeled disabilities. Sharon Duchesneau and Candy McCullough are a lesbian couple who wanted a sperm donor so they could conceive a kid. Both of them are deaf. They consider their deafness to be part of who they are rather than something to be cured, and they wanted a child who would be part of their cultural identity. So they advertised for a sperm donor who was congenitally deaf. They found one, and now they have a deaf child.

A story about the couple in the *Washington Post* caused them to be condemned by some people for inflicting a disability on a child.[7] But they were applauded in the deaf community. Which was the right response? Should they be criticized for making sure their child had a disability, or should they be praised for preserving a subculture that contributes to the diversity and perhaps even the empathy of society? Would it be different if, instead of using a deaf sperm donor, the couple had used preimplantation diagnosis to select an embryo that had the genetic mutation for deafness? What if the embryo was typical, but they edited it to be deaf? Would that be okay? What if they asked a doctor to punch out the child's eardrums after birth?

In some cases when formulating a moral argument, it helps to do a reversal test. The Harvard philosopher Michael Sandel uses this thought experiment: Suppose a parent comes to a doctor and says, "My child is going to be born deaf, but I want you to do something to make her able to hear." The doctor should try, right? But now suppose a parent says, "My child is going to be born able to hear, but I want you to do something to her to make sure she is born deaf." I think most of us would recoil if the doctor agreed. Our natural instinct is to consider deafness a disability.

How do we distinguish between traits that are true disabilities and ones that are disabilities mainly because society is not good at adapting for them? Take the case of the deaf lesbian couple, for example. Some people may consider both the fact that they are deaf and the fact that they are lesbian as disadvantages. What if they wanted a genetic procedure that would make their child more likely to be straight? Suppose they chose the reverse and wanted to make it more likely their child would be gay? (This is a thought experiment. There is no simple gay gene.) Likewise, being born Black in America could be considered a disadvantage. A single gene, *SLC24A5*, has a major influence on determining skin color. What if a set of Black parents considers their race to be a social handicap and wants to edit that gene to produce light-skinned babies?

Such questions prompt us to look at "disabilities" and ask to what extent they are inherently disabling and to what extent the disadvantage is due to our social constructs and prejudices. The disadvantages from being deaf, for a human or any other animal, are very real. In contrast, any disadvantages to being gay or Black are due to social attitudes that can and should be changed. That is why we can make a moral distinction between using genetic techniques to prevent deafness and using these techniques to influence such things as skin color and sexual orientation.

Muscles and sports

Now let's do some thought experiments to see if we might want to cross the blurry line between gene editing that is done to treat true disabilities and gene editing that is done to enhance the traits of our children. The *MSTN* gene produces a protein that curtails the growth of muscles when they reach a normal level. Suppressing the gene takes off the brakes. Researchers have already done this to produce "mighty mice" and cattle with "double muscling." It is what our biohacker Josiah Zayner used to make his kits that produce super-frogs and for the CRISPR he injected into himself.

Among those interested in these types of gene edits, other than cattle breeders, are athletic directors. Pushy parents who want champion children are sure to follow. Especially by using germline editing, they might produce a whole new breed of athletes with bigger bones and stronger muscles.

Add to this mix a rare gene mutation that was discovered in the Olympic champion skier Eero Mäntyranta. Initially accused of doping, he was found to have a gene that increased his number of red blood cells by more than 25 percent, which naturally improved his stamina and ability to use oxygen.

So what do we say to parents who want to use gene editing to produce bigger, more muscular kids with greater stamina? Ones who can run marathons, break tackles, and bend steel with their bare hands? And what does that do to our concept of athletics? Do we go from admiring the diligence of the athlete to admiring instead the wizardry of their genetic engineers? It's easy to put an asterisk next to the home run tallies of José Canseco or Mark McGwire when they admit that they were on steroids. But what do we do if athletes' extra muscles come from genes they were born with? And does it matter if those genes were paid for by their parents rather than bestowed by a random natural lottery?

The role of sports, at least since the first Olympics in 776 BC, is to celebrate two things: natural talent combined with disciplined effort. Enhancements would shift that balance, making human effort less of a component of victory. Therefore the achievement becomes a little less praiseworthy and inspiring. There is a whiff of cheating if an athlete succeeds by obtaining some physical advantages through medical engineering.

But there's a problem with this fairness argument. Most successful athletes have *always* been people who happened to have better athletic genes than the rest of us. Personal effort is a component, but it helps to be born with the genes for good muscles, blood, coordination, and other innate advantages.

For example, almost every champion runner has what is known as the R allele of the *ACTN3* gene. It produces a protein that builds fast-twitch muscle fibers, and it is also associated with improving strength and recovery from muscle injury.[8] Someday it may be possible to edit this variation of the *ACTN3* gene into the DNA of your kids. Would that be unfair? Is it unfair that some kids are born with it naturally? Why is one more unfair than the other?

Height

One way to think through the fairness of using gene editing for physical enhancements is by looking at height. A condition called IMAGe syndrome, which severely curtails size, is caused by a mutation in the

CDKN1C gene. Should it be permissible to genetically edit out this defect so that these kids will grow to an average height? Most of us would think so.

Now let's take the case of parents who just happen to be short. Should they be permitted to edit the genes of their kids so they will grow to average height? If not, what's the moral difference between these two cases?

Suppose there was a genetic edit that could add eight inches to a kid's height. Would it be proper to use it on a boy who would otherwise be under five feet tall to turn him into someone of average height? What about using it on a boy who would otherwise be average height to make him six-foot-five?

A way to wrestle with these questions is by making a distinction between "treatments" and "enhancements." For various traits—height, eyesight, hearing, muscular coordination, and so on—we could use a statistical method to define "typical species functioning." A significant variation below that would be defined as a disability.[9] Using that standard, we might approve of treating a kid who would be less than five feet tall but reject the idea of enhancing a kid who would otherwise be of average height.

By pondering the question of height, we can make another distinction that is useful: the difference between an absolute improvement and a positional improvement. In the first category are enhancements that are beneficial to you even if everyone else gets them. Imagine there was a way to improve your memory or your resistance to virus infections. You'd be better off with it, even if others got the same enhancement. In fact, as the coronavirus pandemic shows, you would be better off *especially* if others had that enhancement as well.

But the advantages of increased height are more positional. Let's call it the standing-on-tiptoes problem. You're in the middle of a crowded room. To see what's going on in the front, you stand on your tiptoes. It works! But then everyone else around you tries it. They all get two inches higher. Then nobody in the room, including you, sees any better than the people in the front row.

Likewise, suppose I'm an average height. If I were enhanced by eight inches, I'd be way taller than most people, and that could be a benefit to me. But if everyone else got the same eight-inch enhancement I did, then I would get no real benefit. The enhancement wouldn't make me or society as a whole better off, especially given the legroom of airline seats these days. The only sure beneficiaries would be carpenters who specialized in raising door frames. So enhanced height is a *positional* good, while enhanced resistance to viruses is an *absolute* good.[10]

That doesn't answer the question of whether we should allow genetic enhancements. But as we grope for a set of principles to include in our moral calculus, the distinction does point to a factor we should consider: favoring enhancements that would benefit all of society over those that would give the recipient a positional advantage.

Super-enhancements and transhumanism

Perhaps some enhancements will gain broad social acceptance. Then what about super-enhancements? Should we ever want to engineer traits and capacities that exceed what any human has ever had? The golfer Tiger Woods had laser surgery to improve his eyesight to be even better than 20/20. Might we want our kids to have super-eyesight? What about adding the capacity to see infrared light or some new color?

DARPA, the Pentagon's research agency, might someday want to create superior soldiers with night vision. They could also imagine an enhancement that allowed human cells to be more resistant to radiation in case of a nuclear attack. Actually, they aren't just imagining that. DARPA already has a project going, in conjunction with Doudna's lab, to study how to create genetically enhanced soldiers.

One odd result of allowing super-enhancements could be that children will become like iPhones: a new version will come out every few years with better features and apps. Will children as they age feel that they are becoming obsolete? That their eyes don't have the cool triple-lens enhancements that are engineered into the latest version of kids?

Fortunately, these are questions we can ask for amusement but not for an answer. It will be up to our grandchildren to figure these out.

Psychological disorders

Two decades after the completion of the Human Genome Project, we still have little understanding of how human psychology is influenced by genetic dispositions. But eventually, we may isolate genes that contribute to a predisposition to schizophrenia, bipolar disorder, severe depression, and other mental challenges.

Then we will have to decide whether we should allow, or perhaps even encourage, parents to make sure that these genes get edited out of their children. Let's pretend to go back in time. If some of the genetic factors predisposing James Watson's son Rufus to schizophrenia could have been edited out, would that have been a good thing? Should we have allowed his parents to make the decision to do that?

Watson has no doubt what the answer should be. "Of course we should use germline therapy to fix things like schizophrenia that nature got horribly wrong," he says. Doing so would lead to a lot less suffering. Schizophrenia, depression, and bipolar disorder can be brutal, often deadly. No one would want to inflict it on a person or on any person's family.

But even if we agree that we want to rid humanity of schizophrenia and similar disorders, we should consider whether there might be some cost to society, even to civilization. Vincent van Gogh had either schizophrenia or bipolar disorder. So did the mathematician John Nash. (And also Charles Manson and John Hinckley.) People with bipolar disorder include Ernest Hemingway, Mariah Carey, Francis Ford Coppola, Carrie Fisher, Graham Greene, Julian Huxley (the eugenicist), Gustav Mahler, Lou Reed, Franz Schubert, Sylvia Plath, Edgar Allan Poe, Jane Pauley, and hundreds of other artists and creators. The number of creative artists with major depressive disorder is in the thousands. A study by schizophrenia-research pioneer Nancy Andreasen of thirty prominent contemporary authors showed that twenty-four had experienced at least one episode of major depression or mood disorder, and twelve were diagnosed with bipolar disorder.[11]

To what extent does dealing with mood swings, fantasies, delusions, compulsions, mania, and deep depression help spur, in some people, creativity and artistry? Is it harder to be a great artist without having some compulsive or even manic traits? Would you cure your own child from being schizophrenic if you knew that, if you didn't, he would become a Vincent van Gogh and transform the world of art? (Don't forget: Van Gogh committed suicide.)

At this point in our deliberations, we have to face the potential conflict between what is desired by the individual versus what is good for human civilization. A reduction in mood disorders would be seen as a benefit by most of the afflicted individuals, parents, and families. They would desire it. But does the issue look different when asked from society's vantage point? As we learn to treat mood disorders with drugs and eventually with genetic editing, will we have more happiness but fewer Hemingways? Do we wish to live in a world in which there are no Van Goghs?

This question of engineering away mood disorders gets to an even more fundamental question: What is the aim or purpose of life? Is it happiness? Contentment? Lack of pain or bad moods? If so, that may be easy. A painless life was engineered by the overlords in *Brave New World*, who made sure the masses had soma, a drug that enhanced their sense of joy and allowed them to avoid discomfort, sadness, or anger. Suppose we could hook our brains to something the philosopher Robert Nozick called an "experience machine," which allowed us to believe that we were hitting home runs and dancing with movie stars and floating in a beautiful bay.[12] It would make us always feel blissful. Would that be desirable?

Or does the good life have aims that are deeper? Should the goal be that each person can flourish, in a more profound fashion, by using talents and traits in a way that is truly fulfilling? If so, that would require authentic experiences, real accomplishments, and true efforts, rather than engineered ones. Does the good life entail making a contribution to our community, society, and civilization? Has evolution encoded such goals into human nature? That might entail sacrifice, pain, mental discomforts, and challenges that we would not always choose.[13]

Smarts

Now let's deal with the final frontier, the one most promising and frightful: the possibility of improving cognitive skills such as memory, focus, information processing, and perhaps even someday the vaguely defined concept of intelligence. Unlike height, cognitive skills are beneficial in more than just a positional way. If everyone were a bit smarter, it probably would make all of us better off. In fact, even if only a portion of the population became smarter, it might benefit everyone in society.

Memory may be the first mental improvement we will be able to engineer, and fortunately it is a less fraught topic than IQ. It has already been improved in mice, such as by enhancing the genes for NMDA receptors in nerve cells. In humans, enhancing those genes could help prevent memory loss in old age, but it could also enhance memory in younger people as well.[14]

Perhaps we will be able to improve our cognitive skills so that we can keep up with the challenges of using our technology wisely. Ah, but there's the rub: *wisely*. Of all the complex components that go into human intelligence, wisdom may be the most elusive. Understanding the genetic components of wisdom may require us to understand consciousness, and I suspect that's not going to happen in this century. In the meantime, we will have to deploy the finite allocation of wisdom that nature has dealt us as we ponder how to use the gene-editing techniques that we've discovered. Ingenuity without wisdom is dangerous.

Chapter 42

Who Should Decide?

The National Academy's video

The tweet was provocative, a bit more provocative than it was intended to be. It read:

> Dream of being stronger? 💪 Or smarter? 🧠 Do you dream of having a top student or star athlete? Or a child free of inheritable #diseases? 👨‍🏫 ⛹️ Can human #GeneEditing eventually make this and more possible?

It was an attempt by the usually staid National Academy of Sciences in October 2019 to spur a "broad public discussion" of gene editing, just like all of those conferences on the topic had recommended. The tweet linked to a quiz and a video explaining germline gene editing.

The video began with five "everyday people" putting sticky notes onto a diagram of a body and fantasizing about what changes they would make in their genes. "I guess I would like to be taller," said one. Other personal desires included: "I would like to change body fat"; "Let's prevent baldness"; "Take away dyslexia."

Doudna was in the video explaining how CRISPR works. Then it showed people discussing the prospect of designing the genes of their future children. "Create the perfect human being?" one man mused. "That's pretty cool!" Said another, "You want the best qualities to be put into your offspring." A woman chimed in, "If I had the chance to choose the best DNA for my child, I would definitely want her to be smart." Others discussed their own health problems, such as attention-deficit disorder and high blood pressure. "I would take that out, for sure," a man said of his heart disease. "I don't want my kids to deal with it."[1]

Bioethicists immediately erupted on Twitter. "What a mistake," tweeted Paul Knoepfler, a cancer researcher and bioethicist at the University of California, Davis. "Who at National Academy of Sciences' media office is behind this bizarre tweet & page it links to that seems troublingly upbeat about human heritable gene editing & to trivialize idea of designer babies?"

Twitter, unsurprisingly, is not the best forum to discuss bioethics. There is a truism about internet comment boards: any discussion descends to shouting "Nazi!" within seven responses. In the case of the gene-editing threads, it was more like by the third response. "Are we still in 1930s Germany?" one person tweeted. Another added, "How did this read in the original German?"[2]

Within a day, the folks at the National Academy of Sciences had sounded retreat. The tweet was deleted and the video pulled off the web. A spokesperson apologized that they had "left the misimpression that the use of genome editing for the 'enhancement' of human traits is permissible or taken lightly."

The brief tempest showed that the bromide of calling for greater societal discussion about the morals of gene editing was easier preached than practiced. It also raised the question of who should get to decide how gene-editing tools should be used. As we saw in the thought experiments in the previous chapter, many of the difficult questions about gene editing involve not just how to decide the issue, but *who* should decide. As is the case with so many policy issues, the desires of an individual might conflict with the good of the community.

The individual or the community?

On most great moral issues, there are two competing perspectives. One emphasizes individual rights, personal liberty, and a deference to personal choice. Stemming from John Locke and other Enlightenment thinkers of the seventeenth century, this tradition recognizes that people will have different beliefs about what is good for their lives, and it argues that the state should give them a lot of liberty to make their own choices, as long as they do not harm others.

The contrasting perspectives are those that view justice and morality through the lens of what is best for the society and perhaps even (in the case of bioengineering and climate policy) the species. Examples include requirements that schoolkids be vaccinated and that people wear masks during a pandemic. The emphasis on societal benefits rather than individual rights can take the form of John Stuart Mill's utilitarianism, which seeks the greatest amount of happiness in a society even if that means trampling on the liberty of some individuals. Or it can take the form of more complex social contract theories, in which moral obligations arise from the agreements we would make to form the society we want to live in.

These contrasting perspectives form the most basic political divide of our times. On the one side are those who wish to maximize individual liberty, minimize regulations and taxes, and keep the state out of our lives as much as possible. On the other side are those who wish to promote the common good, create benefits for all of society, minimize the harm that an untrammeled free market can do to our work and environment, and restrict selfish behaviors that might harm the community and the planet.

The modern foundations for each of these perspectives was expressed in two influential books written fifty years ago: John Rawls's *A Theory of Justice*, which comes down on the side of favoring the good of the community, and Robert Nozick's *Anarchy, State, and Utopia*, which emphasizes the moral foundation for individual liberty.

Rawls seeks to define the rules that we would agree to if we had gathered to make a compact. In order to make sure things are "fair," he said that we should imagine what rules we would make if we didn't know what place we would each end up occupying in society and what natural abilities we would have. He argues that, from behind this "veil of ignorance," people would decide that inequalities should be permitted only to the extent that they result in benefits for all of society, and specifically for the least advantaged. In his book, this leads Rawls to justify genetic engineering only if it does not increase inequality.[3]

Nozick, whose book was a response to that of his Harvard colleague Rawls, likewise imagined how we might emerge from the anarchy of a state of nature. Instead of a complex social contract, he argues that social rules should arise through the voluntary choices of individuals. His guiding principle is that individuals should not be used to promote a social or moral goal devised by others. This leads him to favor a minimalist state that is limited to functions of public safety and enforcement of contracts but avoids most regulations or redistribution efforts. He addresses, in a footnote, the question of genetic engineering, and he takes a libertarian, free-market view. Instead of centralized control and rules set by regulators, he says that there should be "a genetic supermarket." Doctors should accommodate "the individual specifications (within certain moral limits) of prospective parents."[4] Since he wrote his book, the term "genetic supermarket" has become a catchphrase, used by fans and foes, for leaving genetic engineering decisions to individuals and the free market.[5]

Two science fiction books can also help shape our discussion: George Orwell's *1984* and Aldous Huxley's *Brave New World*.[6]

Orwell conjures up an Orwellian world in which information technology is used by "Big Brother," a leader that is always watching you, to centralize power in a super-state and exert control over a cowed populace. Individual freedom and independent thinking are crushed by electronic surveillance and total information control. Orwell was warning about the danger that a Franco or Stalin would someday control information technology and destroy individual freedom.

It didn't happen. When the real 1984 actually rolled around, Apple introduced an easy-to-use personal computer, the Macintosh, and in the words that Steve Jobs wrote for its ad, "you'll see why 1984 won't be like *1984*." That phrase contained a deep truth. Instead of computers becoming an instrument for centralized repression, the combination of the *personal* computer and the decentralized nature of the internet became a way to devolve more power down to each individual, thus unleashing a gusher of free expression and radically democratized media. Perhaps too much so. The dark side of our new information technology is not that it allows government repression of free speech but just the opposite: it permits anyone to spread, with little risk of being held accountable, any idea, conspiracy, lie, hatred, scam, or scheme, with the result that societies become less civil and governable.

The same may be the case for genetic technologies. In his 1932 novel, Huxley warned of a brave new world of centralized government control of reproductive science. Human embryos are created at a "hatchery and conditioning center" and then sorted to be engineered for different social purposes. Those chosen for the "alpha" class are enhanced physically and mentally to become leaders. At the other end of the spectrum, those in the "epsilon" class are bred to become menial laborers and conditioned for a life of induced blissful stupor.

Huxley said that he wrote the book as a reaction to "the current drift toward totalitarian control of everything."[7] But as was the case with information technology, the danger of genetic technology might not be too much *government* control. Instead, it may be too much *individual* control. The excess of the early twentieth-century eugenics movement in America and then the evil of the Nazi program gave a horrid stench to the idea of state-controlled genetic projects. It gave eugenics, which means "good genes," a bad name. Now, however, we may be ushering in a new eugenics—a liberal or libertarian eugenics, one based on free choice and marketed consumerism.

Huxley may have supported this free-market eugenics. He wrote a little-known utopian novel in 1962, *Island*, in which women voluntarily choose to be inseminated by sperm from men with high IQs and artistic talents. "Most married couples feel that it's more moral to take a shot at having a child of superior quality than to run the risk of slavishly reproducing whatever quirks and defects may happen to run in the husband's family," the main character explains.[8]

Free-market eugenics

In our day and age, decisions about genetic editing are likely to be driven, for better or worse, by consumer choice and the persuasive power of marketing. So what's wrong with that? Why shouldn't we leave decisions about gene editing to individuals and parents, just like we do with other reproductive choices? Why do we have to convene ethics conferences, seek a broad societal consensus, and wring our collective hands? Isn't it best to allow the decisions to be made by me and you and other individuals who want the best prospects for our kids and grandkids?[9]

Let's begin by loosening our minds and avoiding a bias for the status quo by asking the most basic question: What's wrong with genetic improvements? If we can do so safely, why shouldn't we prevent abnormalities, diseases, and disabilities? Why not improve our capabilities and create enhancements? "I don't see why eliminating a disability or giving a kid blue eyes or adding fifteen IQ points is truly a threat to public health or to morality," says Doudna's friend George Church, the Harvard geneticist.[10]

In fact, aren't we morally obligated to look after the welfare of our children and of future humans in general? Almost all species share an evolutionary instinct—encoded in the essence of evolution itself—to use whatever wiles they can muster to maximize the chance that their offspring will thrive.

The foremost philosopher advocating this view is Julian Savulescu, a professor of practical ethics at Oxford. He coined the phrase "procreative beneficence" to make the case that it is moral to choose the best genes for your unborn children. Indeed, he argues, it may be immoral *not* to. "Couples should select embryos or fetuses which are most likely to have the best life," he asserts. He even dismissed the concern that this could allow rich people to buy better genes for their children and thereby create a new class (or even subspecies) of enhanced elites. "We should allow selection for non-disease genes even if this maintains or increases social inequality," he writes, specifically citing "genes for intelligence."[11]

To analyze that point of view, let's do another thought experiment. Imagine a world where genetic engineering is determined mainly by individual free choice, with few government regulations and no pesky bioethics panels telling us what's permissible. You go into a fertility clinic and are given, as if at a genetic supermarket, a list of traits you can buy for your children. Would you eliminate serious genetic diseases, such as Huntington's or sickle cell? Of course you would. I personally would also choose that my kids not have genes leading to blindness. How about avoiding below-average height or above-average weight or a low IQ? We would all probably select those options as well. I might even choose a premium-priced option for extra height and muscles and IQ. Now let's say there were, hypothetically, genes that predisposed a child to more likely be straight rather than gay. You're not prejudiced, so you'd likely resist choosing that option, at least initially. But then, assuming no one was judging you, might you rationalize that you wanted your child to avoid discrimination or be a little bit more likely to produce grandchildren for you? And while you were at it, might you throw in blond hair and blue eyes as well?

Whoa!!! Something just went wrong. It really did turn out to be a slippery slope! Without any gates or flags, we might all go barreling down at uncontrollable speed, taking society's diversity and the human genome along with us.

Although this sounds like a scene from *Gattaca*, a real-world version of this baby-designing service—using preimplantation diagnosis—was launched in 2019 by a New Jersey startup, Genomic Prediction. *In vitro* fertilization clinics can send the company genetic samples of prospective babies. The DNA in cells from days-old embryos is sequenced to come up with a statistical estimate of the chances of developing a long list of conditions. Prospective parents can choose which embryo to implant based on the characteristics they want in their child. The embryos can be screened for single-gene disorders such as cystic fibrosis and sickle cell. The tests can also statistically predict multigene conditions, such as diabetes, heart attack risk, hypertension, and, according to the company's promotional material, "intellectual disability" and "height." Within ten years, the founders say, they are likely to be able to make predictions of IQ so that parents can choose to have very smart children.[12]

So now we can see a problem with simply leaving such decisions to individual choice. A liberal or libertarian genetics of individual choice could eventually lead us—just as surely as government-controlled eugenics—to a society with less diversity and deviation from the norm. That might be pleasing to a parent, but we would end up in a society with a lot less creativity, inspiration, and edge. Diversity is good not only for society but for our species. Like any species, our evolution and resilience are strengthened by a bit of randomness in the gene pool.

The problem is that the value of diversity, as our thought experiments showed, can conflict with the value of individual choice. As a society, we may feel that it is profoundly beneficial to the community to have people who are short and tall, gay and straight, placid and tormented, blind and sighted. But what moral right do we have to require another family to forgo a desired genetic intervention simply for the sake of adding to the diversity of society? Would we want the state to require that of us?

One reason to be open to some kind of limit on individual choice is that gene editing could exacerbate inequality and even permanently encode it into our species. Of course, we already tolerate some inequality based on birth and parental choices. We admire parents who read to their kids, make sure they go to good schools, and coach them in soccer. We even accept, perhaps with a roll of the eyes, those who hire SAT tutors and send their kids to computer camp. Many of these confer the advantages of inherited privilege. But the fact that inequality already exists is not an argument to increase or permanently enshrine it.

Permitting parents to buy the best genes for their kids would represent a true quantum leap in inequality. In other words, it won't be just a big leap, but a leap into a new disconnected orbit. After centuries of reducing aristocratic and caste systems based on birth, most societies have embraced a principle of morality that is also a basic premise of democracy: we believe in equal opportunity. The social bond that arises from this "created equal" creed would be severed if we turn financial inequalities into genetic inequalities.

This does not mean that gene editing is inherently bad. But it does argue against allowing it to be part of a free-market bazaar where the rich can buy the best genes and ingrain them into their families.[13]

Restricting individual choice would be difficult to enforce. Various college admissions scandals show us how far some parents will go and what they will pay to give their kids an advantage. Add to that the natural instinct of scientists to pioneer procedures and make discoveries. If a nation imposes too many restrictions, its scientists will move elsewhere and its wealthy parents will seek clinics in some enterprising Caribbean island or foreign haven.

Despite such objections, it's possible to aim for some social consensus on gene editing rather than simply leaving the issue totally to individual choice. There are practices we cannot fully control, from shoplifting to sex trafficking, that are kept to a minimum by a combination of legal sanctions and social shaming. The Food and Drug Administration, for example, regulates new drugs and procedures. Even though some people score drugs for off-label purposes or travel to places for unconventional treatments, FDA restrictions are pretty effective. Our challenge is to figure out what the norms for gene editing should be. Then we can try to find the regulations and social sanctions that will cause most people to follow them.[14]

Playing God

Another reason we might feel uncomfortable with directing our evolution and designing our babies is that we would be "playing God." Like Prometheus snatching fire, we would be usurping a power that properly resides above our pay grade. In so doing, we'd lose a sense of humility about our place in Creation.

The reluctance to play God can also be understood in a more secular way. As one Catholic theologian said at a National Academy of Medicine panel, "When I hear someone say that we shouldn't play God, I'd guess that ninety percent of the time they are atheists." The argument can simply mean that we should not have the hubris to believe that we should fiddle with the awesome, mysterious, delicately interwoven, and beautiful forces of nature. "Evolution has been working toward optimizing the human genome for 3.85 billion years," says NIH director Francis Collins, who is not an atheist. "Do we really think that some small group of human genome tinkerers could do better without all sorts of unintended consequences?"[15]

Our respect for nature and nature's God should, indeed, instill some humility about meddling with our genes. But should it absolutely forbid it? After all, we *Homo sapiens* are part of nature, no less so than bacteria and sharks and butterflies. Through its infinite wisdom or blind stumbling, nature has endowed our species with an ability to edit our own genes. If it's wrong for us to use CRISPR, the reason cannot merely be that it's unnatural. It's just as natural as all of the tricks that bacteria and viruses use.

For all of history, humans (and every other species) have been battling rather than accepting nature's poisoned offerings. Mother Nature has produced massive suffering and distributed it unequally. Thus we devise ways to combat plagues, cure diseases, fix disabilities, and breed better plants, animals, and children.

Darwin wrote about "the clumsy, wasteful, blundering, low, and horridly cruel works of nature." Evolution, he discovered, bears no fingerprints of an intelligent designer or benevolent God. He made a detailed list of things that evolved in a flawed way, including the path of the urinary tract in male mammals, the poor drainage of the sinuses in primates, and the inability of humans to synthesize vitamin C.

These design flaws are not mere exceptions. They are the natural consequence of the way evolution progresses. It stumbles upon and then cobbles together new features, sort of like what happened during the worst eras of Microsoft Office, rather than proceed with a master plan and end product in mind. Evolution's primary guide is reproductive fitness—what traits might cause an organism to reproduce more—which means it permits, and perhaps even encourages, all sorts of plagues, including coronaviruses and cancers, that afflict an organism once its childbearing use is over. This does not mean that, out of respect for nature, we should quit searching for ways to fight against coronaviruses and cancer.[16]

There is, however, a more profound argument against playing God, best articulated by the Harvard philosopher Michael Sandel. If we humans find ways to rig the natural lottery and engineer the genetic endowments of our children, we will be less likely to view our traits as gifts that we accept. That would undermine the empathy that comes from our sense of "there but for the grace of God go I" toward our fellow humans who are less lucky. "What the drive to mastery misses and may even destroy is an appreciation of the gifted character of human powers and achievements," Sandel writes. "To acknowledge the giftedness of life is to recognize that our talents and powers are not wholly our own doing."[17]

Of course I don't fully believe, nor does Sandel, that we must be reverential about the giftedness of all that nature offers us unbidden. Human history has been a quest—a very natural one—to master challenges that happen to us unbidden, be they pandemics or droughts or storms. Few of us would regard Alzheimer's or Huntington's to be a result of giftedness. When we create chemotherapies to fight cancer or vaccines to fight coronaviruses or gene-editing tools to fight birth defects, we are, quite properly, exercising mastery over nature rather than accepting the unbidden as a gift.

But Sandel's argument should nudge us, I think, toward some humility, especially when it comes to trying to design enhancements and perfections for our children. He makes a profound, beautiful, and even spiritual case for eschewing attempts at complete mastery over the unbidden. We can steer a course that avoids a Promethean quest for controlling our endowments while also avoiding complete submission to the vagaries of a lottery. Wisdom involves finding the right balance.

At the Hong Kong summit

Chapter 43

Doudna's Ethical Journey

When it became clear that the CRISPR-Cas9 tool that she co-invented could be used for editing human genes, Doudna had a "visceral, knee-jerk reaction." The idea of editing a child's genes, she says, felt unnatural and scary for humanity. "In the early days I was instinctively against it."[1]

Her position began to change at the January 2015 conference on gene editing in Napa Valley that she organized. At one of the sessions, during a heated debate over whether germline editing should ever be allowed, a participant leaned forward and said quietly, "Someday we may consider it unethical *not* to use germline editing to alleviate human suffering."

The idea that germline editing was "unnatural" began to recede in her thinking. All medical advances attempt to correct something that happened "naturally," she realized. "Sometimes nature does things that are downright cruel, and there are many mutations that cause enormous suffering, so the idea that germline editing was unnatural began to carry less weight for me," she says. "I am not sure how to make a sharp distinction in medicine between what is natural and what is unnatural, and I think it's dangerous to use that dichotomy to block something that could alleviate suffering and disability."

Once she became famous for her gene-editing discoveries, she began to hear stories from people who had been affected by genetic diseases and were yearning for science to help. "The ones about kids were especially touching to me as a mother," she recalls. One example sticks in her mind. A woman sent beautiful pictures of her new baby boy, bald and cute, which reminded Doudna of when her own son, Andy, was born. The baby had just been diagnosed with a genetic neurodegenerative disease. His nerve cells would soon start dying and eventually he would be unable to walk, speak, then swallow or eat. He was doomed to die an early and painful death. The note was a wrenching plea for help. "How could you not want to make progress on coming up with ways to prevent such a thing?" Doudna asks. "My heart broke." If gene editing could prevent this in the future, it would be immoral not to pursue it, she decided. She answered all such emails. She wrote the mother back and promised that she and other researchers were working diligently to find therapies and preventions for such genetic conditions. "But I also had to tell her that it would be years before something like gene editing would be potentially useful for her," she says. "I didn't want to mislead her in any way."

After appearing at the World Economic Forum in Davos in January 2016, where she shared her ethical qualms about gene editing, Doudna was pulled aside by another woman on the panel, who described how her sister had been born with a degenerative disease. It affected not only her but the lives and finances of her whole family. "She said if we could have done gene editing to avoid that, everyone in her family would be absolutely in favor of it," Doudna recalls. "She was very emotional about the cruelty of those who would prevent germline editing, and she was on the verge of tears. I found it so touching."

Later that year, a man came to see her at Berkeley. His father and grandfather had died of Huntington's. Three of his sisters had been diagnosed with it and faced a slow, agonizing death. Doudna refrained from asking the man if he was also afflicted. But his visit convinced her that if germline editing became a safe and effective way to eliminate Huntington's, she was in favor. Once you've seen the face of someone with a genetic disease, she says, especially one like Huntington's, it's hard to support why we would refrain from gene editing.

Her thinking was also influenced by long conversations with Janet Rossant, the chief of research at the Hospital for Sick Children in Toronto, and George Daley, dean of the Harvard Medical School. "I realized how we were on the verge of being able to correct disease-causing mutations," she says. "How could you not want to do that?" Why should CRISPR be held to a far higher standard than any other medical procedure?

The evolution in her thinking made her more sympathetic to the view that many gene-editing decisions should be left to individual choice rather than to bureaucrats and ethics panels. "I'm an American, and putting a high priority on personal freedom and choice is part of our culture," she says. "I also think that as a parent I feel that I would want to have that choice to make about my own health or own family's health as these new technologies come along."

However, because there are still huge risks that may be unknown, she feels that CRISPR should be used only when it is medically necessary and there are no good alternatives. "That means we have no reason to be doing it yet," she says. "That's why I had a problem with He Jiankui's use of CRISPR to attempt to achieve immunity to HIV. There were other ways of doing that. It wasn't medically necessary."

One moral issue that continues to loom large for her is inequality, especially if the wealthy are able to buy genetic enhancements for their children. "We could create a gene gap that would get wider with each new generation," she says. "If you think we face inequalities now, imagine what it would be like if society became genetically tiered along economic lines and we transcribed our financial inequality into our genetic code."

By limiting gene edits to those that are truly "medically necessary," she says, we can make it less likely that parents could seek to "enhance" their children, which she feels is morally and socially wrong. The line between medical treatment and enhancement can be blurry, she acknowledges, but it is not totally meaningless. We know the difference between correcting a very harmful gene variant and adding some genetic trait that is not medically necessary. "As long as we are correcting genetic mutations by restoring the 'normal' version of the gene—not inventing some wholly new enhancement not seen in the average human genome—we're likely to be on the safe side."

She is confident that the good that can come from CRISPR will eventually outweigh the dangers. "Science doesn't move backwards, and we can't unlearn this knowledge, so we need to find a prudent path forward," she says, reprising the phrase in the title of the report she wrote after her 2015 Napa Valley meeting. "We've never seen anything like this before. We now have the power to control our genetic future, which is awesome and terrifying. So we must move forward cautiously and with respect for the power we've gained."

Part Eight

Dispatches from the Front

Here's to the crazy ones. The misfits. The rebels. The troublemakers. The round pegs in the square holes. The ones who see things differently. They're not fond of rules. And they have no respect for the status quo. You can quote them, disagree with them, glorify or vilify them. About the only thing you can't do is ignore them. Because they change things. They push the human race forward. And while some may see them as the crazy ones, we see genius. Because the people who are crazy enough to think they can change the world are the ones who do.

—Steve Jobs, Apple's "Think Different" ad, 1997

Samuel Sternberg

CHAPTER 44

Quebec

Jumping genes

While attending the 2019 CRISPR Conference in Quebec, I am struck by the realization that biology has become the new tech. The meeting has the same vibe as those of the Homebrew Computer Club and the West Coast Computer Faire in the late 1970s, except that the young innovators are buzzing about genetic code rather than computer code. The atmosphere is charged with the catalytic combination of competition and cooperation reminiscent of when Bill Gates and Steve Jobs frequented the early personal computer shows, except this time the rock stars are Jennifer Doudna and Feng Zhang.

The biotech nerds, I realize, are no longer the outsiders. The CRISPR revolution and coronavirus crisis have turned them into the cool kids on the edge, just as happened to the awkward pioneers who once populated the cyber-frontier. As I wandered around reporting dispatches from the front lines of their revolution, I noticed that even as they pursue their new discoveries they feel tugged, sooner than the digital techies did, to engage in a moral reckoning about the new age they are creating.

The buzz in Quebec is about a fascinating breakthrough that reignited the tension between Doudna's realm and that of Zhang. It involves dueling discoveries of an efficient way to add new sequences into DNA. Instead of making a cut in the double-stranded DNA, the newly discovered CRISPR system would insert a new chunk of DNA by harnessing transposons, known as "jumping genes," which are big segments of DNA that can hop from one place to another on chromosomes.

Sam Sternberg, the whip-smart biochemist who studied under Doudna and then was recruited to open his own lab at Columbia, has just published in *Nature* his first major paper as an assistant professor. It describes a CRISPR-guided system that inserts a tailored jumping gene into a desired DNA location. But to Sternberg's surprise, Zhang was able to get a similar paper of his published online in *Science* a few days earlier.[1]

Sternberg seems deflated when he arrives in Quebec, and his friends, including Doudna, are angry. He had submitted his paper to *Nature* on March 15, and word of his discovery began to spread after one of his graduate students gave a talk about it. "Feng then quietly raced to get his paper published first," Martin Jinek tells me at the conference. To Doudna, it was typical of Zhang: "His network of people will tell him about a paper and he will rush ahead."[2]

She and Eric Lander had both conceded to me, when recalling the 2012 race, that rushing a paper into print when you sense competition is fair play. Nevertheless, Zhang's publication on transposons causes resentment. He had submitted his paper to *Science* on May 4, seven weeks after Sternberg had submitted his, but Zhang's was published online on June 6 and Sternberg's did not appear until June 12.

I find it hard to share the Doudna camp's outrage about Zhang. The two papers both involve harnessing jumping genes, but they differ in important ways and each makes a distinctive contribution to the progress of CRISPR. I happened to be visiting Zhang at his Broad Institute lab the day after his paper went online, which was ten days before the Quebec conference, and he described to me the research he had done on transposons. His paper was not a rush job. It had been in the works for a long time. But when he heard footsteps, he pushed *Science* to get it reviewed and online expeditiously—just as Doudna had done with the seminal 2012 paper she coauthored with Charpentier when she heard the footsteps of Virginijus Šikšnys and others.[3]

On the first day of the Quebec conference, Sternberg's friends, including Doudna, both celebrate and commiserate with him in the hotel lobby bar over Romeo's gin, a fragrant Canadian product. His personality is so naturally ebullient that he seems to get over his annoyance by the time he does his presentation the next day, following one given by Zhang. After all, his discovery is an important triumph and step in his career, one not diminished by Zhang's complementary finding. So Sternberg is gracious in his talk. "We heard from Feng earlier today about how CRISPR-Cas12 can mobilize transposable elements," he says. "What I am going to tell you about is a recently published work on type-one systems that work in similar but also different ways to mobilize these bacterial transposons." He makes sure to heap credit on the PhD student in his Columbia lab, Sanne Klompe, who carried out the main experiments.

"Is there any field that is more cutthroat and competitive than biological research?" one of the participants asks me after Zhang and Sternberg give their dueling talks. Well, yes, I think, almost every field can be, from business to journalism. What distinguishes biological research is the collaboration that is woven in. The camaraderie of being rival warriors in a common quest suffuses the Quebec conference. The desire to win prizes and patents tends to create competition, which spurs the pace of discoveries. But equally motivating, I think, is the passion to uncover what Leonardo da Vinci called the "infinite wonders of nature," especially when it comes to something so breathtakingly beautiful as the inner workings of a living cell. "The jumping gene discoveries show just how fun biology is," Doudna says.

Seared bison

When the first day of presentations is over, Doudna and Sternberg go to a casual restaurant in Old Quebec City, but I accept an invitation from Feng Zhang to join him and a small group of his friends for dinner. Not only do I want to hear his perspective, but I also want to check out the inventive new restaurant he has chosen, Chez Boulay, which features crispy seal meatloaf, huge raw scallops, Arctic char, seared bison, and cabbage blood sausage. Our group of a dozen diners includes Kira Makarova of the U.S. National Center for Biotechnology Information, who was a coauthor of Zhang's jumping-gene paper; the CRISPR pioneer Erik Sontheimer, who was Luciano Marraffini's mentor but has stayed above the personal rivalries of the CRISPR world; and April Pawluk, who had been a postdoc in Doudna's lab and was now an editor at *Cell*, a peer-reviewed journal that competes with *Science* and *Nature*. There is a symbiotic relationship between top researchers, who want to make sure their papers get speedy and favorable treatment, and smart journal editors such as Pawluk, who want to publish the most important new discoveries.

Sontheimer orders the wine, which comes from Quebec and is unexpectedly good, and we drink a toast to transposons. When the talk turns from science to the ethical issues hovering over CRISPR, most of the diners agree that, when it's safe and practical, genetic editing—even making inheritable edits in the human germline—ought to be used if necessary to fix bad single-gene mutations, such as Huntington's disease and sickle-cell anemia. But they recoil at the idea of using gene editing for human enhancements, such as trying to give our kids more muscle mass or height or perhaps someday higher IQ and cognitive skills.

The problem is that the distinction is difficult to define and even more difficult to enforce. "There's a blurry line between fixing abnormalities and making enhancements," Zhang says. So I ask him, "What is wrong with making enhancements?" He pauses for a long time. "I just don't like it," he says. "It's messing with nature. And from a longer term population perspective, you may be reducing diversity." He took the famous Harvard course on moral justice taught by the philosopher Michael Sandel, and he has clearly wrestled with these issues in a profound way. But like the rest of us, he hasn't found easy answers.

A looming ethical issue, everyone at the table agrees, is that gene editing could exacerbate, and even encode, inequality in society. "Should rich people be allowed to buy the best genes they can afford?" Sontheimer asks. It is true, of course, that all of society's benefits, including medical ones, are unequally distributed, but creating a marketplace for inheritable genetic enhancements would kick that issue into an entirely new orbit. "Look at what parents are willing to do to get kids in college," Zhang says. "Some people will surely pay for genetic enhancement. In a world in which there are people who don't get access to eyeglasses, it's hard to imagine how we will find a way to have equal access to gene enhancements. Imagine what that will do to our species."

Gavin Knott showing how to edit

Chapter 45

I Learn to Edit

Gavin Knott

Now that I had become immersed in the world of CRISPR pioneers, I decided that I should, in my own small way, be initiated into the club. I should learn how to edit DNA using CRISPR.

So I arrange to spend a few days in Doudna's open-space lab amid the dozens of workspaces, cluttered with centrifuges and pipettes and Petri dishes, where her students and postdocs perform their experiments. I want to replicate the major advances I've recounted: using CRISPR-Cas9 to edit DNA in a test tube, like Doudna and Charpentier described in June 2012, and then using it to make an edit in a human cell, as Zhang, Church, Doudna, and others described in January 2013.

For the first, I am helped by Gavin Knott, a young postdoc from western Australia with a trim beard and easygoing manner. As a graduate student, he decided he wanted to find CRISPR-associated enzymes that attack RNA rather than DNA, and he wrote a letter to Doudna proposing that he come to her lab to do that. Doudna's team was already on the case, working with an enzyme known as Cas13. "She had her finger on the pulse much more than I did," Knott says. But she invited him to be a postdoc in her lab anyway. Among other duties, he became part of the group working on the Safe Genes project for DARPA.[1]

When we go into the secure part of the Doudna Lab where experiments are done, I put on my lab coat and goggles, spray my gloved hands to sterilize them, and instantly feel like a pro. Knott takes me to one of the hoods, a tabletop workspace that is partially enclosed by plastic sides and specially ventilated. Just before we begin work, Doudna buzzes through, wearing a white lab coat over jeans and a black Innovative Genetics Institute T-shirt. She briefly checks on the experiments being done by each of her students (and me), before heading off to an all-day strategy retreat with the institute's top researchers.

The experiment that Knott walks me through involves a snippet of DNA that contains a gene that can make bacteria resistant to the antibiotic ampicillin. This is not a good thing, especially if you're a person who's been infected by such bacteria. So Knott concocts for me some Cas9 with a guide RNA that is designed to eliminate the gene. The lab had brewed all of this from scratch. "The Cas9 we need is encoded on a piece of DNA, and anyone who can grow bacteria in a lab can produce large quantities of it," he assures me. My look probably conveys that I'm not sure this is part of my skill set. "Don't worry," he says. "If you don't want to make it all from scratch, you can just buy the Cas9 from companies like IDT on the web. You can even buy the guide RNAs. If you want to edit genes, it's easy to order the components online."

(Later, I go online to see. The IDT website advertises "all of the reagents needed for successful genome editing," with kits designed for delivery into human cells beginning at $95. Over at a site called GeneCopoeia, a Cas9 protein with a nuclear location signal starts at $85.)[2]

Some of the vials that Knott prepared are lined up in an old-fashioned chill box, one that uses ice to keep liquids cool. "This chill box has a significant history," he says, turning it around. On the back is etched the name "Martin." It had been Jinek's before he left to start his own lab at the University of Zurich. "I inherited it," Knott says proudly. I feel part of a historic chain. The experiments we are about to do mirror Jinek's from 2012: taking a piece of DNA and incubating it with the Cas9 and guide RNA to cut it in the desired location. It's sweet to be using his chill box.

Knott walks me through a variety of steps, using pipettes to combine the ingredients and then incubating it for ten minutes. We add a dye to help us visualize the results, and then we are able to create an image of what we had done by using a process called electrophoresis, which puts an electric field through a gel to separate DNA molecules of different sizes. The resulting printout shows bands at different locations along the gel, indicating if and how they were cut by Cas9. "Textbook success!" Knott exclaims as he takes the image off the printer. "Look at the differences in these bands."

On the way out of the lab, I run into Jamie Cate, Doudna's husband, by the elevator, and I show him my printouts. He points to blurry bars at the bottom of two of the columns and asks, "What are those?" I actually know the answer (thanks to Knott's tutorial). "It's the RNA," I say. Later that day, Cate sends out a tweet attached to a picture of Knott and me working at the lab bench, saying, "And Walter Isaacson passed my pop quiz!" For just a moment, until I realize that Knott did all of the real work, I feel like a true gene editor.

Jennifer Hamilton

The next challenge is to edit a gene in a human cell. In other words, I want to take the step that the labs of Zhang and Church and Doudna accomplished at the end of 2012.

For that I team up with another postdoc in Doudna's lab, Jennifer Hamilton, a Seattle native who earned her doctorate in microbiology at Mount Sinai Medical Center in New York City. With her big glasses and even bigger smile, Hamilton radiates enthusiasm for harnessing viruses to deliver gene-editing tools into human cells. When Doudna came to give a talk to the Women in Science group at Mount Sinai in 2016, Hamilton served as her student escort. "I felt instantly a connection with her," Hamilton recalls.

Doudna was then beginning to build the Innovative Genomics Institute at Berkeley, which would bring together researchers from around the Bay Area. Part of its mission was to find ways to deliver CRISPR editing tools into human cells for medical treatments. So she recruited Hamilton. "I had skills in engineering viruses, and I wanted to apply them to figuring out delivery methods for getting CRISPR into humans," Hamilton says.[3] It was a specialty that would prove valuable when the lab took on the coronavirus pandemic and needed to find ways to deliver CRISPR-based treatments into human cells.

When we begin our attempt to edit DNA in a human cell, Hamilton stresses that it is more challenging than doing it in a test tube. The strands of DNA that I had edited the day before with Knott contained only 2.1 kilobases (2,100 pairs of DNA base letters) versus the 6.4 *million* kilobases in the cell we plan to use, which was derived from a human kidney cell. "The challenge with human gene editing," she tells me, "is to get your editing tools past the cell's outer plasma membrane and past its nuclear membrane to get to where the DNA is, and then you also have to get your tools to find the location in the genome."

Hamilton's explanation of our planned procedure seems to support, albeit inadvertently, Zhang's argument that it is not a simple step to move from editing DNA in a tube to editing it in a human cell. However, the fact that I was about to do it could be used, I guess, to make the opposite argument.

Our plan, Hamilton says, is to make a double-strand break at a targeted place in the DNA of the human cell. In addition, we will supply a template so that a new gene will be inserted. The human cell we start with has been engineered to have a gene that creates a fluorescent protein that glows blue. In one of our procedures, we will use CRISPR-Cas9 to cut the gene and thus deactivate it. This means that the cell should no longer glow. In another sample, we will supply a template that the cell will then incorporate, changing three base pairs of the cell's DNA in order to make the fluorescent protein change from blue to green.

The method we use to get the CRISPR-Cas9 and the template into the nucleus of the cell is called nucleofection. It employs electrical pulses to make the cell's membranes more permeable. At the end of the full editing process, I am able to look through a fluorescent microscope and see the results. The control group still glows blue. A group that had been cut with CRISPR-Cas9 but not supplied with a replacement template doesn't glow at all. And finally, there is the group that we had cut and then edited. I look into the microscope and see them glowing green! I have edited—well, Hamilton has actually edited, with me as an eager copilot—a human cell and changed one of its genes.

Before you get too frightened by what I may have wrought, rest assured: we take everything I did, mix it with chlorine bleach, and wash it down a sink. But I did learn how relatively easy the process can be for a student or rogue scientist who has some skill at a lab bench.

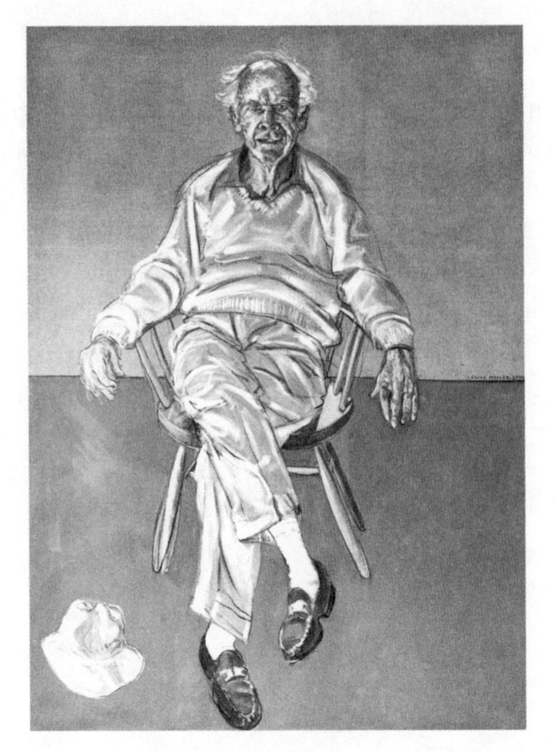

The oil portrait at Cold Spring Harbor, by Lewis Miller

James Watson and his son Rufus in the PBS documentary *Decoding Watson*

CHAPTER 46

Watson Revisited

Intelligence

Cold Spring Harbor Laboratory, where James Watson launched an influential series of annual meetings on the human genome in 1986, decided to add a new series focused on CRISPR gene editing beginning in the fall of 2015. Among the speakers that first year were our four main characters: Jennifer Doudna, Emmanuelle Charpentier, George Church, and Feng Zhang.

Watson attended that initial meeting of the CRISPR group, as he did most meetings at Cold Spring Harbor, and he sat in the front row of the auditorium, underneath a grand oil portrait of himself, to hear Doudna's talk. It was a reprise of her first visit there as a graduate student in the summer of 1987, when Watson also sat up front as she presented, with youthful nervousness, a paper on how some RNAs could replicate themselves. After Doudna's CRISPR talk, he came up to say a few words of praise, just as he had done almost thirty years earlier. It was important, he said, to push the science of making gene edits in humans, including enhancing intelligence. For some in attendance, it felt historic. Stanford biology professor David Kingsley took a picture of Watson and Doudna talking.[1]

But when I show up at the 2019 meeting, Watson is not in his usual seat in the front row. After fifty years, he has been banished from meetings, and the oil portrait of him removed. He is now sentenced to internal exile, living with his wife, Elizabeth, in elegant but tortured isolation at the northern end of the campus in a pale Palladian-style mansion called Ballybung.

His troubles began in 2003, when he marked the fiftieth anniversary of his co-discovery of DNA's structure by giving an interview for a documentary on PBS and the BBC. Genetic engineering should someday be used to "cure" people who have low intelligence, he said. "If you really are stupid, I would call that a disease." It reflected his deep belief, perhaps fostered by pride in his seminal scientific discovery as well as the daily angst of living with his schizophrenic son, Rufus, in the power of DNA to explain human nature. "The lower ten percent who really have difficulty, even in elementary school, what's the cause of it?" Watson asked. "A lot of people would like to say, 'Well, poverty, things like that.' It probably isn't. So I'd like to get rid of that, to help the lower ten percent." As if to make sure that he stoked enough controversy, Watson added that gene-editing could also be used to enhance people's looks. "People say it would be terrible if we made all girls pretty. I think it would be great."[2]

Watson considered himself a political progressive. He supported Democrats from Franklin Roosevelt to Bernie Sanders. His advocacy for gene editing, he insisted, was because he wanted to improve the lot of the less fortunate. But as the Harvard philosopher Michael Sandel noted, "Watson's language contains more than a whiff of the old eugenic sensibility."[3] It was a whiff that was particularly odious wafting from Cold Spring Harbor, given the lab's long history of fomenting that eugenic sensibility.

Watson's comments about intelligence were controversial, but in 2007 he crossed a line by connecting it to race. That year he published another memoir, *Avoid Boring People*, a phrase that he meant to be read in two ways, with "boring" as both a verb and an adjective. Naturally averse, perhaps congenitally so, to boring people, Watson relished mumbling unfiltered and provocative comments, often accompanying them with a windy snort and impish grin. This proved combustible when, as part of the publicity campaign for the book, he gave a series of interviews to Charlotte Hunt-Grubbe, a freelance science journalist who was writing a profile of him for the *Sunday Times* of London. Always unguarded, he was in this case even more so because she was a former student and tennis partner of his who had lived with the Watsons in Cold Spring Harbor for a year.

The result was a languid feature story in which Hunt-Grubbe followed Watson from the library of his house to a local diner to the lawn tennis courts of the Piping Rock Club. After a match, he reflected on his current life. "I'm still thinking," he said, "can we find the genes for mental disease while I'm still alive, and will we have stopped cancer in ten years, and will my tennis serve improve?"[4]

Near the end of her four-thousand-word piece, she offhandedly described him offering some ruminations about race:

> He says that he is "inherently gloomy about the prospect of Africa" because "all our social policies are based on the fact that their intelligence is the same as ours—whereas all the testing says not really, and I know that this hot potato is going to be difficult to address." His hope is that everyone is equal, but he counters that "people who have to deal with black employees find this not true."

The article set off an explosion, and Watson was forced to resign as chancellor of Cold Spring Harbor. But for the time being, he was allowed to wander down the hill from his house at the top of the campus whenever he wanted to attend meetings.

Watson tried to walk his comments back, saying that he was "mortified" about having implied that Africans were "somehow genetically inferior." In a prepared statement released by the lab, he added, "That is not what I meant. More importantly from my point of view, there is no scientific basis for such a belief."[5] There was one problem with his apology: it actually *was* what he meant, and being the type of person he was, he would inevitably have trouble in the future not saying so.

Watson's ninetieth birthday

By the time Watson turned ninety in 2018, the controversy surrounding him seemed to have subsided. His birthday, along with the fiftieth anniversary of his arrival at Cold Spring Harbor Laboratory and his marriage to Elizabeth, was celebrated in the campus auditorium with a concert that featured pianist Emanuel Ax playing Mozart followed by a gala dinner. The benefit raised $750,000 toward an endowed professorship at the laboratory in his honor.

Watson's friends and colleagues tried to sustain a delicate balance. He was honored for being one of the most influential thinkers in modern science, tolerated for being saucy in his writings and conversations, and condemned for his comments on racial intelligence. That balance was sometimes difficult to sustain. A few weeks after the birthday celebration, at a genetics meeting on campus, Eric Lander was asked to make a toast to Watson, who was sitting in the audience. Lander noted that Watson was "flawed," but in his ebullient way added gracious comments about his leadership of the Human Genome Project and for "pushing all of us to explore the frontiers of science for the benefit of humankind."

The toast prompted a backlash, especially on Twitter. Lander, already burned by blasts for minimizing the roles of Doudna and Charpentier in his "Heroes of CRISPR" article, apologized. "I was wrong to toast, and I'm sorry," he wrote in a note to his Broad colleagues that he made public. "I reject his views as despicable. They have no place in science, which must welcome everyone." He added a cryptic comment, which referred to a conversation he once had with Watson about Jewish donors to their respective institutions. "As someone who has been on the receiving end of his abhorrent remarks, I should have been sensitive to the damage caused by recognizing him in any way."[6]

Watson was infuriated by Lander's assertion that it was wrong to be "recognizing him in any way" and the insinuation that he was anti-Semitic. "Lander is regarded as a joke," Watson exploded. "My life has been dominated by, first, my father's love for Jews, and all my good friends in America have been Jewish." He went on to emphasize to me, in a way that would not have mollified his critics, his view that Ashkenazi Jews, who lived for centuries in northern Europe, were genetically more intelligent than other ethnic groups, a point he supported by rattling off those who had won Nobel Prizes.[7]

An American Master

When the *American Masters* series on PBS decided to do a documentary on Watson in 2018, it set out to produce a balanced, intimate, complex, and nuanced look at both his scientific triumphs and his controversial views. He cooperated fully, allowing the cameras to follow him around his elegant home and the Cold Spring Harbor campus. The documentary covered his whole life, including his intellectual bromance with Francis Crick, the controversy over his unauthorized use of Rosalind Franklin's DNA images, and his late-career quest to find genetic treatments for cancer. Most poignant were the scenes of him with his wife and their son Rufus, still living at home at age forty-eight while coping with schizophrenia.[8]

It also dealt with the controversy over his remarks on race. Joseph Graves, the first African American to get a PhD in evolutionary biology, gave a studied rebuttal to those views. "We know a great deal about human genetic variation and how it is apportioned around the world," he said, "and there is absolutely no evidence that there are genetic differences that favor intelligence in any sub-population of human beings." Then the interviewer gave Watson the opportunity to—almost prodded him to—renounce or abandon some of his previous controversial statements.

He didn't. Caught close up on camera, he seemed to pause and even tremble slightly like an aged schoolkid who was unable to say what he was supposed to. It was as if he was congenitally incapable of sugarcoating his thoughts or biting his tongue. "I would like for them to have changed, that there be new knowledge that says that your nurture is much more important than nature," he said as the cameras rolled. "But I haven't seen any knowledge. And there's a difference on the average between blacks and whites on I.Q. tests. I would say the difference is, it's genetic." Later, there was a moment of self-awareness. "It should be no surprise that someone who won the race to find the double helix should think that genes are important."

The documentary aired the first week of January 2019, and Amy Harmon of the *New York Times* wrote a story about his remarks. "James Watson Had a Chance to Salvage His Reputation on Race" was the headline. "He Made Things Worse."[9] She noted that there were complex debates over the relationship between race and IQ, then she quoted Francis Collins, the director of the National Institutes of Health and Watson's successor as head of the Human Genome Project, giving the consensus view. Experts on intelligence, he said, "consider any black-white differences in I.Q. testing to arise primarily from environmental, not genetic, differences."[10]

The board of Cold Spring Harbor Laboratory finally decided it had to cut almost all of its remaining ties to Watson. Calling his comments "reprehensible and unsupported by science," it stripped him of his honorary titles and removed the large and casually elegant oil portrait of him from its main auditorium. He was, however, allowed to remain in his bayfront manor house on the campus.[11]

The Jefferson Conundrum

Watson thus presents historians with what could be called the Jefferson Conundrum: To what extent can you respect a person for great achievements ("We hold these truths") when they are accompanied by reprehensible failings ("are created equal")?

One question raised by the conundrum relates, at least metaphorically, to gene editing. Cutting out a gene for an unwanted trait (sickle-cell anemia or HIV receptivity) might change some existing desirable trait (resistance to malaria or the West Nile virus). The issue is not simply whether we can balance a respect for a person's achievements with a contempt for their flaws. The more complex issue is whether the achievements and the flaws are interwoven. If Steve Jobs had been kinder and gentler, would he have had the passion that allowed him to bend reality and push people to realize their full potential? Did Watson have a congenital tendency to be heretical and provocative, and did that help him push the frontiers of science when he was right and lead him into a dark abyss of prejudice when he was wrong?

I believe that people's flaws cannot be excused by saying they are interwoven with their greatness. But Watson is an important part of the story I am writing—this book begins with Doudna picking up his seminal *The Double Helix* and deciding to become a biochemist—and his views on genetics and human enhancement are an undercurrent of the policy debates over gene editing. So I decide to go visit him right before the summer 2019 CRISPR meeting at Cold Spring Harbor.

A visit with Watson

I have known James Watson since the early 1990s, when I was at *Time* before he was so controversial, and we covered his work on the Human Genome Project, commissioned essays by him, and selected him for our list of the hundred most influential people of the twentieth century. At the 1999 dinner celebrating what we called "the *Time* 100," I asked him to give a toast to the late Linus Pauling, whom he had beaten in the race to discover the structure of DNA. "Failure hovers uncomfortably close to greatness," he said of Pauling. "What matters now are his perfections, not his past imperfections."[12] Perhaps people may say that of Watson someday, but in 2019 he was an outcast.

When I arrive at his house on the Cold Spring Harbor campus, Watson settles into a chintz-covered armchair looking very frail. A few months earlier, he had come back from a trip to China and, with no car provided by the lab to pick him up at the airport, had driven himself in the dark. He ended up veering off the road and into the bay by his home, leading to a long hospitalization. But his mind is still sharp and he is still focused on deploying CRISPR in an equitable way. "If it's only used to solve the problems and desires of the top ten percent, that will be horrible," he says. "We have evolved more and more in the past few decades into an inequitable society, and this would make it much worse."[13]

One step that might help a little, he suggests, is to not allow patents for genetic engineering techniques. There would still likely be a lot of funding for finding safe ways to fix maladies that are devastating, such as Huntington's and sickle-cell anemia. But if there were no patents, there might be less payoff for racing to be the first to devise methods of enhancements, and those that did get invented might be cheaper and more widely available if anyone could copy them. "I would accept some slowdown in the science in return for making it more equitable," he says.

When he made an assertion he knew might shock, he gave his short snort of a laugh and grinned like a scamp who's just done something naughty. "I think my blunt and contrary nature helps my science, because I don't simply accept things just because other people believe it," he says. "My strength is not that I am smarter, it's that I'm more willing to offend the crowd." Sometimes, he admits, he has been "too honest" in order to push an idea. "You have to exaggerate."

Was that the case, I ask, with his comments on race and intelligence? As is his nature, he is able to seem regretful but not repentant. "The PBS documentary on me was actually very good, but I wish they didn't emphasize my old comments on race," he replies. "I don't say anything publicly about that anymore."

But then, as if compelled, he starts to veer off into that realm again. "I couldn't deny what I believed," he tells me. He begins talking about various historical measurements of IQ, the effect of climate, and what he had been taught as an undergraduate at the University of Chicago by Louis Leon Thurstone about factor analysis in intelligence.

Why, I ask him, does he feel the need to say such things? "I haven't given one interview on race since I talked to that girl from the *Sunday Times*," he says. "She had lived in Africa and knew. The only time I repeated it was to this television interviewer, because I couldn't help myself." I suggest that he could help himself if he wanted to. "I always follow my father's advice of saying the truth," he replies. "Someone has to say the truth."

But it's not the truth. Most experts, I tell him, say your views are wrong.

He doesn't engage, so I ask him what other advice his father gave him. "Always be kind," he answers.

Has he heeded that advice?

"I wish I had been better on that one," he admits. "I wish I had worked harder on always being kind."

He badly wanted to be in the audience again for the annual CRISPR meeting at Cold Spring Harbor, which was a week later, but the lab was unwilling to lift its ban. So he requested that I bring Doudna up the hill from the meeting so he could talk to her.

Rufus

Sitting in the kitchen during my visit with Watson was his son Rufus. He didn't join us, but he was listening to every word.

As a boy, Rufus looked like his father did when young: lanky, tousled hair, an easy grin, and an angular face often slightly tilted as if in curiosity. Like father, like son. Heritage and breeding. But now Rufus was in his late forties, pudgy and somewhat disheveled. He has lost the ability to laugh casually. He is acutely aware of his own condition—and also of his father's. Volatile, sensitive, brilliant, disheveled, unfiltered, prone to spouting off, brutally candid, attentive to every conversation, and also gentle—these are all traits that mark Rufus's schizophrenia. To a different degree and in some form, each and every one of these traits can be ascribed to his father as well. Perhaps, someday, the deciphering of the human genome will be able to explain that. Or maybe it won't.

"My dad will say, 'My son Rufus, he's bright but he's mentally ill,'" Rufus told the *American Masters* interviewer. "Whereas I think of it as the opposite. I think I'm dim but not mentally ill." He feels that he has let his father down. "It wasn't until I became aware of how dim I was that I thought this was strange, because my dad's not dim," he says. "Then I thought that I'm a burden on my parents because he's successful, and he deserves to have a successful child. He's worked hard, and if you believe in karma he should have earned himself a successful son."[14]

At one point during my conversations with James Watson, when he veers toward the issue of race, Rufus bursts in from the kitchen shouting. "If you are going to let him say these things, then I am going to have to ask you to leave." Watson merely shrugs and says nothing to his son, but he quits talking about race.[15]

I can sense the intense protectiveness that Rufus feels toward his father. These outbursts also reveal in him a wisdom that his father often lacks. "My dad's statements might make him out to be a bigot and discriminatory," he once said. "They just represent his rather narrow interpretation of genetic destiny." He's right. In many ways he is wiser than his father.[16]

Chapter 47

Doudna Pays a Visit

Careful conversation

As Watson requested, I ask Doudna if she would be willing to visit him during the meeting he was barred from attending. When the two of us enter his house, he asks to see the conference book with the abstracts of the scientific papers being presented. I am reluctant to show it to him because the cover of the book is Rosalind Franklin's "photograph 51" X-ray diffraction image that helped Watson discover the structure of DNA. But he seems amused rather than upset. "Ah, that picture, it will always haunt me," he says, then pauses and smiles his impish grin. "But she never figured out it was a helix."[1]

Watson, wearing a peach-colored sweater in the sun-dappled sitting room, points out some of the art he has collected over the years. Tellingly, his most prominent pieces are modernist and abstract depictions of human faces contorted in emotion. These include paintings and drawings by John Graham, André Derain, Wifredo Lam, Duilio Barnabé, Paul Klee, Henry Moore, and Joan Miró, as well as a drawing of Watson's own slightly contorted and emotionally pensive face by David Hockney. Classical music plays in the background. Elizabeth Watson sits in the corner reading a book, and Rufus hovers out of sight in the kitchen, listening. Everyone tries to be careful in the conversation—even Watson, for the most part.

Doudna talking with James Watson under his portrait

"The reason that CRISPR is the most important discovery since DNA's structure," he tells Doudna, "is that it not only describes the world, as we did with the double helix, but makes it easy to change the world." He and Doudna discuss the Watsons' other son, Duncan, who lives in Berkeley near Doudna. "We were just there visiting him," Watson says. "The students at Berkeley are the pits, they are so progressive. These progressive kids are even dumber than Republicans." Elizabeth chimes in to change the subject.

Doudna reminisces about the first meeting Watson had convened on genome editing at Cold Spring Harbor five years earlier, and how he had asked her a question from the audience. "I was enthusiastic about the use of it," he says. "People who cannot think well enough will be able to be made immensely better." Elizabeth again chimes in on a different topic.

The complexity of human life

It was a short visit, and as we walk back down the hill from Watson's home, I ask Doudna her thoughts. "I was thinking back to when I was twelve and began reading the dog-eared copy of *The Double Helix*," she says. "It would have been wild to know that years later I would be visiting with him in his home having that conversation."

She doesn't say much else that day, but the visit resonated. Over the next few months, we would return to it in our conversations. "It was a poignant and sad visit," she says. "He is clearly someone who has had a huge impact on biology and genetics, but he's expressing views that are quite abhorrent."

She admits that she had mixed feelings about agreeing to go to see him. "But I agreed to because of his influence on biology and on my own life. Here's a person who had this incredible career, and had this potential to be a real figure of respect in the field, and it was all squandered because of these views that he holds. Some people may say you shouldn't have met with him. But for me it's not so simple."

Doudna recalls one aspect of her father's personality that used to upset her. Martin Doudna tended to categorize people as good or bad, with little respect for the shadings that most people contain. "He had people that he revered and thought were wonderful and they could do no wrong, and then he had people who were horrible and he disagreed with them on everything, and they could do no right." Reacting to that, Doudna worked hard to see people in all of their complexity. "I felt like the world is kind of grayscale. There are people who have great qualities, but they also have flaws."

I mention "mosaic," a term often used in biology. "That's a better description than grayscale," she says. "And frankly that's true for all of us. All of us, if we're honest with ourselves, know that we have things that we're great at and things that we're not so great at."

That indirect admission that we all have our flaws intrigued me. I tried to tease more out of her, asking how that applied to herself. "If I have a regret, it's that I don't really feel proud of the way I, in some cases, interacted with my dad," she responds. "I got frustrated with him because he viewed people with a black-and-white lens."

Does that influence, I ask, how she tries to view James Watson? "I don't want to do what my father did and come to simple judgments," she answers. "I try to grapple with people who do great things, but who I also completely disagree with on some things." Watson is a prime example, she says. "He has said some really bad things, but every time I see him, I am brought back to that day when I read *The Double Helix* and first started thinking, 'Gee, I wonder if I could do that kind of science someday.'"[2]

Part Nine

Coronavirus

I have no idea what's awaiting me, or what will happen when this all ends. For the moment I know this: there are sick people and they need curing.

—Albert Camus, *The Plague*, 1947

CHAPTER 48

Call to Arms

Innovative Genomics Institute

At the end of February 2020, Doudna was scheduled to travel from Berkeley to Houston for a seminar. Life in the United States had not yet been disrupted by the looming coronavirus pandemic. There had been no officially reported deaths. But red flags were flying. There were already 2,835 deaths in China, and the stock market was beginning to take notice. The Dow fell more than a thousand points on February 27. "I was nervous," Doudna recalls. "I talked with Jamie about whether or not to go. But at the time everyone I knew was carrying on as usual, and so I went to Houston." She took with her a supply of hand wipes.

When she returned, she began thinking about what she and her colleagues should be doing to fight the pandemic. Having turned CRISPR into a gene-editing tool, she had a profound feel for the molecular mechanisms that could be used by humans to detect and destroy viruses. More important, she had become a maestro of collaboration. It became clear to her that battling coronavirus would require putting together teams that spanned many specialties.

Fortunately, she had a base from which she could build such an effort. She had become the executive director of the Innovative Genomics Institute (IGI), a joint research partnership between Berkeley and the University of California, San Francisco, with a spacious five-story modern building on the northwest corner of the Berkeley campus. (It was originally going to be called the Center for Genetic Engineering, but the university began to worry that the name might unnerve people.)[1] One of the institute's core principles is to foster collaboration between different fields, which is why its building houses plant scientists, microbial researchers, and biomedical specialists. Among the researchers who have their labs in the facility are her husband, Jamie; her original CRISPR collaborator Jillian Banfield; her former postdoc Ross Wilson; and the biochemist Dave Savage, who was using CRISPR to improve how bacteria in ponds convert carbon from the atmosphere into organic compounds.[2]

Doudna had been talking to Savage, whose office is next to hers, for almost a year about launching some project at IGI that would become a model for cross-disciplinary teamwork. One genesis for the plan came from her son, Andy, who had a summer internship at a local biotech company. His day there began with a check-in where leaders from different divisions shared what they were doing to further the company's projects. Hearing this, Doudna had laughed and told Andy she couldn't imagine running an academic lab that way. "Why not?" he asked. She explained that academic researchers get comfortable in their silos and too protective of their independence. It started a long-running conversation in their house about teams, innovation, and how to create a work environment that stimulates creativity.

She kicked around ideas with Savage in late 2019 at a Japanese noodle house in Berkeley. How could you combine, she asked, the best features of a corporate team culture with academic autonomy? They wondered if it would be possible to find a project that would coalesce researchers from a variety of labs around a single goal. They nicknamed the idea "Wigits," for Workshop for IGI Team Science, and they joked that they would all join hands and build wigits together.

When they floated the idea at one of the institute's Friday happy hours, it met with enthusiasm from some of the students but not from most of the professors. "In industry everyone focuses on achieving agreed-upon common goals," says Gavin Knott, one of the students eager to see this happen. "But in academia, everyone functions in their own bubble. We all work on our own research interests and we collaborate only when it's necessary." So with no source of funding and little faculty enthusiasm, the idea remained in limbo.[3]

Then coronavirus came along. Savage's students had been texting him to ask what Berkeley was doing to address the crisis, and he realized it could be the focus of the type of team approach they had discussed. When he wandered into Doudna's office with the idea, he found that she had been thinking along the same lines.

They agreed that she should call a meeting of their IGI colleagues and other Bay Area associates who might be interested in joining a coronavirus effort. That meeting, which is the one described in the introduction of this book, was at 2 p.m. on Friday, March 13—the day after Doudna and her husband made their predawn drive to Fresno to retrieve their son from his robotics competition.

SARS-CoV-2

The rapidly spreading new coronavirus had by then been given an official name: severe acute respiratory syndrome coronavirus 2, or SARS-CoV-2. It was so named because it was similar in its symptoms to the SARS coronavirus that spread out of China in 2003, infecting more than eight thousand people worldwide. The disease caused by the new virus was named COVID-19.

Viruses are deceptively simple little capsules of bad news.* They are just a tiny bit of genetic material, either DNA or RNA, inside a protein shell. When they worm their way into a cell of an organism, they can hijack its machinery in order to replicate themselves. In the case of coronaviruses, the genetic material is RNA, Doudna's specialty. In SARS-CoV-2, the RNA is about 29,900 base letters long, compared to more than three billion in human DNA. The viral sequence provides the code for making a mere twenty-nine proteins.[4]

Here is a sample snippet of the letters in the coronavirus's RNA: CCUCGGCGGGCACGUAGUGUAGCUAGUCAAUCCAU-

*Yes, the world is filled with some very useful and necessary viruses, but they are for a different book.

CAUUGCCUACACUAUGUCACUUGGUGCAGAAAAUUC. That sequence is part of a string that codes for making a protein that sits on the outside of the virus shell. The protein looks like a spike, which gives the virus, when viewed through an electron microscope, the appearance of a crown, hence *corona*. This spike is like a key that can fit into specific receptors on the surface of human cells. Notably, the first twelve letters of the sequence above allow the spike to bind very tightly to one specific receptor on human cells. This evolution of this short sequence explains how the virus could have jumped from bats to other animals to us.

For the SARS-CoV-2 coronavirus, the human receptor is a protein known as ACE2. It plays a role that is similar to the one played for HIV by the CCR5 protein, which the rogue Chinese doctor He Jiankui edited out of his CRISPR twins. Because the ACE2 protein has functions other than just being a receptor, it's probably not a good idea to try to edit it out of our species.

The new coronavirus jumped into humans sometime in late 2019. The first officially certified death was reported on January 9, 2020. Also on that day, Chinese researchers publicly posted the full genetic sequence of the virus. Using cryo-electron microscopy, which fires electrons at proteins that have been frozen in a liquid, structural biologists were able to create a precise model, atom by atom and twist by twist, of the coronavirus and its spikes. With the sequencing information and structural data in hand, molecular biologists began racing to find treatments and vaccines that would block the ability of the virus to latch on to human cells.[5]

The order of battle

The March 13 meeting that Doudna summoned drew far more participants than she and Savage expected. A dozen key lab leaders and students gathered that Friday afternoon in the ground-floor conference room of the IGI building just as the rest of the campus was being locked down. Another fifty researchers from the Bay Area joined by Zoom. "Without planning it or imagining how it would come about," Doudna says, "our idea from the noodle house became reality."[6]

As Doudna discovered, there is an advantage to being part of large organizations such as UC Berkeley and the IGI. Innovation often happens in garages and dorm rooms, but it is sustained by institutions. An infrastructure is needed to handle the logistics required for complex projects. This is especially true during a pandemic. "Having the IGI in place was incredibly useful," Doudna says, "because there were teams of people who could help with things like writing proposals, setting up Slack channels, sending out group emails, arranging Zoom meetings, and coordinating equipment."

Berkeley's legal team came up with a policy for sharing discoveries freely with other coronavirus researchers while protecting the underlying intellectual property. At one of the first meetings, a university lawyer laid out a template for royalty-free licensing. "We will allow non-exclusive no-fee licensing of any of the work that's coming out of this effort," she said. "We still want to file for patent protection for anything discovered, but then we will make it available for this purpose." Doudna had a slide presentation on this for the group's second Zoom meeting, held on March 18. She summarized its message succinctly: "It's not about making money here."

By the time of this second meeting, Doudna also had a slide listing ten projects they had decided to pursue, with the names of the team leaders. Some of the planned tasks made use of the latest CRISPR technology, including developing a CRISPR-based diagnostic test and finding ways to deliver safely into the lungs a CRISPR-based system that could target and destroy the genetic material of the virus.

When the ideas first started rolling in, one of the wise hands in the room, a professor named Robert Tjian, had interjected a note of clarity. "Let's split this in two parts," he said. There are new things we could try to invent, "but first there's the fire-on-my-ass problem." There was a pause for a moment, then he explained. They had to deal with the urgent need for public testing before they could sit at their lab benches and come up with biotechnologies for the future. So the first team Doudna launched was given the mission of converting a space on the ground floor of the building, near where they were seated, into a state-of-the-art, high-speed, automated coronavirus testing lab.

Fyodor Urnov getting the first test samples from Dori Tieu of the Berkeley Fire Department as Dirk Hockemeyer watches

Chapter 49

Testing

America's failure

The first official guidance to local health officials in the U.S. about testing for the new coronavirus came in a conference call on January 15, 2020, led by Stephen Lindstrom, a microbiologist at the Centers for Disease Control (CDC). The CDC had developed a test for the new coronavirus, he said, but it could not make it available to state health departments until the Food and Drug Administration (FDA) approved it. That should be soon, Lindstrom promised, but until then, doctors would have to send samples to the CDC in Atlanta for testing.

The next day, a Seattle doctor sent the CDC a nose-swab sample from a thirty-five-year-old man who had returned from a visit to Wuhan and come down with flu-like symptoms. He became the first person in the U.S. to test positive.[1]

On January 31, Health and Human Services Secretary Alex Azar, whose department oversees the FDA, declared a public health emergency. The declaration gave the FDA the right to speed up approvals for coronavirus tests. But it had a weird unintended consequence. In normal circumstances, hospitals and university labs can devise their own tests to use at their facilities, as long as they do not market them. But a declaration of a public health emergency imposes the requirement that such tests not be used until they get an "emergency use authorization." The intent is to avoid the use of unproven tests during a health crisis. As a result, Azar's declaration triggered new restrictions on academic labs and hospitals. That would have been fine if the CDC's test was widely available. But the FDA had still not approved it.

That approval finally came on February 4, and the next day the CDC began sending test kits to state and local labs. The way the test works, or was supposed to work, is that a long swab is inserted into the back of a patient's nasal passage. The lab uses some of the chemical mixtures in the kit to extract any RNA that is in the mucus. The RNA is then "reverse-transcribed" to turn it into DNA. The DNA strands are amplified into millions of copies using a well-known process called a polymerase chain reaction (PCR), which most college biology students learn how to do.

The PCR process was invented in 1983 by Kary Mullis, a chemist at a biotech company. Driving in his car one night, Mullis crafted a way to tag a sequence of DNA and use enzymes to duplicate it through repeated cycles of heating and cooling known as thermocycling. "Beginning with a single molecule of the DNA, the PCR can generate 100 billion similar molecules in an afternoon," he wrote.[2] These days the process is usually done using a machine the size of a microwave that raises and lowers the temperature of the mixture. If the genetic material of the coronavirus is present in the mucus, the PCR process amplifies it so that it can be detected.

When state health officials received the test kits from the CDC, they set about verifying that they worked by trying them on patient samples that were already known to be positive or negative. "Early on Feb. 8, one of the first CDC test kits arrived in a Federal Express package at a public health laboratory on the east side of Manhattan," the *Washington Post* reported. "For hours, lab technicians struggled to verify that the test worked." When they ran the tests on samples known to contain the virus, they got a positive result. That was good. Unfortunately, when they ran the test on purified water, they also got a positive result. One of the chemical compounds in the CDC test kits was defective. It had been contaminated during the manufacturing process. "Oh, shit," said Jennifer Rakeman, an assistant commissioner of the city's health department. "What are we going to do now?"[3]

Adding to the disgrace was the fact that the World Health Organization had delivered 250,000 diagnostic tests that worked just fine to countries around the world. The U.S. could have gotten some of those tests or replicated them, but it had refused.

A university steps in

The University of Washington, at the epicenter of one of the first COVID outbreaks in the U.S., was the first to rush into this minefield. At the beginning of January, after seeing the reports from China, Alex Greninger, a round-faced young assistant director of the virology lab at the university's medical center, talked to his boss, Keith Jerome, about developing their own test. "We're probably going to be wasting some money on this," Jerome said. "It's probably not going to come over here. But you've got to be ready."[4]

Within two weeks, Greninger had a working test, which, under normal regulations, they could use in their own hospital system. But then HHS Secretary Azar issued his emergency declaration, which made regulations more strict. So Greninger submitted a formal application to the FDA for an "Emergency Use Authorization." It took him close to one hundred hours to fill out all of the forms. Then came an astonishing bureaucratic snafu. He got a response from the FDA on February 20 informing him that, in addition to sending his application electronically, he had to mail in a printed copy along with a copy burned onto a compact disc (remember what those were?) to FDA headquarters in Maryland. In an email he wrote to a friend that day describing the FDA's bizarre approach, Greninger vented, "Repeat after me, emergency."

A few days later, the FDA responded by requiring him to do more trials to see if the test he was using inadvertently detected the MERS and SARS viruses, even though they had been dormant for years and he had no samples of those viruses to test. When he called the CDC to see if he could get a sample of the old SARS virus, it refused. "That's when I thought, 'Huh, maybe the FDA and the CDC haven't talked about this at all,'" Greninger told reporter Julia Ioffe. "I realized, Oh, wow, this is going to take a while."[5]

Others had similar problems. The Mayo Clinic had created a crisis team to deal with the pandemic. Of its fifteen members, five were tasked to deal full time with the FDA's paperwork requirements. By late February, there were dozens of hospitals and academic labs, including at Stanford and the Broad Institute of MIT and Harvard, that had developed testing capabilities, but none had managed to win FDA authorization.

At that point Anthony Fauci, the National Institutes of Health infectious disease chief who had become a national superstar, stepped in. On February 27, he spoke to HHS Secretary Azar's chief of staff, Brian Harrison, and urged that the FDA allow universities, hospitals, and private testing services to start using their own tests while waiting for Emergency Use Authorizations. Harrison held a conference call with the relevant agencies and told them, using strong language, that before the end of the meeting they had to come up with such a plan.[6]

The FDA finally relented on Saturday, February 29, and announced that it would allow non-government labs to use their own tests as they waited to get Emergency Use Authorizations. That Monday, Greninger's lab tested thirty patients. Within a few weeks, it would be testing more than 2,500 a day.

Eric Lander's Broad Institute also jumped into the fray. Deborah Hung, the codirector of the Broad's infectious diseases program, also worked as a physician at Brigham and Women's Hospital in Boston. On the evening of March 9, when confirmed cases of COVID in the state had risen to forty-one, it struck her how bad the virus was going to be. She called her colleague Stacey Gabriel, the director of the Broad Institute's genomics sequencing facility, which is a few blocks from the Broad headquarters in a former warehouse that stored beer and popcorn for Fenway Park. Could she turn the lab into a facility for testing for the coronavirus? Gabriel said yes, then called Lander to see if that was okay. Lander was, as always, eager to deploy science in the public interest and rightly proud of the teammates he had assembled who shared that instinct. "The call was kind of irrelevant," Lander says. "I of course said yes, but she was going to do it anyway, as well she should." The lab went into full operation on March 24, receiving samples from hospitals across the Boston area.[7] With the failure of the Trump administration to carry out widespread testing, university research labs began taking on a role that has normally been performed by the government.

Enrique Lin Shiao and Jennifer Hamilton

CHAPTER 50

The Berkeley Lab

The volunteer army

When Doudna and her colleagues at Berkeley's Innovative Genomics Institute decided at their March 13 meeting to focus on building their own coronavirus testing lab, there was a discussion about what technology to use. Should it be the cumbersome but reliable process of amplifying the genetic material from test swabs using a polymerase chain reaction (PCR), as described earlier? Or should they try to invent a new type of test, one that used CRISPR technology to directly detect the RNA of the virus?

They decided to do both, but they would initially scramble to do the first approach. "We need to walk before we run," Doudna said at the conclusion of the discussion. "Let's use current technology right away, then we can innovate."[1] By having its own testing lab, the IGI would have the data and patient samples to try out new approaches.

After the meeting, the institute sent out a tweet:

> Innovative Genomics Institute @igisci: We are working as hard as possible to establish clinical #COVID19 testing capability at @UCBerkeley campus. We will update this page often to ask for reagents, equipment, and volunteers.

Within two days, more than 860 people had responded and the volunteer list had to be cut off.

The team that Doudna put together reflected the diversity of her lab and of the biotech field in general. To command the operation, she turned to Fyodor Urnov, a gene-editing wizard who had been leading IGI's efforts to develop affordable methods to cure sickle-cell anemia.

Born in 1968 in the heart of Moscow, Urnov learned English from his mother, Julia Palievsky, who was a professor, and his father, Dmitry Urnov, a distinguished literary critic and Shakespeare scholar, William Faulkner fan, and biographer of Daniel Defoe. I asked Fyodor whether the coronavirus had led him to ask his father, who now lives near him in Berkeley, about Defoe's 1722 book, *A Journal of the Plague Year*. "Yes," he said, "I'm going to get him to give me and our daughter who lives in Paris a Zoom lecture on the book."[2]

Like Doudna, Urnov read Watson's *The Double Helix* when he was about thirteen and decided to become a biologist. "Jennifer and I joke about the fact that we both read *The Double Helix* at about the same age," he says. "For all of Watson's shortcomings as a human being, which are substantial, he produced a ripping good yarn that makes the hunt for the mechanisms of life seem very exciting."

At eighteen, Urnov, a bit of a rebel, was drafted into the Soviet military and his head shaved. "I survived unscathed," he says, after which he left for the United States. "In August of 1990, I found myself landing in Boston's Logan Airport, having been accepted to Brown, and a year later my mom got a Fulbright to be a visiting scholar at the University of Virginia." Soon he was happily pursuing his doctorate at Brown, buried in test tubes. "I realized that I was not going back to Russia."

Urnov is among those researchers comfortable with having one foot in academia and the other in industry. For sixteen years, while teaching at Berkeley, he was a team leader at Sangamo Therapeutics, which translates scientific discoveries into medical treatments. His Russian roots and literary parentage instilled in him a dramatic flair, which he earnestly combines with a passion for America's can-do spirit. When he got the assignment from Doudna to lead the lab, he sent around a quote from Tolkien's *Lord of the Rings*:

> "I wish it need not have happened in my time," said Frodo.
>
> "So do I," said Gandalf, "and so do all who live to see such times. But that is not for them to decide. All we have to decide is what to do with the time that is given us."

One of his two scientific field marshals was Jennifer Hamilton, the Doudna protégée who a year earlier had spent a day teaching me to edit a human gene using CRISPR. She grew up in Seattle, studied biochemistry and genetics at the University of Washington, and then worked as a lab technician while listening to the podcast *This Week in Virology*. She did her doctorate at Mount Sinai Medical Center in New York, where she turned viruses and virus-like particles into mechanisms for delivering medical treatments, and then joined Doudna's lab as a postdoc. At the 2019 Cold Spring Harbor conference, Doudna watched proudly when Hamilton presented her research on using virus-like particles to deliver CRISPR-Cas9 gene-editing tools into humans.

When the coronavirus crisis hit in early March, Hamilton told Doudna that she wanted to get involved like people at her University of Washington alma mater were. So Doudna tapped her to lead the technical development of the lab. "It felt like a call to arms," Hamilton says. "I simply had to say yes." She never dreamed that her dexterity at optimizing RNA extraction would turn out to be an urgent skill in a global crisis. The real-world deployment also gave her and her fellow academics a taste of the type of project-oriented teamwork that is common in the business world. "It's the first time that I've been a part of a scientific team where so many people with different talents have coalesced around a common goal."[3]

Working with Hamilton to get the testing lab running was Enrique Lin Shiao, born and raised in Costa Rica, the son of Taiwanese immigrants who left everything behind to start over in a very new place. The cloning of Dolly the sheep in 1996 sparked his interest in genetics. After high school, he got a scholarship to the Technical University of Munich, where he researched how to fold DNA into different shapes to build nanotech biology tools. From there he went to Cambridge University to study how DNA folding is important for cell function. For his doctorate, he went to the University of Pennsylvania, where he figured out how non-coding regions of our genome, previously described as "junk DNA," could play a role in disease progression. In other words, like Feng Zhang, Enrique Lin Shiao was a typical American success story from when the nation was a magnet for diverse global talent.

As a postdoc researcher in Doudna's lab, Lin Shiao worked on ways to make new gene-editing tools that could cut and paste long DNA sequences. While sheltering at home in March 2020, he was scrolling through his Twitter feed and saw the tweet from his IGI colleagues seeking volunteers for the planned testing lab. "They were asking for experience in RNA extraction and PCR, which are techniques I routinely perform in the lab," he says. "The next day I got an email from Jennifer asking if I would be interested in co-leading the technical efforts, and I immediately agreed."[4]

The lab

The IGI was fortunate that there was a 2,500-square-foot space on the building's ground floor that was being converted into a gene-editing lab. Doudna's team began moving in new machines and boxes filled with chemicals to turn the space into a coronavirus testing facility. A lab-building project that normally would take months was done in days.[5]

They begged and borrowed and commandeered supplies from labs across campus. One day, when they were ready to start an experiment, they realized that they did not have the right plates to run in one of the PCR machines. Lin Shiao and others went through all the labs in the IGI building and then in two nearby buildings until they found some. "Since campus was largely closed, it felt like a giant scavenger hunt," he says. "Every day felt a bit like a roller coaster, where we discovered a new problem early in the morning, got worried, and then figured it out by the end of the day."

The lab spent about $550,000 on equipment and supplies.[6] One key machine was a contraption to automate the task of extracting the RNA in patient samples. The Hamilton STARlet uses robotic pipettes to suck small amounts from each patient sample and put them onto plates the size of an iPhone with ninety-six little wells. The trays are moved into the chamber of the machine, where each of the samples is doused with reagents to extract the RNA. Using a barcode, the machines keep track of the patient information from each sample, making sure to follow privacy guidelines. It was a new experience for academic researchers. "Usually for bench scientists like ourselves we feel our impact is a bit indirect and it comes in the long term," Lin Shiao says. "This feels so direct and immediate."[7]

Hamilton's grandfather had been an engineer on the NASA Apollo rocket launches, and one day her team paused to watch a clip someone had posted on their Slack channel from the movie *Apollo 13* where the engineers have to figure out how to make a "square peg fit in a round hole" in order to save the astronauts. "Every day we've been facing challenges, but we're solving these problems as they come up because we know that time is short," Hamilton says. "This experience has made me wonder if this is what it was like for my grandfather working at NASA in the 1960s." It was a fitting analogy. COVID and CRISPR were helping to make human cells the next frontier.

Doudna had to figure out what legal liability the university might incur by testing outsiders. That was a process that would normally have taken the lawyers weeks of hand-wringing, so Doudna called the president of the University of California system, Janet Napolitano, a former Homeland Security secretary. In twelve hours, Napolitano had given her approval and brought the system's legal bureaucracy in line. Urnov notes that it was useful to roll out Doudna as a big gun on such occasions. "I jokingly call her the USS *Jennifer Doudna*," he says.

With federal testing still in disarray and commercial labs taking more than a week to return results, there was huge demand for Berkeley's testing. The town's health officer, Lisa Hernandez, asked Urnov for five thousand tests, some of which would be done on the area's poor and homeless. The fire chief, David Brannigan, told Urnov that thirty of his firefighters were quarantined because they couldn't get test results. Doudna and Urnov promised to accommodate them all.

"Thank you, IGI"

The first major challenge for the new lab was making sure that their COVID tests were accurate. Doudna brought a special eye to this task, since she had been an expert at deciphering readouts involving RNA ever since she was a graduate student. As the results came in, researchers would share them on a Zoom screen and then watch online as Doudna leaned forward and looked intensely at the images of inverted blue triangles, green triangles, and squares indicating data points. Sometimes she would just sit and stare, not moving, as others held their breath. "Yes, that looks good," she said during one session as she pointed a cursor to a part of an RNA detection test. Then her expression changed for all on Zoom to see as she pointed to another place and muttered, "Nope, nope, nope."

Finally, early in April, she looked at the latest data that Lin Shiao had gathered and pronounced it "awesome." The tests were ready to go live.

On Monday, April 6, at 8 a.m., a fire department van pulled up to the door of the IGI and an officer named Dori Tieu delivered a box filled with samples. Urnov, wearing white gloves and a blue mask, accepted the Styrofoam cooler as his colleague Dirk Hockemeyer watched. They promised that they would have results the next morning.

As they were making the final preparations to get the lab into operation, Urnov went to get a takeout meal for his parents, who live nearby. When he arrived back at the IGI building, he saw a sheet of paper taped to the big glass door. On it was written, "Thank you, IGI! Sincerely, the people of Berkeley and the World."

Fyodor Urnov's reflection as he photographs the note

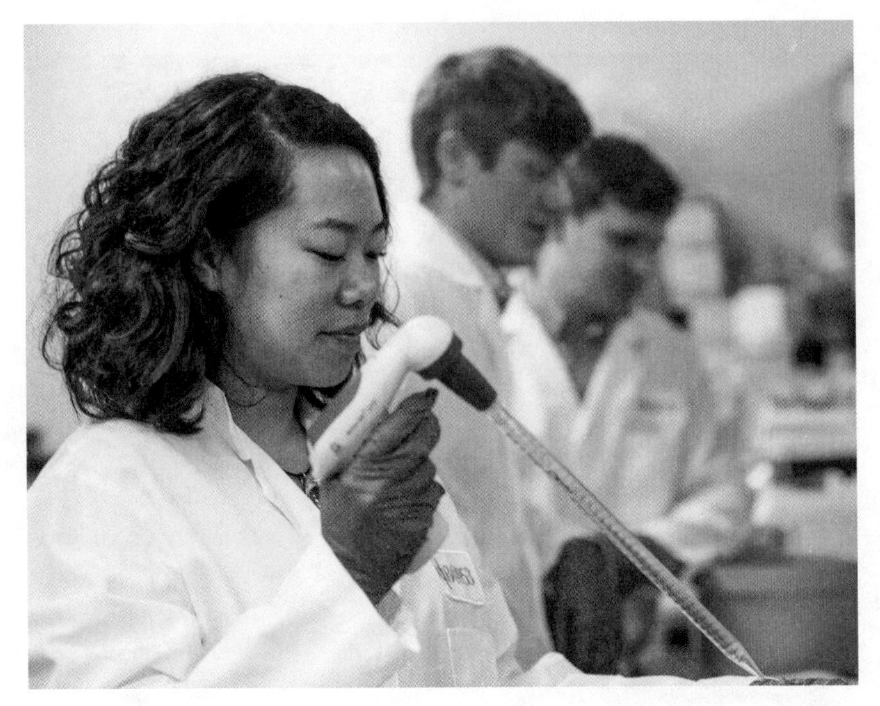

Janice Chen and Lucas Harrington

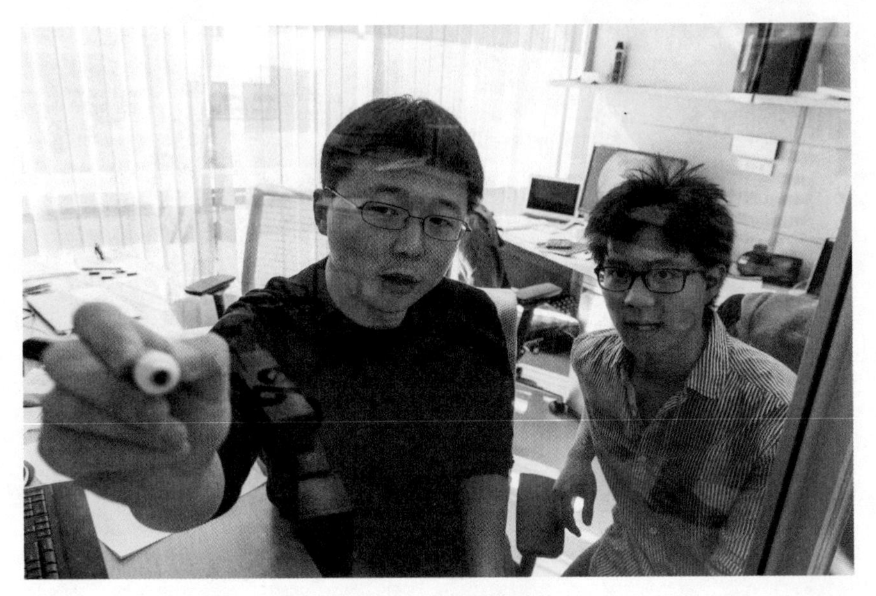

Feng Zhang with Patrick Hsu

CHAPTER 51

Mammoth and Sherlock

CRISPR as a detection tool

At the March 13 meeting that Doudna convened to address the coronavirus, she decided that a top priority was to create a high-speed conventional PCR testing lab. But during the discussion, Fyodor Urnov suggested that they also consider a more innovative idea: using CRISPR to detect the RNA of the coronavirus, similar to how bacteria use CRISPR to detect attacking viruses.

"There's a paper that just came out on that," a participant interjected.

Urnov showed a slight flash of impatience and interrupted, for he knew the paper well. "Yes, from Janice Chen, formerly of the Doudna Lab."

There were actually two similar papers that had just come out. One was from former members of the Doudna Lab who had formed a company to use CRISPR as a detection tool. The other, not surprisingly, sprang from Feng Zhang of the Broad Institute. Once again, the two realms were competing. This time, however, it was not a race to patent methods for editing human genes. In this new race, the goal was to help save humanity from the novel coronavirus, and their discoveries were being shared for free.

Cas12 and Mammoth

Back in 2017, Janice Chen and Lucas Harrington were doctoral students working in Doudna's lab exploring newly discovered CRISPR-associated enzymes. Specifically, they were analyzing one that became known as Cas12a, which had a special property. It could be targeted, like Cas9, to find and cut a specified sequence of DNA. But it didn't stop there. Once it cleaved the double-stranded DNA target, it went into an indiscriminate cutting frenzy, chopping up any single-stranded DNA that was nearby. "We started to see this very weird behavior," Harrington says.[1]

Over breakfast one day, Doudna's husband, Jamie Cate, suggested that this property could be harnessed to create a diagnostic tool. Chen and Harrington had the same idea. They combined a CRISPR-Cas12 system with a "reporter" molecule, which was a fluorescent signal connected to a bit of DNA. When the CRISPR-Cas12 system found a targeted sequence of DNA, it would also chop up the reporter molecules and cause a glowing signal. The result was a diagnostic tool that could detect whether the patient had a particular virus or bacteria or cancer. Chen and Harrington dubbed it the "DNA endonuclease targeted CRISPR trans reporter," a very clunky phrase that was crafted in order to create the CRISPR-like acronym DETECTR.

When Chen, Harrington, and Doudna submitted their findings in an article to *Science* in November 2017, the editors requested that they write more about how to turn the discovery into a diagnostic test. Even the traditional scientific journals were now showing greater interest in connecting basic science to potential applications. "If a journal tells you to do something like that," Harrington says, "you start working on it very hard." So over Christmas break of 2017, he and Chen collaborated with a researcher at UC San Francisco to show how their CRISPR-Cas12 tool could detect human papillomavirus (HPV), a sexually transmitted infection. "We were going back and forth with a giant piece of lab equipment in an Uber, testing different patient samples," he says.

Doudna prodded *Science* to expedite publication as part of its fast-track program. They resubmitted their article in January 2018 with the data the editors had requested showing that DETECTR detected HPV infections; it was accepted, and a version went online in February.

Ever since Watson and Crick ended their famous DNA paper by saying, "It has not escaped our notice that the specific pairing we have postulated immediately suggests a possible copying mechanism for the genetic material," it has become standard to end journal papers with an understated but important forward-looking sentence. Chen, Harrington, and Doudna ended their paper by saying that the CRISPR-Cas12 system "offers a new strategy to improve the speed, sensitivity and specificity of nucleic acid detection for point-of-care diagnostic applications." In other words, it might be used to create a simple test to detect virus infections quickly, at home or in a hospital.[2]

Even though Harrington and Chen had not yet gotten their doctorates, Doudna encouraged them to form a company. She was now a strong believer that basic research should be combined with translational research, moving discoveries from bench to bedside. "A lot of other technologies that we had discovered were bought as a defensive strategy by big companies that then didn't develop them," Harrington says. "So that motivated us to start our own company." Mammoth Biosciences launched officially in April 2018 with Doudna as chair of its scientific advisory board.

Cas13 and SHERLOCK

As was often the case, Doudna and her team were in a competition with her cross-country rival, the Broad Institute's Feng Zhang. Working with the CRISPR pioneer Eugene Koonin of the NIH, Zhang had used computational biology to sort through the genomes of thousands of microbes, and in October 2015 they reported on their discovery of many new CRISPR-associated enzymes. In addition to the previously known Cas9 and Cas12 enzymes that target DNA, Zhang and Koonin found a class of enzymes that target RNA.[3] They became known as Cas13.

Cas13 had the same odd trait as Cas12: when it found its target, it went into a cutting frenzy. The Cas13 not only cut its targeted RNA, it then proceeded to cut up any other nearby RNA.

At first Zhang assumed this was a mistake. "We thought that Cas13 would cleave the RNA just the way that Cas9 cleaved DNA," he says. "But whenever we did a reaction with Cas13, the RNA got shredded in many different places." He asked his lab team whether they were sure they had been purifying the enzyme correctly; maybe it was contaminated. They painstakingly eliminated all possible sources of contamination, but the indiscriminate cleavage kept happening. Zhang speculated that it was an evolutionary method to have the cell commit suicide if it got too infected by an invading virus, thus preventing the virus from spreading as fast.[4]

Doudna's lab then contributed to the study of precisely how Cas13 works. In a paper in October 2016, Doudna and her coauthors—including her husband, Jamie Cate, and Alexandra East-Seletsky, the graduate student who had done some of the key 2012 experiments on CRISPR in human cells—explained the different functions that Cas13 performs, including being able to indiscriminately chop up thousands of other nearby RNAs once it reaches its target. This promiscuous chopping makes it possible to use Cas13 with fluorescent reporters (as was done with Cas12) to be a detection tool for a specified RNA sequence, such as that of a coronavirus.[5]

Zhang and his colleagues at the Broad were able to create such a detection tool in April 2017, which they named "specific high sensitivity enzymatic reporter unlocking," which was reverse-engineered (though not very well) to produce the acronym SHERLOCK. The game was afoot! They showed that SHERLOCK could detect specific strains of Zika and Dengue viruses.[6] Over the next year, they made a version that combined Cas13 and Cas12 to detect multiple targets in one reaction. Then they were able to simplify the system and make it possible for the detection to be reported on paper lateral flow strips, similar to pregnancy tests.[7]

Zhang decided to start a diagnostics company to commercialize SHERLOCK, just like Chen and Harrington had launched Mammoth. Zhang's cofounders included the two graduate students who were the lead authors on many of the papers from his lab describing CRISPR-Cas13: Omar Abudayyeh and Jonathan Gootenberg. Gootenberg recalls that they almost decided not to publish a paper when they first discovered the tendency of Cas13 to go into a frenzy of indiscriminate RNA cutting. It seemed like a useless quirk of nature. But once Zhang figured out how to harness that quirk to create a virus-detection technology, Gootenberg realized how discoveries in basic science can turn out to have unexpected real-world applications. "You know, nature's got a ton of amazing secrets in it," he says.[8]

It took a while to get Sherlock Biosciences funded and launched because Zhang and his two graduate students did not want profit to be the main goal of the company. They wanted the technologies to be affordable in the developing world. So the company was structured in a way that allowed it to profit on its innovations while still taking a nonprofit approach in places where there was great need.

Unlike the Doudna-Zhang competition for the patents, the one involving diagnostic companies was not very contentious. Both sides knew that the technologies had enormous potential to do good. Whenever there was a new epidemic, Mammoth and Sherlock could quickly reprogram their diagnostic tools to target the novel virus and produce testing kits. The Broad team, for example, sent a team with SHERLOCK to Nigeria in 2019 to help detect victims of an outbreak of Lassa fever, a virus in the same family as Ebola.[9]

At the time, using CRISPR as a diagnostic tool seemed to be a worthy endeavor, though not a particularly exciting one. It did not get as much buzz as using CRISPR to treat diseases or edit human genes. But then, at the beginning of 2020, the world suddenly changed. The ability to quickly detect an attacking virus became critical. And the best way to do it faster and cheaper than the conventional PCR tests, which required a lot of mixing steps and temperature cycles, was to deploy RNA-guided enzymes that had been programmed to detect the genetic material of the virus—in other words, adapt the CRISPR system that bacteria had been deploying for millions of years.

Feng Zhang (*top left*) with Omar Abudayyeh (*top right*) and Jonathan Gootenberg (*middle right*) at a Zoom meeting on COVID detection

CHAPTER 52

Coronavirus Tests

Feng Zhang

In early January 2020, Feng Zhang started getting emails about coronavirus written in Chinese. Some were from Chinese academics he had met, but he also got an unexpected one from the science officer at China's consulate in New York City. "Even though you are American and not living in China," it said, "this is really a problem that's important for humanity." It quoted an old Chinese saying: *When one place is in trouble, assistance comes from all quarters*. "So we hope that you can think about it and see what you can do," the email urged.[1]

Zhang knew little about the novel coronavirus other than what he had read in a *New York Times* article describing the situation in Wuhan, but the emails "gave me a sense of urgency about the situation," he says. This was especially true of the exchange he had with the Chinese consulate. "I usually don't have any interactions from them," says Zhang, who had immigrated to Iowa with his parents when he was eleven.

I asked him whether Chinese authorities think of him as a Chinese scientist. "Yeah, probably," he says after a pause. "I think they probably think of all Chinese people as Chinese. But that's irrelevant because the world is so connected now, especially in a pandemic."

Zhang decided to reconfigure the SHERLOCK detection tool so that it could test for the new coronavirus. Unfortunately, he didn't have anyone in his lab to handle the necessary experiments. So he resolved to go to his bench and do the experiments himself. He also enlisted his two former graduate students, Omar Abudayyeh and Jonathan Gootenberg. They had moved on to open their own lab at MIT's McGovern Institute, a block from the Broad, and they agreed to collaborate with him again.

Zhang did not initially have access to samples of the coronavirus from human patients, so he made a synthetic version of it. Using the SHERLOCK process, he and his team devised a detection test that took only three steps and could be done in an hour without fancy equipment. All it required was a small device to keep the temperature constant while the genetic material from the samples was amplified through a chemical process that was simpler than PCR. The results could be read using a paper dipstick.

On February 14, well before most of the U.S. had focused on the novel coronavirus, Zhang's lab posted a white paper describing the test and inviting any lab to use or adapt the process freely. "Today we are sharing a research protocol for SHERLOCK-based COVID-19 #coronavirus detection, and hope it will help others who are working to combat the outbreak," Zhang tweeted. "We will continue to update this as we make further progress."[2]

The company he had founded, Sherlock Biosciences, quickly began work on turning the process into a commercial testing device that could be used in hospitals and doctors' offices. When the CEO, Rahul Dhanda, told his team that he wanted the company to focus on COVID, the researchers literally swung their chairs back to their workbenches to take on the mission. "When we say a pivot, there was a literal pivot of chairs at the same time there was a pivot of the company towards a new goal," he says. By the end of 2020, the company was working with manufacturing partners to turn out small machines that could be used to get results in less than an hour.[3]

Chen and Harrington

Around the time that Zhang began working on his coronavirus test, Janice Chen got a call from a researcher on the scientific advisory board of the company she had founded with Doudna and Lucas Harrington, Mammoth Biosciences. "What do you think about developing a CRISPR-based diagnostic to detect the SARS-CoV-2 virus?" he asked. She agreed that they should try. As a result, she and Harrington became part of yet another cross-country competition between Doudna's circle and Zhang's.[4]

Within two weeks, the Mammoth team was able to reconfigure its CRISPR-based DETECTR tool so that it would detect SARS-CoV-2. One benefit of collaborating with UC San Francisco, which has its own hospital, was that they could test on real human samples, drawn from thirty-six COVID patients, unlike the Broad, which initially had to use synthetic viruses.

The Mammoth test relied on the CRISPR-associated enzyme that Chen and Harrington had studied in Doudna's lab, Cas12, which targets DNA. That would seem to make it less suited than SHERLOCK's Cas13, which targets RNA, the genetic material of the coronavirus. However, both detection techniques need to convert the RNA of the coronavirus into DNA in order for it to be amplified. In the SHERLOCK test, it has to be transcribed back into RNA to be detected, thus adding a small step to the process.

Chen and Harrington rushed to get a white paper online with the details of their Mammoth test. In many ways it was similar to the SHERLOCK process. All that was necessary was a heating block, the reagents, and paper flow strips to give a readout of the results. Like Zhang, the Mammoth team decided to put what they had devised into the public domain, to be shared freely.

On February 14, while they were preparing to put their white paper online, Chen and Harrington saw a message pop up on the Slack channel they were using. Someone posted the tweet that Zhang had just sent out announcing that he had just published his white paper on how to use the SHERLOCK protocol for detecting the coronavirus.

"We were like, 'Oh, shoot'" Chen recalls of that Friday afternoon. But after a few minutes, they realized that having both papers appear was a good thing. They appended a postscript to the paper they were just about to post. "While we were preparing this whitepaper, another protocol for SARS-CoV-2 detection using CRISPR diagnostics (SHERLOCK, v.20200214) was published," it said. They then included a useful chart comparing the workflows of the two techniques.[5]

Zhang was gracious, though it was easy for him to be since he had beaten the Mammoth team by a day. "Check out the resource provided by Mammoth," he tweeted, including a link to its white paper. "Glad that scientists are working together and sharing openly. #coronavirus."

That tweet reflected a welcome new trend in the CRISPR world. The passionate competition for patents and prizes had led to secrecy about research and the formation of competing CRISPR companies. But the urgency that Doudna and Zhang and their colleagues felt about defeating the coronavirus pushed them to be more open and willing to share their work. Competition was still an important, and useful, part of the equation. There continued to be a race between Doudna's world and Zhang's to publish papers and make advances on the new COVID tests. "I'm not going to sugarcoat it," Doudna says. "There's definitely competition going on. It makes people feel an urgency to move ahead or, if they don't, other people are going to get to something first." But coronavirus made the rivalry less cutthroat, because patents were not a paramount concern. "The awesomely good thing about this terrible situation is that all the intellectual property questions have been put aside, and everyone's really intent on just finding solutions," says Chen. "People are focused on getting something out there that works, rather than on the business aspect of it."

At-home tests

The CRISPR-based tests developed by Mammoth and Sherlock are cheaper and faster than conventional PCR tests. They also have an advantage over antigen tests, such as the one developed by Abbott Labs that was approved in August of the plague year. The CRISPR-based tests can detect the presence of the RNA of a virus as soon as a person has been infected. But the antigen tests, which detect the presence of proteins that exist on the surface of the virus, are most accurate only after a patient has become highly infectious to others.

The ultimate goal for all of these methods was to create a CRISPR-based coronavirus test that would be like a home pregnancy test: cheap, disposable, fast, and simple, which you could buy at the corner drugstore and use in the privacy of your bathroom.

Harrington and Chen of the Mammoth team unveiled their concept for such a device in May 2020 and announced a partnership with the London-based multinational pharmaceutical company GlaxoSmithKline (maker of Excedrin and Tums) to manufacture it. It would provide accurate results in twenty minutes and require no special equipment.

Likewise, Zhang's lab that same month developed a way to simplify the SHERLOCK detection system, which originally required two steps, into a process that required just a single-step reaction. The only equipment necessary was a pot to keep the system heated at a steady 140 degrees Fahrenheit. Zhang named it STOP, for SHERLOCK Testing in One Pot.[6] "Let me show you what it will look like," Zhang says to me with his boyish enthusiasm as he shares slides and renderings on a Zoom call. "You just put a nasal or saliva sample into this cartridge, slide it into the device, break one blister to release a solution that will extract the virus RNA, and then break another blister that will release some freeze-dried CRISPR for a reaction in the amplification chamber."

Zhang named the device STOP-COVID. But the platform can be easily adapted to detect any virus. "That's why we chose the STOP name, which can be paired with any target," he says. "We could create a STOP-flu or a STOP-HIV or have many detection targets on the same platform. The device is agnostic about what virus it's looking for."[7]

Mammoth has the same vision of making it easy to reprogram its own tool to detect any new virus that comes along. "The beauty of CRISPR is that once you have the platform, then it's just a matter of reconfiguring your chemistry to detect a different virus," Chen explains. "It can be used for the next pandemic or any virus. It can also be used against any bacteria or anything that has a genetic sequence, even cancer."[8]

Biology hits home

The development of home testing kits has a potential impact beyond the fight against COVID: bringing biology into the home, the way that personal computers in the 1970s brought digital products and services—and an awareness of microchips and software code—into people's daily lives and consciousness.

Personal computers and then smartphones became platforms on which waves of innovators could build neat products. In addition, they helped make the digital revolution into something *personal*, which caused people to develop some understanding of the technology.

When Zhang was growing up, his parents emphasized that he should use his computer as a tool to build things on. After his attention turned from microchips to microbes, he wondered why biology did not have the same involvement in people's daily lives as computers did. There were no simple biology devices or platforms that innovators could build things on or that people could use in their homes. "As I was doing molecular biology experiments, I thought, 'This is so cool and it's so robust, but why hasn't it impacted people's lives in ways that a software app does?'"

He was still asking that question when he got to graduate school. "Can you think of how we can bring molecular biology into the kitchen or into people's homes?" he would ask his classmates. As he was working on developing his at-home CRISPR tests for viruses, he realized that they could be the way to do that. Home testing kits could become the platform, operating system, and form factor that will allow us to weave the wonders of molecular biology more into our daily lives.

Developers and entrepreneurs may someday be able to use CRISPR-based home testing kits as platforms on which to build a variety of biomedical apps: virus detection, disease diagnosis, cancer screening, nutritional analyses, microbiome assessments, and genetic tests. "We can get people in their homes to check if they have the flu or just a cold," says Zhang. "If their kids have a sore throat, they can determine if it's strep throat." In the process, it might give us all a deeper appreciation for how molecular biology works. The inner workings of molecules may remain, for most people, as mysterious as those of microchips, but at least all of us will be a bit more aware of the beauty and power of both.

Dariia Dantseva, Josiah Zayner, and David Ishee injecting their own vaccine

Chapter 53

Vaccines

My shot

"Look me in the eyes," the doctor ordered, staring at me from behind her plastic face guard. Her eyes were vividly blue, almost as blue as her hospital mask. Yet after a moment, I started to turn to the doctor on my left, who was jabbing a long needle deep into the muscle of my upper arm. "No!" the first doctor snapped. "Look at me!"

Then she explained. Because I was part of a double-blind clinical trial of an experimental COVID vaccine,[1] they had to make sure that I didn't get any clues about whether I was being injected with a real dose or merely a placebo made of saline solution. Would I really be able to tell just by looking at the syringe? "Probably not," she answered, "but we want to be careful."

It was early August of the plague year, and I had enlisted as a participant in the clinical trial for the COVID vaccine that was being developed by Pfizer with the German company BioNTech. It was a new type of vaccine that had never before been deployed. Instead of delivering deactivated components of the targeted virus, like traditional vaccines do, it injects into humans a snippet of RNA.

As you know by now, RNA is the strand that runs throughout Doudna's career and this book. In the 1990s, while other scientists were focused on DNA, her Harvard professor Jack Szostak turned her on to its less-celebrated but harder-working sibling that oversaw the making of proteins, acted as a guide for enzymes, could replicate itself, and was probably the root of all life on earth. "I never, ever got over my fascination about how RNA can do so many things," she says when I tell her of my participation in the RNA vaccine trial. "It's the genetic material of the coronavirus and, in a very interesting way, could be the basis for vaccines and cures."[2]

Traditional vaccines

Vaccines work by stimulating a person's immune system. A substance that resembles a dangerous virus (or any other pathogen)* is delivered into a person's body. That substance could be a deactivated version of the virus or a safe fragment of the virus or genetic instructions to make that fragment. This is intended to kick the person's immune system into gear. When it works, the body produces antibodies that will, sometimes for many years, fend off any infection if the real virus ever attacks.

Vaccinations were pioneered in the 1790s by an English doctor named Edward Jenner who noticed that many milkmaids were immune to smallpox. They had all been infected by a form of pox that afflicts cows but is harmless to humans, and Jenner surmised that the cowpox had given the milkmaids immunity to smallpox. So he took some pus from a cowpox blister, rubbed it into scratches he made in the arm of his gardener's eight-year-old son, and then (this was in the days before bioethics panels) exposed the kid to smallpox. He didn't become ill.

Vaccines use a variety of methods to try to stimulate the human immune system. One traditional approach is to inject a weakened and safe (attenuated) version of the virus. These can be good teachers, because they look very much like the real thing. The body responds by

*A "pathogen," commonly referred to as a "germ," is any microorganism that causes disease or infection. The most common are viruses, bacteria, fungi, and protozoa.

making antibodies for fighting them, and the immunity can last a lifetime. Albert Sabin used this approach for the oral polio vaccine in the 1950s, and that's the way we now fend off measles, mumps, rubella, and chicken pox. It takes a long time to develop and cultivate these vaccines (the viruses have to be incubated in chicken eggs), but some companies in 2020 were using this method as a long-term option for attacking COVID.

When Sabin was trying to develop a weakened polio virus for a vaccination, Jonas Salk succeeded with an approach that seemed somewhat safer: using a killed virus. This type of vaccine can still teach a person's immune system how to fight off the live virus. The Beijing-based company Sinovac used this approach to devise an early COVID vaccine.

Another traditional approach is to inject a subunit of the virus, such as one of the proteins that are on the virus's coat. The immune system will then remember these, allowing the body to mount a quick and robust response when it encounters the actual virus. The vaccine against the hepatitis B virus, for example, works this way. Using only a fragment of the virus means that they are safer to inject into a patient and easier to produce, but they are usually not as good at producing long-term immunity. Many companies pursued this approach in the 2020 race for a COVID vaccine by developing ways to introduce into human cells the spike protein that is on the surface of the coronavirus.

Genetic vaccines

The plague year of 2020 is likely to be remembered as the time when these traditional vaccines began to be supplanted by genetic vaccines. Instead of injecting a weakened or partial version of the dangerous virus into humans, these new vaccines deliver a gene or piece of genetic coding that will guide human cells to produce, on their own, components of the virus. The goal is for these components to stimulate the patient's immune system.

One method for doing this is by taking a harmless virus and engineering into it a gene that will make the desired component. As we all now know, viruses are very good at worming their way into human cells. That is why safe viruses can be used as a delivery system, or vector, to transport material into the cells of patients.

This approach led to one of the earliest COVID vaccine candidates, which was developed at the aptly named Jenner Institute of Oxford University. Scientists there genetically reengineered a safe virus—an adenovirus that causes flu in chimpanzees—by editing into it the gene to make the spike protein of the coronavirus. Similar vaccines developed by other companies in 2020 used a human version of the adenovirus. The vaccine created by Johnson & Johnson, for example, used a human adenovirus as the delivery mechanism to carry a gene that codes for making part of the spike protein. But the Oxford team decided that using one from a chimpanzee was better, because patients who previously had cold infections might have an immunity to the human version.

The idea behind both the Oxford and the Johnson & Johnson vaccines was that the reengineered adenovirus would make its way into human cells, where it would cause the cells to make lots of these spike proteins. That in turn would stimulate the person's immune system to make antibodies. As a result, the person's immune system would be primed to respond rapidly if the real coronavirus struck.

The lead researcher at Oxford was Sarah Gilbert.[3] In 1998, when she had triplets who were born prematurely, her husband took time off from his job so that she could return to her lab. In 2014, she worked on developing a vaccine for Middle East respiratory syndrome (MERS), using a chimp adenovirus edited to contain the gene for a spike protein. That epidemic died away before her vaccine could be deployed, but it gave her a head start when COVID struck. She already knew that the chimp adenovirus had successfully delivered into humans the gene for the spike protein of MERS. As soon as the Chinese published the genetic sequence of the new coronavirus in January 2020, she began engineering its spike protein gene into the chimp virus, waking each day at 4 a.m.

By then her triplets were twenty-one, and all were studying biochemistry. They volunteered to be early testers, getting the vaccine and seeing if they developed antibodies. (They did.) Trials in monkeys conducted at a Montana primate center in March also produced promising results.

The Bill and Melinda Gates Foundation provided early funding. Bill Gates also pushed Oxford to team up with a major company that could manufacture and distribute the vaccine if it worked. So Oxford forged a partnership with AstraZeneca, the British-Swedish pharmaceutical company.

DNA vaccines

There is another way to get genetic material into a human cell and cause it to produce the components of a virus that can stimulate the immune system. Instead of engineering the gene for the component into a virus, you can just deliver the genetic code for the component—as DNA or RNA—into human cells. The cells thus become a vaccine-manufacturing facility.

Let's start with DNA vaccines. Although no DNA vaccine had ever been approved before the COVID plague, the concept seemed promising. Researchers at Inovio Pharmaceuticals and a handful of other companies in 2020 created a little circle of DNA that coded for parts of the coronavirus spike protein. The idea was that if it could get inside the nucleus of a cell, the DNA could very efficiently churn out many strands of messenger RNA to go forth and oversee the production of the spike protein parts, which serve to stimulate the immune system. DNA is cheap to produce and do not require dealing with live viruses and incubating them in chicken eggs.

The big challenge facing a DNA vaccine is delivery. How can you get the little ring of engineered DNA not only into a human cell but into the nucleus of the cell? Injecting a lot of the DNA vaccine into a patient's arm will cause some of the DNA to get into cells, but it's not very efficient.

Some of the developers of DNA vaccines, including Inovio, tried to facilitate the delivery into human cells through a method called electroporation, which delivers electrical shock pulses to the patient at the site of the injection. That opens pores in the cell membranes and allows the DNA to get in. The electric pulse guns have lots of tiny needles and are unnerving to behold. It's not hard to see why this technique is unpopular, especially with those on the receiving end.

One of the teams that Doudna organized at the beginning of the coronavirus crisis in March 2020 focused on these delivery challenges facing DNA vaccines. It was led by her former student Ross Wilson, who now runs his own lab down the hall from her at Berkeley, and Alex Marson of the University of California, San Francisco. At one of Doudna's regular Zoom meetings, Wilson showed a slide of the Inovio electric zapper. "They actually shoot the patient in the muscle with one of these guns," he said. "About the only visible advance they've made in ten years is now they have a little plastic thing to hide the tiny needles so they don't frighten the patient as much."

Marson and Wilson devised a way to address the DNA vaccine delivery problem using CRISPR-Cas9. They put together a Cas9 protein, a guide RNA, and a nuclear localization signal that helps the complex get into the nucleus. The result was a "shuttle" that could get the DNA vaccine into cells. The DNA then directs the cells to make coronavirus spike proteins and thus stimulate the immune system to fend off the real coronavirus.[4] It's a brilliant idea that could have uses for many treatments in the future, but it has been difficult to make work. By the beginning of 2021, Wilson and Marson were still trying to prove it could be effective.

RNA vaccines

That leads us back to our favorite molecule, the biochemical star of this book: RNA.

The vaccine that was tested in my clinical trial makes use of the most basic function that RNA performs in the central dogma of biology: serving as a messenger RNA (mRNA) that carries genetic instructions from DNA, which is bunkered inside a cell's nucleus, to the manufacturing region of the cell, where it directs what protein to make. In the case of the COVID vaccine, the mRNA instructs cells to make part of the spike protein that is on the surface of a coronavirus.[5]

RNA vaccines deliver their payloads inside tiny oily capsules, known as lipid nanoparticles, that are injected by a long syringe into the muscles of the upper arm. My muscle hurt for days.

An RNA vaccine has certain advantages over a DNA vaccine. Most notably, the RNA does not need to get into the nucleus of the cell, where DNA is headquartered. The RNA does its work in the outer region of cells, the cytoplasm, which is where proteins are constructed. So an RNA vaccine simply needs to deliver its payload into this outer region.

In 2020, two innovative young pharmaceutical companies produced RNA vaccines for COVID: Moderna, based in Cambridge, Massachusetts, and the German company BioNTech, which formed a partnership with the American company Pfizer. My clinical trial was for BioNTech/Pfizer.

BioNTech was founded in 2008 by the husband-and-wife research team of Uğur Şahin and Özlem Türeci with the goal of creating cancer immunotherapies, which stimulate the immune system to fight cancerous cells. It soon also became a leader in devising medicines that use mRNA as vaccines against viruses. In January 2020, when Şahin read a medical journal article on the new coronavirus in China, he sent an email to the BioNTech board saying that it was wrong to believe that this virus would come and go as easily as MERS and SARS. "This time it is different," he told them.[6]

BioNTech launched what they dubbed Project Lightspeed to devise a vaccine based on RNA sequences that would cause human cells to make versions of the coronavirus's spike protein. Once it looked promising, Şahin called Kathrin Jansen, the head of vaccine research and development at Pfizer. The two companies had been working together since 2018 to develop flu vaccines using mRNA technology, and he asked her whether Pfizer would want to enter a similar partnership for a COVID vaccine. Jansen said she had been about to call and propose the same thing. The deal was signed in March.[7]

By then, a similar RNA vaccine was being developed by Moderna, a much-smaller company with only eight hundred employees. Its chair and cofounder, Noubar Afeyan, a Beirut-born Armenian who immigrated to the United States, became fascinated in 2005 by the prospect that mRNA could be inserted into human cells to direct the production of a desired protein. So he hired some young graduates from the Harvard lab of Jack Szostak, who had been Jennifer Doudna's PhD adviser and turned her on to the wonders of RNA. The company mainly focused on using mRNA to try to develop personalized cancer treatments, but it also had begun experimenting with using the technique to make vaccines against viruses.

In January 2020, Afeyan was celebrating the birthday of one of his daughters at a Cambridge restaurant when he got an urgent text message from the CEO of his company, Stéphane Bancel, in Switzerland. So he stepped outside in the freezing temperature to call him back. Bancel said that he wanted to launch a project to use mRNA to attempt a vaccine against the new coronavirus. At that point, Moderna had twenty drugs in development but none had been approved or even reached the final stage of clinical trials. Afeyan instantly authorized him to start work without waiting for full board approval. Lacking Pfizer's resources, Moderna had to depend on funding from the U.S. government. Anthony Fauci, the government's infectious disease expert, was supportive. "Go for it," he declared. "Whatever it costs, don't worry about it." It took Moderna only two days to create the desired RNA sequences that would produce the spike protein, and thirty-eight days later it shipped the first box of vials to the NIH to begin early-stage trials. Afeyan keeps a picture of that box on his cell phone.

As with CRISPR therapies, a difficult part of the vaccine development was creating the delivery mechanism into the cell. Moderna had been working for ten years to perfect lipid nanoparticles, the tiny synthetic capsules that can carry molecules into a human cell. This gave it one advantage over BioNTech/Pfizer: its particles were more stable and did not have to be stored at extremely low temperatures. Moderna is also using this technology to deliver CRISPR into human cells.[8]

Our biohacker steps in

At this point Josiah Zayner, the garage scientist who injected himself with CRISPR, came back onstage to play Puck again. As others were eagerly awaiting results for the genetic vaccines that went into clinical trials in the summer of 2020, Zayner brought his wise-fool spirit to the battle, enlisting a couple of like-minded biohackers in the cause. His plan was to produce and then inject himself with one of the many potential coronavirus vaccines that were being developed. Then he would see whether (a) he survived and (b) he developed antibodies to protect against COVID. "You can call it a stunt if you want, but it's really about people taking control of science and moving it fucking faster," he told me.[9]

Specifically, he decided to make and test a potential vaccine that had been described that May in a *Science* paper by researchers at Harvard. The vaccine was just beginning human trials.[10] It was a DNA vaccine that included the genetic code for the spike of the coronavirus. The paper described precisely how to make it. With the recipe in hand, Zayner ordered the ingredients and went to work.

From his garage lab in Oakland, just seven miles south of Doudna's at Berkeley, Zayner launched a YouTube streaming course—named Project McAfee, after the anti-virus software—so that others could follow along and perform the experiments on themselves. "Biohackers can be like the test pilots of the modern world by doing the slightly crazy shit that needs to be done," he declared.

He had two copilots. David Ishee is a ponytailed rural Mississippi dog breeder who uses CRISPR to edit the genes of Dalmatians and mastiffs to try to make them healthier, stronger, and in one offbeat experiment glow in the dark. He joined by Skype from a wooden shed in his backyard crammed with lab equipment. When Zayner said that they would be streaming their experiments for the next two months, Ishee took a sip of a Monster energy drink and interjected in his languid honeysuckle-scented drawl, "Or at least until the authorities come for us." Also Skyping in was Dariia Dantseva, a student in Dnipro, Ukraine, who created her country's first biohacking lab. "Ukraine is pretty easy about regulating biohacking, because the state literally does not exist," she says. "I believe that knowledge is not just for the elites, it's for all of us. That's why we do this."

The experiments that Zayner performed through the summer of 2020 were not just a showy stunt, like when he injected CRISPR into his arm at the San Francisco conference. "We could just inject this shit," he said of the DNA vaccine described by the Harvard researchers. "But I don't think anyone would get anything out of that. We want to add a lot more value." Instead, he and his copilots carefully, week after week, did a livestream demonstration to teach people how to make the code for the spike proteins of the coronavirus. That way they could get dozens, perhaps hundreds, of people to test it, thus gathering useful data about its effectiveness. "If a bunch of scrubs like us can do this, hundreds of people could be doing it and moving science forward more quickly," he says. "We want everyone to have the opportunity to create this DNA vaccine and test if it creates antibodies in human cells."

I asked him why he thought a DNA vaccine would work with just a simple injection rather than the electroporation shocks and other techniques that some researchers said were needed to assure that the DNA got into the nucleus of human cells. "We wanted to follow the Harvard paper as closely as possible, and they did not use any special techniques like electroporation," he replied. "DNA is easy to produce, so if some delivery method doubles the efficiency you can get the same results just by doubling or so the amount of DNA you inject."

On Sunday, August 9, the three biohackers appeared together—from California, Mississippi, and Ukraine—in a live video-stream to inject into their arms the vaccines they had been concocting over the past two months. "We three tried to push science forward by showing what people are capable of doing in a do-it-yourself environment," Zayner explained as the video began. "So anyway, here we go! We're doing it!" Then Zayner, wearing a Michael Jordan red tank-top jersey, plunged a long needle into his arm as Dantseva and Ishee followed suit. He offered a bit of reassurance to his audience: "For all of you who signed in to see us die, it's not going to happen."

He was right. They didn't die. They simply winced a lot. And in the end, there was evidence that the vaccine may have worked. Because his experiment did not include any special method for getting the DNA into the nucleus of human cells, the results were not totally clear or convincing. But when he tested his blood in September, streaming it live on the internet for everyone to watch, Zayner found evidence that he had developed neutralizing antibodies to fight the coronavirus. He called it a "mild success," but noted that biology often produces murky results. It gave him a greater appreciation for careful clinical trials.

Some of the scientific researchers I talked to were appalled by what Zayner did. But I found myself rooting for him. If his shadow has offended, think but this and all is mended: *More citizen involvement in science is a good thing*. Genetic coding will never become as crowd-sourced and democratized as software coding, but biology should not remain the exclusive realm of a gospel-guarding priesthood. When Zayner kindly sent me a dose of his homemade vaccine, I decided not to inject it. But I admired him and his two other musketeers for doing so. It made me want to get involved in testing vaccines, though in a more authorized way.[11]

My clinical trial

My own involvement in citizen science was to sign up for a clinical trial of the Pfizer/BioNTech mRNA vaccine. As noted in the opening of this chapter, it was a double-blind study, meaning that neither I nor the researchers were told who got a real vaccine and who got a placebo.

When I volunteered at Ochsner Hospital in New Orleans, I was told that the study could last up to two years. That raised a few questions in my mind. What would happen, I asked the coordinator, if the vaccine got approved before then? She told me that I would then be "unblinded," meaning that they would tell me if I had gotten the placebo and, if so, give me the real vaccine.

What would happen if some other vaccines got approved while our trial was still underway? I could drop out whenever I wanted, she said, and seek to get the approved vaccine. Then I asked a more difficult question: If I dropped out, would I then be unblinded? She paused. She called her supervisor, who paused as well. Finally I was told, "That's not been decided."[12]

So I went to the top. I posed these questions to Francis Collins at the National Institutes of Health, which was overseeing the vaccine studies. (There is an advantage to being a book writer.) "You have asked a question that is currently engaging the members of the Vaccines Working Group in serious debate," he replied. Just a few days earlier, a "consultation report" on this issue had been prepared by the Department of Bioethics at NIH headquarters in Bethesda, Maryland.[13] Even before reading the five-page report, I was impressed and comforted that the NIH had something called a Department of Bioethics.

The report was thoughtful. For a variety of scenarios, the scientific value that could come from continuing a blinded study was balanced against the health of the trial participants. In the case that the vaccine got FDA approval, the advice was: "There will be an obligation to inform participants so that they can decide whether to obtain the vaccine."

After digesting all of this, I decided to quit asking questions and enroll. It might aid the science a little bit, and I would learn first-hand, or first arm, about RNA vaccines. Some people are very skeptical about vaccines and clinical trials. I err on the side of being trusting.

RNA victorious

In December of 2020, with COVID once again resurging throughout much of the world, the two RNA vaccines were the first to be authorized in the United States and became the vanguard of the biotech battle to beat back the pandemic. The plucky little RNA molecule, which had spawned life on our planet and then plagued us in the form of coronaviruses, rode to our rescue. Jennifer Doudna and her colleagues had employed RNA in a tool to edit our genes and then as a method to detect coronaviruses. Now scientists had found a way to enlist RNA's most basic biological function in order to turn our cells into manufacturing plants for the spike protein that would stimulate our immunity to the coronavirus.

Look at the halo of letters—GCACGUAGUGU . . .—on the

cover of this book. It is a snippet of the RNA that creates the part of the spike protein that binds to human cells, and these letters became part of the code used in the new vaccines. Never before had an RNA vaccine been approved for use. But a year after the novel coronavirus was first identified, both Pfizer/BioNTech and Moderna had devised these new genetic vaccines and tested them in large clinical trials, involving people like me, where they proved more than 90 percent effective. When the CEO of Pfizer, Albert Bourla, was informed of the results on a conference call, even he was stunned. "Repeat it," he asked. "Did you say 19 or 90?"[14]

Throughout human history, we have been subjected to wave after wave of viral and bacterial plagues. The first known one was the Babylon flu epidemic around 1200 BC. The plague of Athens in 429 BC killed close to 100,000 people, the Antonine plague in the second century killed ten million, the plague of Justinian in the sixth century killed fifty million, and the Black Death of the fourteenth century took almost 200 million lives, close to half of Europe's population.

The COVID pandemic that killed more than 1.5 million people in 2020 will not be the final plague. However, thanks to the new RNA vaccine technology, our defenses against most future viruses are likely to be immensely faster and more effective. "It was a bad day for viruses," Moderna's chair Afeyan says about the Sunday in November 2020 when he got the first word of the clinical trial results. "There was a sudden shift in the evolutionary balance between what human technology can do and what viruses can do. We may never have a pandemic again."

The invention of easily reprogrammable RNA vaccines was a lightning-fast triumph of human ingenuity, but it was based on decades of curiosity-driven research into one of the most fundamental aspects of life on planet earth: how genes encoded by DNA are transcribed into snippets of RNA that tell cells what proteins to assemble. Likewise, CRISPR gene-editing technology came from understanding the way that bacteria use snippets of RNA to guide enzymes to chop up dangerous viruses. Great inventions come from understanding basic science. Nature is beautiful that way.

Stanley Qi

Nathan and Cameron Myhrvold

Chapter 54

CRISPR Cures

The development of vaccines—both the conventional sort and those employing RNA—would eventually help to beat back the coronavirus pandemic. But they are not a perfect solution. They rely on stimulating a person's immune system, always a risky thing to do. (Most deaths from COVID-19 came from organ inflammation due to unwanted immune-system responses.)[1] As vaccine makers have repeatedly discovered, the multilayered human immune system is very tricky to control. In it lurk mysteries. It contains no simple on-off switches, but instead works through the interaction of complicated molecules that are not easy to calibrate.[2]

The use of antibodies from the blood plasma of recovering patients or made synthetically also helped fight the COVID plague. But these treatments are, likewise, not a perfect long-term solution for each new wave of virus. Convalescent plasma is difficult to harvest from donors in large quantities, and lab-made monoclonal antibodies are hard to manufacture.

The long-range solution to our fight against viruses is the same as the one bacteria found: using CRISPR to guide a scissors-like enzyme to chop up the genetic material of a virus, without having to enlist the patient's immune system. Once again, the circles of scientists around Doudna and Zhang found themselves in competition as they raced to adapt CRISPR to this urgent mission.

Cameron Myhrvold and CARVER

Cameron Myhrvold straddles the world of digital coding and genetic coding, which is not surprising given his heritage and breeding. The lookalike son of Nathan Myhrvold (pronounced MEER-vold), who was the longtime chief technology officer and sparky genius at Microsoft, he has his father's gleeful eyes, chipmunk-cheeked round face, effervescent laugh, and free-range curiosity. People of my generation were awed by his father's brilliance not only in the digital realm but also in fields ranging from food science to asteroid tracking to the speed at which dinosaurs could whip their tails. Cameron shares his father's facility with computer coding, but like many in his generation he focused more on genetic coding and the wonders of biology.

As a Princeton undergraduate, he studied molecular and computational biology, then he got his doctorate from Harvard's Systems, Synthetic, and Quantitative Biology Program, which combines biology and computer science. He loved the intellectual challenge but worried that his work on nano-engineering of organisms was so cutting-edge that it would have little practical impact in the foreseeable future.[3]

So after he got his PhD, he took time off to hike the Colorado Trail. "I was really trying to figure out where to go scientifically," he says. On one leg of his hike, he met a guy who asked him a lot of earnest questions about science. "During that conversation," Myhrvold says, "it became apparent to me that I liked working on problems that were directly relevant to human health."

That led him to decide to become a postdoc in the lab of Pardis Sabeti, a Harvard biologist who uses computer algorithms to explain the evolution of disease. She was born in Tehran and as a child fled with her family to America during the Iranian Revolution. A member of the Broad Institute, she collaborates closely with Feng Zhang. "Joining

Pardis's lab and working with Feng Zhang seemed like a really great way to take on the problem of fighting viruses," Myhrvold says. As a result, Myhrvold became part of the Boston-area orbit around Zhang and eventually a player in its CRISPR star wars with the Berkeley-area orbit of Jennifer Doudna.

While studying for his doctorate at Harvard, Myhrvold became friends with Jonathan Gootenberg and Omar Abudayyeh, the two grad students who worked with Zhang on CRISPR-Cas13. Myhrvold would often kick around ideas with them when he visited Zhang's lab to use its gene-sequencing machine. "That's when I realized, wow, like those two guys were a really special pair," Myhrvold says. "We came up with ways to use Cas13 to detect different RNA sequences, and I thought it would be a really cool opportunity."

When Myhrvold suggested to Sabeti that they should collaborate with Zhang's lab, she was enthusiastic because there was a lot of synergy between the two teams. It resulted in a made-for-the-movies diverse American platoon: Gootenberg, Abudayyeh, Zhang, Myhrvold, Sabeti.

They worked together on Zhang's 2017 paper describing the SHERLOCK system for detecting RNA viruses.[4] The following year, they collaborated on a paper showing how to make the SHERLOCK process even simpler.[5] It appeared in the same issue of *Science* as the paper from Doudna's lab describing the virus-detection tool developed by Chen and Harrington.

In addition to using CRISPR-Cas13 to detect viruses, Myhrvold became interested in turning it into a therapeutic treatment, one that could get rid of viruses. "There are hundreds of viruses that can infect people, but there's only a handful that have available drugs," he says. "That's in part because viruses are so different from each other. What if we could come up with a system that we could program to treat different viruses?"[6]

Most of the viruses that cause human problems, including the coronavirus, have RNA as their genetic material. "They are precisely the type of virus for which you would want a CRISPR enzyme that targets RNA, such as Cas13," he says. So he came up with a way to use CRISPR-Cas13 to do for humans what it does for bacteria: target a dangerous virus and chop it up. Continuing the tradition of reverse-engineering clever acronyms for CRISPR-based inventions, he dubbed the proposed system CARVER, for "Cas13-assisted restriction of viral expression and readout."

In December 2016, shortly after he joined Sabeti's lab as a postdoc, Myhrvold sent her an email reporting on some initial experiments using CARVER to target a virus that causes the symptoms of meningitis or encephalitis. His data showed that it reduced the levels of the virus significantly.[7]

Sabeti was able to get a DARPA grant to study the CARVER system as a way to destroy viruses in humans.[8] Myhrvold and others in her lab did a computer analysis of more than 350 genomes from RNA viruses that infected humans and identified what are known as "conserved sequences," meaning those that are the same in many viruses. These sequences have been preserved unchanged by evolution, and thus are not likely to mutate away anytime soon. His team engineered an arsenal of guide RNAs designed to target these sequences. He then tested Cas13's ability to stop three viruses, including the type that causes severe flu. In cell cultures in a lab, the CARVER system was able too significantly reduce the level of viruses.[9]

Their paper was published online in October 2019. "Our results demonstrate that Cas13 can be harnessed to target a wide range of single-stranded RNA viruses," they wrote. "A programmable antiviral technology would allow for the rapid development of antivirals that can target existing or newly identified pathogens."[10]

A few weeks after the CARVER paper came out, the first cases of COVID-19 were detected in China. "It was one of these moments when you realize the stuff you've been working on for a long time might be a lot more relevant than you thought," Myhrvold says. He started a new computer folder labeled nCov, for "novel coronavirus," since it had not yet been given an official name.

By late January, he and his colleagues had studied the sequence of the coronavirus genome and begun work on CRISPR-based tests for detecting it. The result was a burst of papers in the spring of 2020 for improving CRISPR-based detection technologies for viruses. These included a system known as CARMEN, designed to detect 169 viruses at one time,[11] and a process that combined SHERLOCK's detection capability with an RNA extraction method called HUDSON to create a single-step detection technique he named SHINE.[12] In addition to its CRISPR wizardry, the Broad was a master at devising acronyms.

Myhrvold decided that his time could best be used in developing tools that could detect viruses rather than working on treatments like CARVER, designed to destroy viruses. He was in the process of moving his lab to Princeton, where he had accepted a position beginning in 2021. "I think in the longer term we need treatments," he says, "but I decided that diagnostics were something that we could actually deliver on quickly."

In the West Coast orbit of Jennifer Doudna, however, there was a team that was pushing forward with a coronavirus treatment. Similar to the CARVER system that Myhrvold had invented, it would use CRISPR to seek and destroy viruses.

Stanley Qi and PAC-MAN

Stanley Qi grew up in what he calls a small city in China: Weifang, on the coast about three hundred miles south of Beijing. Its urban core is actually home to more than 2.6 million people, about the same as Chicago, "but that is regarded as small in China," he says. It is bustling with factories but does not have a world-class university, so Qi (pronounced "tshee") went to Tsinghua University in Beijing, where he majored in math and physics. He applied to Berkeley to do graduate work in physics, but he found himself increasingly attracted to biology. "It seemed to have more application for helping the world," he says, "so I decided to switch from physics to bioengineering after my second year at Berkeley."[13]

There he gravitated to the lab of Doudna, who became one of his two advisors. Instead of focusing on gene editing, he developed new ways to use CRISPR to interfere with the expression of genes. "I was surprised at how she spent time to discuss science with me, not on the superficial level but down to the deep level and including key technical details," he says. His interest in viruses increased in 2019 when he was funded (as Myhrvold and Doudna were) by DARPA's program for preparing against pandemics. "We started with a focus on finding a CRISPR method to fight influenza," he says. Then coronavirus struck. In late January 2020, after reading a story about the situation in China, Qi called together his team and shifted his focus from influenza to COVID.

Qi's approach was similar to that pursued by Myhrvold. He wanted to use a guided enzyme to target and then cleave the RNA of the invading virus. Like Zhang and Myhrvold, he decided to use a version of Cas13. The discovery of Cas13a and Cas13b was done at the Broad by Zhang. But another Cas13 variation had been discovered by a brilliant bioengineer in Doudna's orbit, Patrick Hsu, who had experience in both the Broad and Berkeley camps.[14]

Born in Taiwan, Hsu had gotten his undergraduate degree at Berkeley and his doctorate from Harvard, where he worked in the Zhang Lab when Zhang was racing Doudna to make CRISPR work in human cells. Hsu then spent two years as a scientist at Editas, the CRISPR-based company that Zhang had cofounded and Doudna had quit. From there he went to the Salk Institute in Southern California, where he discovered the enzyme that became known as Cas13d. In 2019, he became an assistant professor at Berkeley and one of the team leaders in Doudna's efforts to tackle COVID.

Because of its small size and highly specific targeting capability, the Cas13d that Hsu discovered was chosen by Qi as the best enzyme to target the coronavirus in human lung cells. In the competition to come up with good acronyms, Qi scored high. He dubbed his system PAC-MAN, which he had extracted from "prophylactic antiviral CRISPR in human cells." The name was that of the chomping character in the once popular video game. "I like video games," Qi told *Wired*'s Steven Levy. "The Pac-Man tries to eat cookies, and it is chased by a ghost.

But when it encounters a specific kind of cookie called the power cookie—in our case a CRISPR-Cas13 design—suddenly it turns itself to be so powerful. It can start eating the ghost and start cleaning up the whole battlefield."[15]

Qi and his team tested PAC-MAN on synthesized fragments of the coronavirus. In mid-February, his doctoral student Tim Abbott ran experiments showing that PAC-MAN in a lab setting reduced the amount of the coronavirus by 90 percent. "We demonstrated that Cas13d-based genetic targeting can effectively target and cleave the RNA sequences of SARS-CoV-2 fragments," Qi and his collaborators wrote. "PAC-MAN is a promising strategy to combat not only coronaviruses, including that causing COVID-19, but also a broad range of other viruses."[16]

The paper went online March 14, 2020, the day after Doudna's initial meeting of Bay Area researchers who had enlisted in the coronavirus fight. Qi emailed her a link, and within an hour she had replied, inviting him to join the group and present at their second weekly online meeting. "I told her we needed some resources to develop the PAC-MAN idea, get access to live coronavirus samples, and figure out delivery systems that might get it into the lung cells of patients," he says. "She was super supportive."[17]

Delivery

The concept behind CARVER and PAC-MAN was a brilliant one, although in fairness I should note that bacteria had thought of it more than a billion years ago. The RNA-cleaving Cas13 enzymes could chomp up coronaviruses in human cells. If they could be made to work, CARVER and PAC-MAN would act more efficiently than a vaccine that produces an immune response. By directly targeting the invading virus, these CRISPR-based technologies avoid having to rely on the body's erratic immune response.

The challenge was delivery: How could you get it to the right cells in a human patient and then through the membranes of those cells? That is a very difficult challenge, especially when it involves getting into lung cells, which is why CARVER and PAC-MAN were still not ready for deployment in humans in 2021.

At her March 22 weekly meeting, Doudna introduced Qi and showed a slide describing the group he would lead in their coronavirus war.[18] She teamed him up with researchers in her lab who were working on novel delivery methods, and she worked with him to prepare a white paper pitching the project to potential funders. "We use a variant of CRISPR, Cas13d, to target viral RNA sequences for cleavage and destruction," they wrote. "Our work offers a new strategy that could potentially be used as a genetic vaccine and treatment for COVID-19."[19]

The traditional way to deliver CRISPR and other genetic therapies is by using safe viruses—such as adeno-associated viruses, which don't cause any disease or provoke severe immune responses—as "viral vectors" that can deliver genetic material into cells. Or they can create synthetic virus-like particles to do the delivery, which is the specialty of Jennifer Hamilton and other researchers in Doudna's lab. Another method, electroporation, works by applying an electric field onto a cell's membrane to make it more permeable. All of these approaches have their drawbacks. The small size of viral vectors often limits the types of CRISPR proteins and number of guide RNAs that are deliverable. In search of a safe and effective delivery mechanism, the IGI would need to live up to its name and innovate.

To work with Qi on delivery systems, Doudna put him in touch with Ross Wilson, her former postdoc. Wilson, who now has a lab next to hers at Berkeley, is an expert in new ways to deliver material into the cells of patients. As noted earlier, he is working with Alex Marson to devise a delivery system for a DNA vaccine.[20]

Wilson fears that delivering PAC-MAN or CARVER into cells will be difficult. Qi is nevertheless hopeful that these CRISPR-based therapies can be deployed in the next few years. One method that is proving promising is to encase the CRISPR-Cas13 complex inside of synthetic molecules called lipitoids, which are about the size of a virus. He has been working with the Biological Nanostructures Facility at Lawrence Berkeley National Lab, a sprawling government complex on a hill above Berkeley's campus, to create lipitoids that can deliver PAC-MAN into lung cells.[21]

One way this could work, Qi says, is by delivering PAC-MAN treatments through a nasal spray or some other form of nebulizer. "My son has asthma," he says, "so as a little kid playing football he used a nebulizer as a preventive measure. People use these regularly to prepare the lung to be less allergic if they are exposed to something." The same could be done during a coronavirus pandemic; people could use a nasal spray so that PAC-MAN or another CRISPR-Cas13 prophylactic treatment will protect them.

Once the delivery mechanisms are worked out, CRISPR-based systems such as PAC-MAN and CARVER will be able to treat and protect people without having to activate the body's own immune system, which can be quirky and delicate. They can also be programmed to target essential sequences in the virus's genetic code so they cannot be easily evaded by the virus mutating. And they are simple to reprogram when a new virus emerges.

This concept of reprogramming is also apt in a larger sense. The CRISPR treatments come from reprogramming a system that we humans found in nature. "That gives me hope," Myhrvold says, "that when we face other great medical challenges, we will be able to find other such technologies in nature and put them to use." It is a reminder of the value of curiosity-driven basic research into what Leonardo da Vinci liked to refer to as the infinite wonders of nature. "You never know," Myhrvold says, "when some obscure thing you're studying is going to have important implications for human health." As Doudna likes to put it, "Nature is beautiful that way."

Cold Spring Harbor Laboratory

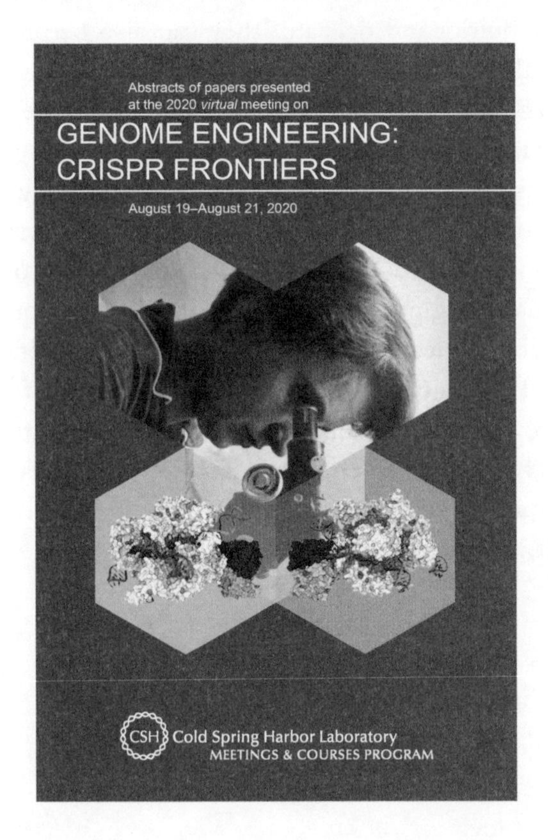

Chapter 55

Cold Spring Harbor Virtual

CRISPR and COVID

The stories of CRISPR and COVID wove together at the Cold Spring Harbor Laboratory's annual CRISPR conference in August 2020. A primary topic was how CRISPR was being used to fight the coronavirus, featuring talks by Jennifer Doudna and Feng Zhang as well as some of the COVID warriors in their rival orbits. Instead of gathering on the rolling campus overlooking an inlet of Long Island Sound, participants convened by Zoom and Slack, looking a bit bleary from months of interacting with boxed faces on their computer screens.

The meeting also wove in another strand of this book. It celebrated the hundredth anniversary of the birth of Rosalind Franklin, whose pioneering work on the structure of DNA inspired Doudna, when she read *The Double Helix* as a young girl, to believe that women could do science. The cover of the meeting's program featured a colorized photograph of Franklin peering into a microscope.

Fyodor Urnov, who directed the COVID testing lab that Doudna created at Berkeley, gave the opening tribute to Franklin. I expected him to deliver it with his usual dramatic flair, but instead he made it, properly, a serious look at her scientific work, including her research into the location of RNA in tobacco mosaic viruses. The only flourish came at the end when he showed a picture of Franklin's empty lab bench after her death. "The best way to honor her is to remember that the structural sexism she faced remains with us today," he said, his voice choking up a bit. "Rosalind is the godmother of gene editing."

Doudna's talk began with a reminder of the natural connection between CRISPR and COVID. "CRISPR is a fabulous way that evolution has dealt with the problem of viral infection," she said. "We can learn from it in this pandemic." Zhang followed with an update on his STOP technology for easy-to-use portable testing machines. As he finished, I sent him a message asking when they would become available at airports and schools, and he texted back seconds later with pictures of the latest prototypes, which had been delivered that week. "We are working hard to make it available this fall," he said. Cameron Myhrvold, speaking animatedly with both hands like his father does, gave a description of how his CARMEN system could be programmed to detect multiple viruses at once. Doudna's former student Janice Chen followed with a presentation about the DETECTR platform that she and Lucas Harrington had created at Mammoth. Patrick Hsu reported on the work being done with Doudna's team to create better methods for amplifying genetic material so it could be detected. And Stanley Qi described how his PAC-MAN system could be used not only to detect coronaviruses but also destroy them.

I was invited to moderate a panel about COVID, and I began by asking Zhang and Doudna about the possibility that the pandemic might create greater public interest in biology. When at-home testing kits become low-cost and easy to use, Zhang replied, they will democratize and decentralize medicine. The most important next steps will be innovations in "microfluidics," which involves channeling tiny amounts of liquid in a device, and then connecting the information to our cell phones. That will allow us all, in the privacy of our homes, to test our saliva and blood for hundreds of medical indicators, monitor our health conditions on our phones, and share the data with doctors and researchers. Doudna added that the pandemic had accelerated the convergence of science with other fields. "The engagement of non-scientists in our work will help achieve an incredibly interesting biotechnology revolution," she predicted. This was molecular biology's moment.

Near the end of the panel, an audience member named Kevin Bishop electronically raised his hand.[1] He worked at the National Institutes of Health and wanted to ask why there were so few African Americans like himself enrolled in the clinical trials for COVID vaccines. That led to a discussion of the distrust Blacks have about medical trials because of historical horrors such as the Tuskegee experiments, in which placebos were given to some sharecroppers suffering from syphilis who thought they were getting real medical treatments. A few of the conference attendees questioned whether it was important to have racial diversity in the COVID vaccine trials. (The consensus: yes, for medical and moral reasons.) Bishop suggested enlisting African American churches and colleges into the effort of enrolling volunteers.

The diversity issue, it struck me, involves far more than just clinical trials. Judging from the list of attendees at the meeting, women are becoming well represented in the field of biological research. But there were very few African Americans, either at the conference or on the benches in the various labs I had visited. In that regard, the new life-sciences revolution resembles, unfortunately, the digital revolution. If there are not efforts at outreach and mentorship, biotechnology will be yet another revolution that leaves most Blacks behind.

CRISPR marches on

The conference presentations on how CRISPR was being deployed to fight COVID were impressive, but equally so were the reports on the discoveries that were pushing CRISPR gene-editing forward. The most important were those made by one of Doudna's co-organizers of the conference, Harvard's soft-spoken superstar David Liu. He has a foot in both the Cambridge and Berkeley camps. After graduating first in his class from Harvard, he got his doctorate from Berkeley and then returned to teach at Harvard, where he became Zhang's colleague at the Broad Institute and cofounder with him of Beam Therapeutics.

With his disarming gentility and friendly intellect, he has remained close to both Doudna and Zhang.

Beginning in 2016, Liu began developing a technique known as "base editing," which can make a precise change in a single letter in DNA without cutting a break in the strands. It's like a very sharp pencil for editing. At the 2019 Cold Spring Harbor meeting, he announced a further advance called "prime editing," in which a guide RNA can carry a long sequence to be edited into a targeted segment of DNA. It requires making only a tiny nick in the DNA rather than a double-strand break. Edits of up to eighty letters are possible.[2] "If CRISPR-Cas9 is like scissors and base editors are like pencils, then you can think of prime editors as like word processors," Liu explained.[3]

Dozens of the presentations at the 2020 meeting involved young researchers who had found clever new ways to use base editing and prime editing. Liu himself described his latest discovery of how to deploy base-editing tools into the energy-producing region of cells.[4] In addition, he was a co-author of a paper that described a user-friendly web app that could be used to design prime-editing experiments.[5] COVID had not slowed the CRISPR revolution.

The importance of base editing was highlighted on the cover of the conference book. Just below the colorized picture of Rosalind Franklin was a beautiful 3-D image of a base editor attached to a purple RNA guide and a blue DNA target. Using some of the structural biology and imaging techniques that Franklin pioneered, the image had been published a month earlier by the labs of Doudna and Liu, with much of the work done by Gavin Knott, the postdoc who had taught me how to edit DNA using CRISPR.[6]

The Blackford Bar

In the dining hall on the Cold Spring Harbor campus, there is a wood-paneled lounge, known as the Blackford Bar, that manages to be both spacious and cozy. Old photographs line the walls, multiple ales and lagers are on tap, TV sets broadcast both scientific lectures and Yankees baseball games, and an outdoor deck overlooks the tranquil harbor. There you can find, on most summer evenings, conference attendees, researchers from nearby lab buildings, and the occasional groundskeeper or campus worker. During previous CRISPR conferences, it was filled with talk of impending discoveries, fanciful ideas, potential job openings, and high and low gossip.

In 2020, the conference organizers tried to re-create the scene with a Slack channel and Zoom room called #virtual-bar. Its purpose, they said, was to "simulate the serendipitous introductions you would've experienced at the Blackford Bar." So I decided to give it a try. About forty others showed up the first night. People introduced themselves in a stilted way, like at a real cocktail reception. Then a moderator broke us into groups of six and sent us to breakout Zoom rooms. After twenty minutes, each breakout session ended, and we were assigned randomly to a different group. Oddly, the format worked rather well when the conversations drilled down on specific scientific questions. There were interesting discussions of such topics as protein synthesis techniques and the hardware being built at Synthego to do automated cell editing. But there was none of the ordinary social chat that lubricates real life and nurtures emotional connections. There was no Yankees game in the background nor sunset to share while sitting on the deck. I left after two rounds.

Cold Spring Harbor Laboratory was founded in 1890 based on a belief in the magic of in-person meetings. The formula is to attract interesting people to an idyllic locale and provide them with opportunities to interact, including at a nice bar. The beauty of nature and the joy that comes from unstructured human engagement is a powerful combination. Even when they don't interact—such as when an awed young Jennifer Doudna passes the aging icon Barbara McClintock on a path through the Cold Spring Harbor campus—people benefit from an atmosphere that is charged in a way that sparks creativity.

One of the transformations wrought by the coronavirus pandemic is that more meetings in the future will be done virtually. It's a shame. If COVID doesn't kill us, Zoom will. As Steve Jobs emphasized when he built a headquarters for Pixar and planned a new Apple campus, new ideas are born out of serendipitous encounters. In-person interactions are especially important in the initial brainstorming of new ideas and the forging of personal bonds. As Aristotle taught, we are a social animal, an instinct that cannot fully be satisfied online.

Nevertheless, there will be an upside to the fact that the coronavirus has expanded how we work together and share ideas. By hastening the Age of Zoom, the pandemic will broaden the horizons of scientific collaboration, allowing it to be even more global and crowdsourced. A walk along the cobblestone streets of San Juan was the catalyst for the collaboration between Doudna and Charpentier, but the technology of Skype and Dropbox allowed them and their two postdocs to work together for six months in three countries to decipher CRISPR-Cas9. Because people have now become comfortable meeting in boxes on a computer screen, teamwork will be more efficient. A balance, I hope, will be struck: the reward for our efficient virtual meetings will be the chance to hang out together in person in places like the campus of Cold Spring Harbor.

Charpentier, remotely

At the end of Doudna's scientific presentation at the conference, a young researcher asked a personal question: "What inspired you to work on CRISPR-Cas9 the very first time?" Doudna paused for a moment, since it was not the type of question that scientific researchers usually ask after a technical presentation. "It started as a wonderful collaboration with Emmanuelle Charpentier," she replied. "I am forever indebted to her for the work we did together."

It was an interesting answer, because a few days earlier Doudna had talked to me about her sorrow that she and Charpentier had drifted apart, personally as well as scientifically. She lamented that she continued to detect a frostiness and asked me if, in my conversations with Charpentier, I had picked up more clues about why. "One of the things I am saddest about in the CRISPR tale is the fact that I really like Emmanuelle, but our relationship fell apart," she said. Doudna had studied French in high school and college, even at one point considering switching her major from chemistry to French. "I always had this fantasy of myself as a French girl, and Emmanuelle in some ways reminded me of that. I just adore her on a certain level. I wish we could have continued to have a wonderful close professional and personal connection and could have enjoyed the science and all of the things that came afterwards as friends."

When she told me this, I suggested that she invite Charpentier to speak at the Cold Spring Harbor virtual conference. Doudna immediately seized on the idea and asked her, through the conference's co-organizer Maria Jasin, to give the tribute to Rosalind Franklin or speak on any other topic. I followed up with Charpentier to encourage her to accept.

She hesitated at first, then replied that she had another meeting to attend remotely during that period. Jasin and Doudna offered to be flexible about the time and date, but Charpentier declined. Sensing her reticence, I tried a different approach: I invited her to join me and Doudna by Zoom the day after the conference for a private chat. I told her that I wanted to include their reminiscences at the end of this book. She surprised me by embracing the idea. She even emailed Doudna to say that she was looking forward to it.

As a result, we met online the Sunday after the conference. I had prepared a list of questions to ask. But as soon as Doudna and Charpentier came online, they began talking to each other and catching up, at first in the slightly stilted manner of people who have not seen each other for a while, and then, after a few minutes, more animatedly. Doudna began referring to Charpentier by her nickname, Manue, and soon they were both laughing. I turned off my video camera so I could leave the screen to them while I just listened.

Doudna talked about how tall her teenage son, Andy, had grown, shared a picture that Martin Jinek had sent of his new baby, and joked about an awards event she and Charpentier did with the American Cancer Society in 2018 at which Joe Biden told them that he did not plan to run for president. Doudna congratulated Charpentier on the success of her CRISPR Therapeutics company in curing sickle-cell anemia in its Nashville trial. "We published our paper in 2012, and here we are in 2020 and someone has already been cured of a disease," Doudna said. Charpentier nodded and laughed. "We can be very happy at how fast things happened," she said.

The talk gradually turned more personal. Charpentier recalled the beginning of their collaboration, when they had lunch at the conference in Puerto Rico, walked the cobblestone streets together, and ended up in a bar for a drink. Many times when you meet another scientist, she said, you know that you could never work with them. But their meeting was the opposite. "I knew we would be good collaborators," she told Doudna. Then they swapped memories about working around the clock by Skype and Dropbox in their six-month race to decode CRISPR-Cas9. Charpentier confessed that she worried whenever she sent Doudna some writing for the paper they produced jointly. "I thought you would have to correct my English," she said. Doudna replied, "Your English is great, and I remember you had to correct some of my own mistakes. It was a lot of fun to write that paper together, because we have different ways of thinking about things."

Finally, when their exchanges began to lag, I turned on my video camera to ask a question. Over the past few years, you've drifted apart, both scientifically and personally, I said. Do you miss the friendship you had?

Charpentier jumped in, eager to explain what had happened. "We were on the road a lot because of the prize ceremonies and other things," she said. "People were overloading our schedule, and we did not have any time to enjoy the in between. So part of the problem was the simple fact that we both became terribly busy." She spoke wistfully of the week they had spent together in Berkeley in June 2012 when they were finishing their paper. "There is this picture of us, with me with a funny haircut, in front of your institute," she said, referring to the picture that is at the beginning of chapter 17 of this book. It was the last time they had been relaxed together, Charpentier said. "After that, it was crazy because of the impact our paper had. We had little time for ourselves."

Charpentier's words made Doudna smile, and she opened up even more. "I enjoyed our friendship as much as doing the science," she said. "I love your delightful manner. I always had this fantasy, ever since I studied French in school, of living in Paris. And Manue, you embodied that for me."

The conversation ended with talk about working together again someday. Charpentier said she had a fellowship to do research in the U.S. Doudna had previously made plans, which COVID scuttled, to spend the Spring 2021 semester on sabbatical at Columbia. They agreed that they should coordinate sabbaticals. "Maybe in the Spring of 2022 in New York," Doudna suggested. "I would very much like that, to be there with you," Charpentier replied. "We could collaborate again."

Celebrating with Andy and Jamie in their kitchen just after the Nobel Prize announcement

CHAPTER 56

The Nobel Prize

"Rewriting the Code of Life"

Doudna was sound asleep when, at 2:53 a.m. on October 9, 2020, she was awakened by the persistent buzz of her cell phone, which she had put on vibrate mode. She was alone in a hotel room in Palo Alto, where she had gone to be part of a small meeting on the biology of aging, the first such in-person event she had attended in the seven months since the onset of the coronavirus crisis. The call was from a reporter for *Nature*. "I hate to bother you so early," she said, "but I wanted your comment on the Nobel."

"Who won?" Doudna asked, sounding slightly irritated.

"You mean you haven't heard?!" the reporter said. "You and Emmanuelle Charpentier!"

Doudna looked at her phone and saw a bunch of missed calls that indeed seemed to have come from Stockholm. After pausing for a moment to absorb the news, she said, "Let me call you back."[1]

The awarding of the 2020 Nobel Prize in Chemistry to Doudna and Charpentier was not a complete surprise, but the recognition came with historic swiftness. Their CRISPR discovery was merely eight years old. The day before, Sir Roger Penrose had shared the Nobel in physics for a discovery about black holes he had made more than fifty years earlier. There was also a sense that this chemistry award was historic. More than just recognizing an achievement, it seemed to herald the advent of a new era. "This year's prize is about rewriting the code of life," the secretary general of the Royal Swedish Academy proclaimed in making the announcement. "These genetic scissors have taken the life sciences into a new epoch."

Also noteworthy was that the prize went only to two people, rather than the usual three. Given the ongoing patent dispute over who first discovered CRISPR as a gene-editing tool, the third slot could have gone to Feng Zhang, although that would have left out George Church, who published similar findings at the same time. In addition, there were many other worthy candidates, including Francisco Mojica, Rodolphe Barrangou, Phillipe Horvath, Eric Sontheimer, Luciano Marraffini, and Virginijus Šikšnys.

There was also the historic significance of the prize going to two women. One could sense a tight smile on the face of Rosalind Franklin's ghost. Although she made the images that helped James Watson and Francis Crick discover the structure of DNA, she became just a minor character in the early histories, and she died before they got their 1962 Nobel Prize. Even if she had lived, it is unlikely she would have displaced Maurice Wilkins as that year's third honoree. Until 2020, only five women, beginning with Marie Curie in 1911, had won a Nobel for chemistry, out of 184 honorees.

When Doudna called the Stockholm number that had been left on her voicemail, she got an answering machine. But after a few minutes, she was able to connect and officially receive the news. After taking a few more calls, including from Martin Jinek and the persistent reporter from *Nature*, she threw her clothes into her bag and jumped in her car for the hourlong drive back to Berkeley. On the way she telephoned Jamie, who said that a communications team from the university was already setting up on their patio. When she arrived home at 4:30 a.m., she texted her neighbors to apologize for the commotion and camera lights.

For a few minutes, she got a chance to celebrate the news over coffee with Jamie and Andy. Then she made a few remarks to the camera team on her patio before heading to Berkeley for a hastily assembled virtual global press conference. On the ride over, she spoke to her colleague Jillian Banfield, who in 2006 had called her out of the blue and asked to meet at the Free Speech Movement Café on campus to discuss some clustered repeated sequences that she kept finding in the DNA of bacteria. "I am so grateful to have you as a collaborator and friend," she told Banfield. "It's been so much fun."

Many of the questions at the press conference focused on how the awards represented a breakthrough for women. "I'm proud of my gender!" Doudna said with a big laugh. "It's great, especially for younger women. For many women there's a feeling that, whatever they do, their work may not be as recognized as it might be if they were a man. I would like to see that change, and this is a step in the right direction." Later, she reflected on her days as a schoolgirl. "I was told more than a few times that girls don't do chemistry or girls don't do science. Fortunately, I ignored that."

As she spoke, Charpentier was holding her own press conference in Berlin, where it was midafternoon. I had reached her a few hours earlier, right after she got the official phone call from Stockholm, and she was unusually emotive. "I had been told that this might someday come," she told me, "but when I received the call I became very moved, very emotional." It took her back, she said, to her early childhood and deciding, while walking past the Pasteur Institute in her native Paris, that she someday would be a scientist. But by the time of her press conference, her emotions were well hidden behind her Mona Lisa smile. Carrying a glass of white wine, she came into the lobby of her institute, posed next to a bust of its namesake, Max Planck, and then answered questions in a way that managed to be both lighthearted and earnest. As happened in Berkeley, most of the focus was on what the award meant for women. "The fact that Jennifer and I were awarded this prize today can provide a very strong message for young girls," she said. "It can show them that women can also be awarded prizes."

That afternoon, their rival Eric Lander sent out a tweet from his perch at the Broad Institute: "Huge congratulations to Drs. Charpentier and Doudna on the @NobelPrize for their contributions to the amazing science of CRISPR! It's exciting to see the endless frontiers of science continue to expand, with big impacts for patients." In public, Doudna reacted graciously. "I'm deeply grateful for the acknowledgment from Eric Lander, and it's an honor to receive his words," she said. Privately, she wondered whether his use of the word "contributions" was a lawyerly way to subtly minimize their Nobel-certified discoveries. More notable to me were his words about "big impacts for patients" in the future. It led me to hope that Zhang and Church and perhaps David Liu will someday win the Nobel Prize for medicine as a companion to the one that Doudna and Charpentier won for chemistry.

Doudna mentioned at her press conference that she was "waving across the ocean" at Charpentier. But she badly wanted to actually talk to her. She texted Charpentier repeatedly throughout the day and left messages on her cell three times. "Please, please call me," Doudna texted at one point. "I won't take much of your time. I just want to say congratulations on the phone to you." Charpentier finally responded, "I'm really, really exhausted, but I promise I'll call you tomorrow." So it wouldn't be until the next morning that they finally connected for a relaxed and rambling chat.

After her press conference, Doudna went to her lab building for a champagne celebration followed by a Zoom party where she was toasted by a hundred or so friends. Mark Zuckerberg and Priscilla Chan, whose foundation was funding some of her work, made a virtual appearance, as did Jillian Banfield and various Berkeley deans and officials. The nicest toast came from Jack Szostak, the Harvard professor who had turned her on to the wonders of RNA back when she was a graduate student. Szostak, who had won a Nobel in medicine in 2009 (jointly with two women), raised a glass of champagne while sitting in the backyard of his stately brick Boston townhouse. "The only thing better than winning a Nobel Prize," he said, "is having one of your students win one."

She and Jamie cooked Spanish omelets for dinner, then Doudna joined her two sisters on a FaceTime call. They talked about how their late parents would have reacted. "I really wish they could have been around," Doudna said. "Mom would have been so emotional, and Dad would have pretended not to be. Instead, he would have made sure he understood the science, then asked me what I planned to do next."

Transformations

By honoring CRISPR, a virus-fighting system found in nature, in the midst of a virus pandemic, the Nobel committee reminded us how curiosity-driven basic research can end up having very practical applications. CRISPR and COVID are speeding our entry into a life-science era. Molecules are becoming the new microchips.

At the height of the coronavirus crisis, Doudna was asked to write a piece for *The Economist* on the social transformations being wrought. "Like many other aspects of life these days, science and its practice seem to be undergoing rapid and perhaps permanent changes," she wrote. "This will be for the better."[2] The public, she predicted, will have more understanding of biology and the scientific method. Elected officials will better appreciate the value of funding basic science. And there will be enduring changes in how scientists collaborate, compete, and communicate.

Before the pandemic, communication and collaboration between academic researchers had become constrained. Universities created large legal teams dedicated to staking a claim to each new discovery, no matter how small, and guarding against any sharing of information that might jeopardize a patent application. "They've turned every interaction scientists have with each other into an intellectual property transaction," says Berkeley biologist Michael Eisen. "Everything I get from or send to a colleague at another academic institution involves a complex legal agreement whose purpose is not to promote science but to protect the university's ability to profit from hypothetical inventions that might arise from scientists doing what we're supposed to do—share our work with each other."[3]

The race to beat COVID was not run by those rules. Instead, led by Doudna and Zhang, most academic labs declared that their discoveries would be made available to anyone fighting the virus. This allowed greater collaboration between researchers and even between countries. The consortium that Doudna put together of labs in the Bay Area could not have coalesced so quickly if they had to worry about intellectual property arrangements. Likewise, scientists around the world contributed to an open database of coronavirus sequences that, by the end of August 2020, had thirty-six thousand entries.[4]

The sense of urgency about COVID also brushed back the gatekeeper role played by expensive, peer-reviewed, paywall-protected scholarly journals such as *Science* and *Nature*. Instead of waiting months for the editors and reviewers to decide whether to publish a paper, researchers at the height of the coronavirus crisis were posting more than a hundred papers a day on preprint servers, such as *medRxiv* and *bioRxiv*, that were free and open and required a minimal review process. This allowed information to be shared in real time, freely, and even be dissected on social media. Despite the potential danger of spreading research that had not been fully vetted, the rapid and open dissemination worked well: it sped up the process of building on each new finding and allowed the public to follow the advance of science as it happened. On some important papers involving coronavirus, publication on the reprint servers led to crowdsourced vetting and wisdom from experts around the world.[5]

George Church says he had long wondered whether there would ever be a biological event that was catalytic enough to bring science into our daily lives. "COVID is it," he says. "Every now and then a meteor hits, and suddenly the mammals are in charge."[6] Most of us someday will have detection devices in our home that will allow us to check for viruses and many other conditions. We will also have wearables with nanopores and molecular transistors that can monitor all of our biological functions, and they will be networked so that they can share information and create a global bio-weather map showing in real time the spread of biological threats. All of this has made biology an even more exciting field of study; in August 2020, applications to medical school had jumped seventeen percent from the previous year.

The academic world will also change, and not just by the rise of more online classes. Instead of being ivory towers, universities will be engaged in tackling real-world problems, from pandemics to climate change. These projects will be cross-disciplinary, breaking down academic silos and the walls between labs, which have traditionally been independent fiefdoms that fiercely guard their autonomy. Fighting the coronavirus required collaboration across disciplines. In that way, it resembled the effort to develop CRISPR, which involved microbe-hunters working with geneticists, structural biologists, biochemists, and computer geeks. It also resembled the way things operate in innovative businesses, where units work together to pursue a specific project or mission. The nature of the scientific threats we face will accelerate this trend toward project-oriented collaborations among disparate labs.

One fundamental aspect of science will remain the same. It has always been a collaboration across generations, from Darwin and Mendel to Watson and Crick to Doudna and Charpentier. "At the end of the day, the discoveries are what endure," Charpentier says. "We are just passing on this planet for a short time. We do our job, and then we leave and others pick up the work."[7]

All of the scientists I write about in this book say that their main motivation is not money, or even glory, but the chance to unlock the mysteries of nature and use those discoveries to make the world a better place. I believe them. And I think that may be one of the most important legacies of the pandemic: reminding scientists of the nobility of their mission. So, too, might it imprint these values on a new generation of students who, as they contemplate their careers, may be more likely to pursue scientific research now that they have seen how exciting and important it can be.

Mardi Gras, 2020

Epilogue

Royal Street, New Orleans, Fall 2020

The Great Pandemic has temporarily receded, and the earth is beginning to heal. I am sitting on my balcony in the French Quarter, and I can again hear music on the street and smell shrimp being boiled at the corner restaurant.

But I know that more viral waves are likely to come, either from the current coronavirus or novel ones in the future, so we need more than just vaccines. Like bacteria, we need a system that can be easily adapted to destroy each new virus. CRISPR could provide that to us, as it does for bacteria. It could also someday be used to fix genetic problems, defeat cancers, enhance our children, and allow us to hack evolution so that we can steer the future of the human race.

I began this journey thinking that biotechnology was the next great scientific revolution, a subject that was filled with awe-inspiring natural wonders, research rivalries, thrilling discoveries, lifesaving triumphs, and creative pioneers such as Jennifer Doudna, Emmanuelle Charpentier, and Feng Zhang. The Year of the Plague made me realize that I was understating the case.

A few weeks ago, I found my old copy of James Watson's *The Double Helix*. Like Doudna, I got the book as a gift from my father when I was in school. It's a first edition with the pale red jacket, and it might be worth something today on eBay except that my sophomoric pencil notes litter the margins, recording and defining words that were new to me, such as "biochemistry."

Reading the book made me, as it made Doudna, want to become a biochemist. Unlike her, I didn't. If I had to do it all over again—pay attention, you students reading this—I would have focused far more on the life sciences, especially if I was coming of age in the twenty-first century. People of my generation became fascinated by personal computers and the web. We made sure our kids learned how to code. Now we will have to make sure they understand the code of life.

One way to do that is for all of us older kids to realize, as the interwoven tales of CRISPR and COVID show, how useful it is to understand the way life works. It's good that some people have strong opinions about the use of GMOs in food, but it would be even better if more of them knew what a genetically modified organism is (and what the yogurt-makers discovered). It's good to have strong opinions about gene engineering in humans, but it's even better if you know what a gene is.

Fathoming the wonders of life is more than merely useful. It is also inspiring and joyful. That is why we humans are lucky that we are endowed with curiosity.

I am reminded of this by a baby lizard that is crawling around the curves of the wrought iron of my balcony and onto a vine, changing color slightly. I become curious: What *causes* the skin to change color? Also, why in heaven's name has this coronavirus plague been followed by such a profusion of lizards? I have to stop myself from conjuring up medieval explanations. I take a quick diversion online to slake my curiosity, and it's a pleasant experience. It reminds me of my favorite note that Leonardo da Vinci scribbled in the margin of one of his crammed notebook pages: "Describe the tongue of the woodpecker." Who wakes up one morning and decides he needs to know what the tongue of a woodpecker looks like? The passionately and playfully curious Leonardo, that's who.

Curiosity is the key trait of the people who have fascinated me, from Benjamin Franklin and Albert Einstein to Steve Jobs and Leonardo da Vinci. Curiosity drove James Watson and the Phage Group, who wanted to understand the viruses that attack bacteria, and the Spanish graduate student Francisco Mojica, who was intrigued by clustered repeated sequences of DNA, and Jennifer Doudna, who wanted to understand what made the sleeping grass curl up when you touched it. And maybe that instinct—curiosity, pure curiosity—is what will save us.

A year ago, after trips to Berkeley and various conferences, I sat on this balcony and tried to process my thoughts about gene editing. My worry then involved the diversity of our species.

I had returned home in time for the funeral of the beloved grande dame of New Orleans, Leah Chase, who died at ninety-six after running a restaurant in the Tremé neighborhood for almost seven decades. With her wooden spoon, she would stir the roux for her shrimp-and-sausage gumbo (one cup of peanut oil and eight tablespoons of flour) until it was the color of café au lait and could bind together the many diverse ingredients. A Creole of color, her restaurant likewise bound together the diversity of New Orleans life, Black and white and Creole.

The French Quarter was hopping that weekend. There was a naked bicycle race that was intended (oddly enough) to promote traffic safety. There were parades and second lines to celebrate the life of Miss Leah and also of Mac Rebennack, the funk musician known as Dr. John. There was the annual Gay Pride Parade and related block parties. And coexisting quite happily was the French Market Creole Tomato Festival, featuring truck farmers and cooks showing off the many varieties of succulent non–genetically modified local tomatoes.

From my balcony, I marveled at the diversity of the passing humanity. There were people short and tall, gay and straight and trans, fat and skinny, light and dark and café au lait. I saw a cluster wearing Gallaudet University T-shirts excitedly using sign language. The supposed promise of CRISPR is that we may someday be able to pick which of these traits we want in our children and in all of our descendants. We could choose for them to be tall and muscular and blond and blue-eyed and not deaf and not—well, pick your preferences.

As I surveyed the scene with all of its natural variety, I pondered how this promise of CRISPR might also be its peril. It took nature millions of years to weave together three billion base pairs of DNA in a complex and occasionally imperfect way to permit all of the wondrous diversity within our species. Are we right to think we can now come along and edit that genome to eliminate what we see as imperfections? Will we lose our diversity? Our humility and empathy? Will we become less flavorful, like our tomatoes?

On Mardi Gras 2020, marchers in the St. Anne parade strutted past our balcony, a few of them dressed as the coronavirus, with bodysuits that mimicked a Corona beer bottle and hoods that made them look like viral rockets. A few weeks later, our shutdown order came. Doreen Ketchens, the beloved clarinetist who plays with her band in front of our corner grocery, gave a farewell-for-now performance to a near-empty sidewalk. She sang a final rendition of "When the Saints Go Marching In," stressing the verse about "when the sun begins to shine."

The mood now is different than it was last year, as are my thoughts on CRISPR. Like our species, my thinking evolves and adapts with changing situations. I now see the promise of CRISPR more clearly than the peril. If we are wise in how we use it, biotechnology can make us more able to fend off viruses, overcome genetic defects, and protect our bodies and minds.

All creatures large and small use whatever tricks they can to survive, and so should we. It's natural. Bacteria came up with a pretty clever virus-fighting technique, but it took them trillions of life cycles to do so. We can't wait that long. We will have to combine our curiosity with our inventiveness to speed up the process.

After millions of centuries during which the evolution of organisms happened "naturally," we humans now have the ability to hack the code of life and engineer our own genetic future. Or, to flummox those who would label gene editing as "unnatural" and "playing God," let's put it another way: Nature and nature's God, in their infinite wisdom, have evolved a species that is able to modify its own genome, and that species happens to be ours.

Like any evolutionary trait, this new ability may help the species thrive and perhaps even produce successor species. Or it may not. It could be one of those evolutionary traits that, as sometimes happens, leads a species down a path that endangers its survival. Evolution is fickle that way.

That's why it works best as a slow process. Every now and then, a rogue or rebel—He Jiankui, Josiah Zayner—will prod us to go faster. But if we are wise, we can pause and decide to proceed with more caution. Slopes are less slippery that way.

To guide us, we will need not only scientists, but humanists. And most important, we will need people who feel comfortable in both worlds, like Jennifer Doudna. This is why it is useful, I think, for all of us to try to understand this new room that we are about to enter, one that seems mysterious but is rich with hope.

Not everything needs to be decided right away. We can begin by asking what type of world we want to leave for our children. Then we can feel our way together, step by step, preferably hand in hand.

ACKNOWLEDGMENTS

I want to thank Jennifer Doudna for her willingness to put up with me. She sat for dozens of interviews, answered my incessant phone calls and emails, allowed me to spend time in her lab, gave me access to a wide variety of meetings, and even let me lurk in her Slack channels. And her husband, Jamie Cate, he put up with me as well, and he also helped.

Feng Zhang was notably gracious. Although the book focuses on his competitor, he cheerfully hosted me in his lab and gave me multiple interviews. I came to like and admire him, just as I do his colleague Eric Lander, who was likewise generous with his time. One of the joys in reporting this book was getting to spend time in Berlin with Emmanuelle Charpentier, who was *charmante*. Although I'm not sure what that means, I know it when I see it, which I hope comes through in these pages. I likewise got a kick out of hanging around George Church, who is a charming (*charmant*?) gentleman disguised as a mad scientist.

Kevin Doxzen of the Innovative Genomics Institute and Spencer Olesky of Tulane were the scientific vetters of this book. They provided very smart comments and corrections. Max Wendell, Benjamin Bernstein, and Ryan Braun of Tulane also chipped in. All of them were wonderful, so please don't blame them for any mistakes that crept in.

I am also grateful to all of the scientists and their fans who spent time with me, provided insights, gave interviews, and checked facts: Noubar Afeyan, Richard Axel, David Baltimore, Jillian Banfield, Cori Bargmann, Rodolphe Barrangou, Joe Bondy-Denomy, Dana Carroll,

Janice Chen, Francis Collins, Kevin Davies, Meredith DeSalazar, Phil Dormitzer, Sarah Doudna, Kevin Doxzen, Victor Dzau, Eldora Ellison, Sarah Goodwin, Margaret Hamburg, Jennifer Hamilton, Lucas Harrington, Rachel Haurwitz, Christine Heenan, Don Hemmes, Megan Hochstrasser, Patrick Hsu, Maria Jasin, Martin Jinek, Elliot Kirschner, Gavin Knott, Eric Lander, Le Cong, Richard Lifton, Enrique Lin Shiao, David Liu, Luciano Marraffini, Alex Marson, Andy May, Sylvain Moineau, Francisco Mojica, Cameron Myhrvold, Rodger Novak, Val Pakaluk, Duanqing Pei, Matthew Porteus, Stanley Qi, Antonio Regalado, Matt Ridley, Dave Savage, Jacob Sherkow, Virginijus Šikšnys, Erik Sontheimer, Sam Sternberg, Jack Szostak, Fyodor Urnov, Elizabeth Watson, James Watson, Jonathan Weissman, Blake Wiedenheft, Ross Wilson, and Josiah Zayner.

As always, I owe deep thanks to Amanda Urban, my agent for forty years now. She manages to be caring and intellectually honest at the same time, which is very bracing. Priscilla Painton and I worked together at *Time* when we were in our salad days and were neighbors when our kids were in their pre-salad ones. Suddenly, she is now my editor. It's sweet the way the world turns. She did a diligent and smart job both restructuring this book at one point and polishing it line by line.

Science is a collaborative effort. So is producing a book. The joy of being with Simon & Schuster is that I get to work with a great team led by the irrepressible and insightful Jonathan Karp, who seemed to read this manuscript many times and kept suggesting improvements. It includes Stephen Bedford, Dana Canedy, Jonathan Evans, Marie Florio, Kimberly Goldstein, Judith Hoover, Ruth Lee-Mui, Hana Park, Julia Prosser, Richard Rhorer, Elise Ringo, and Jackie Seow. Helen Manders and Peppa Mignone at Curtis Brown did a wonderful job working with international publishers. I also want to thank Lindsey Billups, my assistant, who is smart and wise and very sensible. Her help every day was invaluable.

My greatest thanks as always go to my wife, Cathy, who helped with the research, carefully read my drafts, provided sage counsel, and kept me on an even keel (or tried). Our daughter, Betsy, also read the manuscript and made smart suggestions. They are the foundations of my life.

This book was launched by Alice Mayhew, who was the editor of all of my previous books. In our first discussions, I was amazed at how well she knew the science. She was relentless in insisting that I make this book a journey of discovery. She had edited the classic of the genre, Horace Freeland Judson's *The Eighth Day of Creation*, back in 1979, and forty years later she seemed to remember every passage in it. Over the 2019 Christmas holidays, she read the first half of this book and came back with a torrent of joyful comments and insights. But she didn't live to see it finished. Nor did dear Carolyn Reidy, the CEO of Simon & Schuster, who had always been a mentor, guide, and joy to know. One of life's great pleasures was to make Alice and Carolyn smile. If you'd ever seen their smiles, you'd understand. I hope this book would have. I've dedicated it to their memory.

NOTES

Introduction: Into the Breach

1. Author's interview with Jennifer Doudna. The competition was run by First Robotics, a nationwide program created by the irrepressible Segway inventor Dean Kamen.
2. Interviews, audio and video recordings, notes, and slides provided by Jennifer Doudna, Megan Hochstrasser, and Fyodor Urnov; Walter Isaacson, "Ivory Power," *Air Mail*, Apr. 11, 2020.
3. See chapter 12 on the yogurt makers for a fuller discussion of the iterative process that can occur between basic researchers and technological innovation.

Chapter 1: Hilo

1. Author's interviews with Jennifer Doudna and Sarah Doudna. Other sources for this section include *The Life Scientific*, BBC Radio, Sept. 17, 2017; Andrew Pollack, "Jennifer Doudna, a Pioneer Who Helped Simplify Genome Editing," *New York Times*, May 11, 2015; Claudia Dreifus, "The Joy of the Discovery: An Interview with Jennifer Doudna," *New York Review of Books*, Jan. 24, 2019; Jennifer Doudna interview, National Academy of Sciences, Nov. 11, 2004; Jennifer Doudna, "Why Genome Editing Will Change Our Lives," *Financial Times*, Mar. 14, 2018; Laura Kiessling, "A Conversation with Jennifer Doudna," *ACS Chemical Biology Journal*, Feb. 16, 2018; Melissa Marino, "Biography of Jennifer A. Doudna," *PNAS*, Dec. 7, 2004.
2. Dreifus, "The Joy of the Discovery."
3. Author's interviews with Lisa Twigg-Smith, Jennifer Doudna.
4. Author's interviews with Jennifer Doudna, James Watson.
5. Jennifer Doudna, "How COVID-19 Is Spurring Science to Accelerate," *The Economist*, June 5, 2020.

Chapter 2: The Gene

1. This section on the history of genetics and DNA relies on Siddhartha Mukherjee, *The Gene* (Scribner, 2016); Horace Freeland Judson, *The Eighth Day of Creation* (Touchstone, 1979); Alfred Sturtevant, *A History of Genetics* (Cold Spring Harbor, 2001); Elof Axel Carlson, *Mendel's Legacy* (Cold Spring Harbor, 2004).
2. Janet Browne, *Charles Darwin*, vol. 1 (Knopf, 1995) and vol. 2 (Knopf, 2002); Charles Darwin, *The Journey of the Beagle*, originally published 1839; Darwin, *On the Origin of Species*, originally published 1859. Electronic copies of Darwin's books, letters, writings, and journals can be found at Darwin Online, darwin-online.org.uk.

3. Isaac Asimov, "How Do People Get New Ideas," 1959, reprinted in *MIT Technology Review*, Oct. 20, 2014; Steven Johnson, *Where Good Ideas Come From* (Riverhead, 2010), 81; Charles Darwin, *Autobiography*, describing events of October 1838, Darwin Online, darwin-online.org.uk.
4. In addition to Mukherjee, Judson, and Sturtevant, this section on Mendel also draws from Robin Marantz Henig, *The Monk in the Garden* (Houghton Mifflin Harcourt, 2000).
5 Erwin Chargaff, "Preface to a Grammar of Biology," *Science*, May 14, 1971.

Chapter 3: DNA

1. This section draws from my multiple interviews with James Watson over a period of years and from his book *The Double Helix*, originally published by Atheneum in 1968. I used *The Annotated and Illustrated Double Helix*, compiled by Alexander Gann and Jan Witkowski (Simon & Schuster, 2012), which includes the letters describing the DNA model and other supplemental material. This section also draws from James Watson, *Avoid Boring People* (Oxford, 2007); Brenda Maddox, *Rosalind Franklin: The Dark Lady of DNA* (HarperCollins, 2002); Judson, *The Eighth Day*; Mukherjee, *The Gene*; Sturtevant, *A History of Genetics*.
2. Judson says Watson was turned down at Harvard; Watson said to me and in *Avoid Boring People* that he was accepted but not offered a stipend or financing.
3. The youngest person to win a Nobel Prize is now Malala Yousafzai from Pakistan, who won the Peace Prize. She was shot by the Taliban and became a fighter for girls' education.
4. Mukherjee, *The Gene*, 147.
5. Rosalind Franklin, "The DNA Riddle: King's College, London, 1951–1953," Rosalind Franklin Papers, NIH National Library of Medicine, https://profiles.nlm.nih.gov/spotlight/kr/feature/dna; Nicholas Wade, "Was She or Wasn't She?," *The Scientist*, Apr. 2003; Judson, *The Eighth Day*, 99; Maddox, *Rosalind Franklin*, 163; Mukherjee, *The Gene*, 149.

Chapter 4: The Education of a Biochemist

1. Author's interviews with Jennifer Doudna.
2. Author's interviews with Jennifer Doudna.
3. Author's email interviews with Don Hemmes.
4. Author's interviews with Jennifer Doudna; Jennifer A. Doudna and Samuel H. Sternberg, *A Crack in Creation* (Houghton Mifflin, 2017), 58; Kiessling, "A Conversation with Jennifer Doudna"; Pollack, "Jennifer Doudna."
5. Unless otherwise noted, all Jennifer Doudna quotes in this section are from my interviews with her.
6. Sharon Panasenko, "Methylation of Macromolecules during Development in *Myxococcus xanthus*," *Journal of Bacteriology*, Nov. 1985 (submitted July 1985).

Chapter 5: The Human Genome

1. The Department of Energy launched work on sequencing the human genome in 1986. Official funding for the Human Genome Project was in President Reagan's 1988 budget submission. The Department of Energy and the National Institutes of Health signed a memorandum of understanding to formalize the Human Genome Project in 1990.

2. Daniel Okrent, *The Guarded Gate* (Scribner, 2019).
3. "Decoding Watson," directed and produced by Mark Mannucci, *American Masters*, PBS, Jan. 2, 2019.
4. Author's interviews and meetings with James Watson, Elizabeth Watson, and Rufus Watson; Algis Valiunas, "The Evangelist of Molecular Biology," *The New Atlantis*, Summer 2017; James Watson, *A Passion for DNA* (Oxford, 2003); Philip Sherwell, "DNA Father James Watson's 'Holy Grail' Request," *The Telegraph*, May 10, 2009; Nicholas Wade, "Genome of DNA Discoverer Is Deciphered," *New York Times*, June 1, 2007.
5. Author's interviews with George Church, Eric Lander, and James Watson.
6. Frederic Golden and Michael D. Lemonick, "The Race Is Over," and James Watson, "The Double Helix Revisited," *Time*, July 3, 2000; author's conversations with Al Gore, Craig Venter, James Watson, George Church, and Francis Collins.
7. Author's own notes from the White House ceremony; Nicholas Wade, "Genetic Code of Human Life Is Cracked by Scientists," *New York Times*, June 27, 2000.

Chapter 6: RNA

1. Mukherjee, *The Gene*, 250.
2. Jennifer Doudna, "Hammering Out the Shape of a Ribozyme," *Structure*, Dec. 15, 1994.
3. Jennifer Doudna and Thomas Cech, "The Chemical Repertoire of Natural Ribozymes," *Nature*, July 11, 2002.
4. Author's interviews with Jack Szostak and Jennifer Doudna; Jennifer Doudna, "Towards the Design of an RNA Replicase," PhD thesis, Harvard University, May 1989.
5. Author's interviews with Jack Szostak, Jennifer Doudna.
6. Jeremy Murray and Jennifer Doudna, "Creative Catalysis," *Trends in Biochemical Sciences*, Dec. 2001; Tom Cech, "The RNA Worlds in Context," *Cold Spring Harbor Perspectives in Biology*, July 2012; Francis Crick, "The Origin of the Genetic Code," *Journal of Molecular Biology*, Dec. 28, 1968; Carl Woese, *The Genetic Code* (Harper & Row, 1967), 186; Walter Gilbert, "The RNA World," *Nature*, Feb. 20, 1986.
7. Jack Szostak, "Enzymatic Activity of the Conserved Core of a Group I Self-Splicing Intron," *Nature*, July 3, 1986.
8. Author's interviews with Richard Lifton, Jennifer Doudna, Jack Szostak; Greengard Prize citation for Jennifer Doudna, Oct. 2, 2018; Jennifer Doudna and Jack Szostak, "RNA-Catalysed Synthesis of Complementary-Strand RNA," *Nature*, June 15, 1989; J. Doudna, S. Couture, and J. Szostak, "A Multisubunit Ribozyme That Is a Catalyst of and Template for Complementary Strand RNA Synthesis," *Science*, Mar. 29, 1991; J. Doudna, N. Usman, and J. Szostak, "Ribozyme-Catalyzed Primer Extension by Trinucleotides," *Biochemistry*, Mar. 2, 1993.
9. Jayaraj Rajagopal, Jennifer Doudna, and Jack Szostak, "Stereochemical Course of Catalysis by the Tetrahymena Ribozyme," *Science*, May 12, 1989; Doudna and Szostak, "RNA-Catalysed Synthesis of Complementary-Strand RNA"; J. Doudna, B. P. Cormack, and J. Szostak, "RNA Structure, Not Sequence, Determines the 5' Splice-Site Specificity of a Group I Intron," *PNAS*, Oct. 1989; J. Doudna and J. Szostak, "Miniribozymes, Small Derivatives of the sunY Intron, Are Catalytically Active," *Molecular and Cell Biology*, Dec. 1989.
10. Author's interviews with Jack Szostak.
11. Author's interview with James Watson; James Watson et al., "Evolution of Catalytic Function," Cold Spring Harbor Symposium, vol. 52, 1987.

12. Author's interviews with Jennifer Doudna and James Watson; Jennifer Doudna . . . Jack Szostak, et al., "Genetic Dissection of an RNA Enzyme," Cold Spring Harbor Symposium, 1987, p. 173.

Chapter 7: Twists and Folds

1. Author's interviews with Jack Szostak, Jennifer Doudna.
2. Pollack, "Jennifer Doudna."
3. Author's interview with Lisa Twigg-Smith.
4. Jamie Cate . . . Thomas Cech, Jennifer Doudna, et al., "Crystal Structure of a Group I Ribozyme Domain: Principles of RNA Packing," *Science*, Sept. 20, 1996. For the first major step in the Boulder research, see Jennifer Doudna and Thomas Cech, "Self-Assembly of a Group I Intron Active Site from Its Component Tertiary Structural Domains," *RNA*, Mar. 1995.
5. NewsChannel 8 report, "High Tech Shower International," YouTube, May 29, 2018, https://www.youtube.com/watch?v=FxPFLbfrpNk&feature=share.

Chapter 8: Berkeley

1. Cate et al., "Crystal Structure of a Group I Ribozyme Domain."
2. Author's interviews with Jamie Cate, Jennifer Doudna.
3. Andrew Fire . . . Craig Mello, et al., "Potent and Specific Genetic Interference by Double-Stranded RNA in *Caenorhabditis elegans*," *Nature*, Feb. 19, 1998.
4. Author's interviews with Jennifer Doudna, Martin Jinek, Ross Wilson; Ian MacRae, Kaihong Zhou . . . Jennifer Doudna, et al., "Structural Basis for Double-Stranded RNA Processing by Dicer," *Science*, Jan. 13, 2006; Ian MacRae, Kaihong Zhou, and Jennifer Doudna, "Structural Determinants of RNA Recognition and Cleavage by Dicer," *Natural Structural and Molecular Biology*, Oct. 1, 2007; Ross Wilson and Jennifer Doudna, "Molecular Mechanisms of RNA Interference," *Annual Review of Biophysics*, 2013; Martin Jinek and Jennifer Doudna, "A Three-Dimensional View of the Molecular Machinery of RNA Interference," *Nature*, Jan. 22, 2009.
5. Bryan Cullen, "Viruses and RNA Interference: Issues and Controversies," *Journal of Virology*, Nov. 2014.
6. Ross Wilson and Jennifer Doudna, "Molecular Mechanisms of RNA Interference," *Annual Review of Biophysics*, May 2013.
7. Alesia Levanova and Minna Poranen, "RNA Interference as a Prospective Tool for the Control of Human Viral Infections," *Frontiers of Microbiology*, Sept. 11, 2018; Ruth Williams, "Fighting Viruses with RNAi," *The Scientist*, Oct. 10, 2013; Yang Li . . . Shou-Wei Ding, et al., "RNA Interference Functions as an Antiviral Immunity Mechanism in Mammals," *Science*, Oct. 11, 2013; Pierre Maillard . . . Olivier Voinnet, et al., "Antiviral RNA Interference in Mammalian Cells," *Science*, Oct. 11, 2013.

Chapter 9: Clustered Repeats

1. Yoshizumi Ishino . . . Atsuo Nakata, et al., "Nucleotide Sequence of the iap Gene, Responsible for Alkaline Phosphatase Isozyme Conversion in *Escherichia coli*," *Journal of Bacteriology*, Aug. 22, 1987; Yoshizumi Ishino et al., "History of CRISPR-Cas from Encounter with a Mysterious Repeated Sequence to Genome Editing Technology," *Journal of Bacteriology*, Jan. 22, 2018; Carl Zimmer, "Breakthrough DNA Editor Born of Bacteria," *Quanta*, Feb. 6, 2015.
2. Author's interviews with Francisco Mojica. This section also draws from Kevin Davies,

"Crazy about CRISPR: An Interview with Francisco Mojica," *CRISPR Journal*, Feb. 1, 2018; Heidi Ledford, "Five Big Mysteries about CRISPR's Origins," *Nature*, Jan. 12, 2017; Clara Rodríguez Fernández, "Interview with Francis Mojica, the Spanish Scientist Who Discovered CRISPR," *Labiotech*, Apr. 8, 2019; Veronique Greenwood, "The Unbearable Weirdness of CRISPR," *Nautilus*, Mar. 2017; Francisco Mojica and Lluis Montoliu, "On the Origin of CRISPR-Cas Technology," *Trends in Microbiology*, July 8, 2016; Kevin Davies, *Editing Humanity* (Simon & Schuster, 2020).

3. Francesco Mojica . . . Francisco Rodriguez-Valera, et al., "Long Stretches of Short Tandem Repeats Are Present in the Largest Replicons of the Archaea *Haloferax mediterranei* and *Haloferax volcanii* and Could Be Involved in Replicon Partitioning," *Journal of Molecular Microbiology*, July 1995.
4. Email from Ruud Jansen to Francisco Mojica, Nov. 21, 2001.
5. Ruud Jansen . . . Leo Schouls, et al., "Identification of Genes That Are Associated with DNA Repeats in Prokaryotes," *Molecular Biology*, Apr. 25, 2002.
6. Author's interviews with Francisco Mojica.
7. Sanne Klompe and Samuel Sternberg, "Harnessing 'a Billion Years of Experimentation,'" *CRISPR Journal*, Apr. 1, 2018; Eric Keen, "A Century of Phage Research," *Bioessays*, Jan. 2015; Graham Hatfull and Roger Hendrix, "Bacteriophages and Their Genomes," *Current Opinions in Virology*, Oct. 1, 2011.
8. Rodríguez Fernández, "Interview with Francis Mojica"; Greenwood, "The Unbearable Weirdness of CRISPR."
9. Author's interviews with Francisco Mojica; Rodríguez Fernández, "Interview with Francis Mojica"; Davies, "Crazy about CRISPR."
10. Francisco Mojica . . . Elena Soria, et al., "Intervening Sequences of Regularly Spaced Prokaryotic Repeats Derive from Foreign Genetic Elements," *Journal of Molecular Evolution*, Feb. 2005 (received Feb. 6, 2004; accepted Oct. 1, 2004).
11. Kira Makarova . . . Eugene Koonin, et al., "A Putative RNA-Interference-Based Immune System in Prokaryotes," *Biology Direct*, Mar. 16, 2006.

Chapter 10: The Free Speech Movement Café

1. Author's interviews with Jillian Banfield and Jennifer Doudna; Doudna and Sternberg, *A Crack in Creation*, 39; "Deep Surface Biospheres," Banfield Lab page, Berkeley University website.
2. Author's joint interview with Jillian Banfield and Jennifer Doudna.
3. Author's interview with Jennifer Doudna.

Chapter 11: Jumping In

1. Author's interviews with Blake Wiedenheft and Jennifer Doudna.
2. Kathryn Calkins, "Finding Adventure: Blake Wiedenheft's Path to Gene Editing," National Institute of General Medical Sciences, Apr. 11, 2016.
3. Emily Stifler Wolfe, "Insatiable Curiosity: Blake Wiedenheft Is at the Forefront of CRISPR Research," *Montana State University News*, June 6, 2017.
4. Blake Wiedenheft . . . Mark Young, and Trevor Douglas, "An Archaeal Antioxidant: Characterization of a Dps-Like Protein from *Sulfolobus solfataricus*," *PNAS*, July 26, 2005.
5. Author's interview with Blake Wiedenheft.
6. Author's interview with Blake Wiedenheft.
7. Author's interviews with Martin Jinek, Jennifer Doudna.

8. Kevin Davies, "Interview with Martin Jínek," *CRISPR Journal*, Apr. 2020.
9. Author's interview with Martin Jinek.
10. Jinek and Doudna, "A Three-Dimensional View of the Molecular Machinery of RNA Interference"; Martin Jinek, Scott Coyle, and Jennifer A. Doudna, "Coupled 5' Nucleotide Recognition and Processivity in Xrn1-Mediated mRNA Decay," *Molecular Cell*, Mar. 4, 2011.
11. Author's interviews with Blake Wiedenheft, Martin Jinek, Rachel Haurwitz, Jennifer Doudna.
12. Author's interviews with Blake Wiedenheft, Jennifer Doudna; Blake Wiedenheft, Kaihong Zhou, Martin Jinek . . . Jennifer Doudna, et al., "Structural Basis for DNase Activity of a Conserved Protein Implicated in CRISPR-Mediated Genome Defense," *Structure*, June 10, 2009.
13. Jinek and Doudna, "A Three-Dimensional View of the Molecular Machinery of RNA Interference."
14. Author's interviews with Martin Jinek, Blake Wiedenheft, Jennifer Doudna.
15. Wiedenheft et al., "Structural Basis for DNase Activity of a Conserved Protein."

Chapter 12: The Yogurt Makers

1. Vannevar Bush, "Science, the Endless Frontier," Office of Scientific Research and Development, July 25, 1945.
2. Matt Ridley, *How Innovation Works* (Harper Collins, 2020), 282.
3. Author's interviews with Rodolphe Barrangou.
4. Rodolphe Barrangou and Philippe Horvath, "A Decade of Discovery: CRISPR Functions and Applications," *Nature Microbiology,* June 5, 2017; Prashant Nair, "Interview with Rodolphe Barrangou," *PNAS*, July 11, 2017; author's interviews with Rodolphe Barrangou.
5. Author's interviews with Rodolphe Barrangou.
6. Rodolphe Barrangou . . . Sylvain Moineau . . . Philippe Horvath, et al., "CRISPR Provides Acquired Resistance against Viruses in Prokaryotes," *Science*, Mar. 23, 2007 (submitted Nov. 29, 2006; accepted Feb. 16, 2007).
7. Author's interviews with Sylvain Moineau, Jillian Banfield, and Rodolphe Barrangou. Conference agendas 2008–2012 provided by Banfield.
8. Author's interview with Luciano Marraffini.
9. Author's interview with Erik Sontheimer.
10. Author's interviews with Erik Sontheimer, Luciano Marraffini; Luciano Marraffini and Erik Sontheimer, "CRISPR Interference Limits Horizontal Gene Transfer in Staphylococci by Targeting DNA," *Science*, Dec. 19, 2008; Erik Sontheimer and Luciano Marraffini, "Target DNA Interference with crRNA," U.S. Provisional Patent Application 61/009,317, Sept. 23, 2008; Erik Sontheimer, letter of intent, National Institutes of Health, Dec. 29, 2008.
11. Doudna and Sternberg, *A Crack in Creation*, 62.

Chapter 13: Genentech

1. Author's interviews with Jillian Banfield and Jennifer Doudna.
2. Eugene Russo, "The Birth of Biotechnology," *Nature*, Jan. 23, 2003; Mukherjee, *The Gene,* 230.
3. Rajendra Bera, "The Story of the Cohen-Boyer Patents," *Current Science*, Mar. 25, 2009; US Patent 4,237,224 "Process for Producing Biologically Functional Molecular

Chimeras," Stanley Cohen and Herbert Boyer, filed Nov. 4, 1974; Mukherjee, *The Gene*, 237.
4. Mukherjee, *The Gene*, 238.
5. Frederic Golden, "Shaping Life in the Lab," *Time*, Mar. 9, 1981; Laura Fraser, "Cloning Insulin," Genentech corporate history; *San Francisco Examiner* front page, Oct. 14, 1980.
6. Author's interview with Rachel Haurwitz.
7. Author's interview with Jennifer Doudna.

Chapter 14: The Lab

1. Author's interviews with Rachel Haurwitz, Blake Wiedenheft, Jennifer Doudna.
2. Author's interview with Rachel Haurwitz.
3. Rachel Haurwitz, Martin Jinek, Blake Wiedenheft, Kaihong Zhou, and Jennifer Doudna, "Sequence- and Structure-Specific RNA Processing by a CRISPR Endonuclease," *Science*, Sept. 10, 2010.
4. Samuel Sternberg . . . Ruben L. Gonzalez Jr., et al., "Translation Factors Direct Intrinsic Ribosome Dynamics during Translation Termination and Ribosome Recycling," *Nature Structural and Molecular Biology*, July 13, 2009.
5. Author's interviews with Sam Sternberg.
6. Author's interviews with Sam Sternberg, Jennifer Doudna.
7. Author's interviews with Sam Sternberg, Jennifer Doudna; Sam Sternberg, "Mechanism and Engineering of CRISPR-Associated Endonucleases," PhD thesis, University of California, Berkeley, 2014.
8. Samuel Sternberg, . . . and Jennifer Doudna, "DNA Interrogation by the CRISPR RNA-Guided Endonuclease Cas9," *Nature*, Jan. 29, 2014; Sy Redding, Sam Sternberg . . . Blake Wiedenheft, Jennifer Doudna, Eric Greene, et al., "Surveillance and Processing of Foreign DNA by the *Escherichia coli* CRISPR-Cas System," *Cell*, Nov. 5, 2015.
9. Blake Wiedenheft, Samuel H. Sternberg, and Jennifer A. Doudna, "RNA-Guided Genetic Silencing Systems in Bacteria and Archaea," *Nature*, Feb. 14, 2012.
10. Author's interviews with Sam Sternberg.
11. Author's interviews with Ross Wilson, Martin Jinek.
12. Marc Lerchenmueller, Olav Sorenson, and Anupam Jena, "Gender Differences in How Scientists Present the Importance of Their Research," *BMJ*, Dec. 19, 2019; Olga Khazan, "Carry Yourself with the Confidence of a Male Scientist," *Atlantic*, Dec. 17, 2019.
13. Author's interviews with Blake Wiedenheft, Jennifer Doudna; Blake Wiedenheft, Gabriel C. Lander, Kaihong Zhou, Matthijs M. Jore, Stan J. J. Brouns, John van der Oost, Jennifer A. Doudna, and Eva Nogales, "Structures of the RNA-Guided Surveillance Complex from a Bacterial Immune System," *Nature*, Sept. 21, 2011 (received May 7, 2011; accepted July 27, 2011).

Chapter 15: Caribou

1. Author's interview with Jennifer Doudna and Rachel Haurwitz.
2. Gary Pisano, "Can Science Be a Business?," *Harvard Business Review*, Oct. 2006; Saurabh Bhatia, "History, Scope and Development of Biotechnology," *IPO Science*, May 2018.
3. Author's interviews with Rachel Haurwitz, Jennifer Doudna.
4. Bush, "Science, the Endless Frontier."
5. "Sparking Economic Growth," The Science Coalition, April 2017.

6. "Kit for Global RNP Profiling," NIH award 1R43GM105087-01, for Rachel Haurwitz and Caribou Biosciences, Apr. 15, 2013.
7. Author's interviews with Jennifer Doudna, Rachel Haurwitz; Robert Sanders, "Gates Foundation Awards $100,000 Grants for Novel Global Health Research," *Berkeley News*, May 10, 2010.

Chapter 16: Emmanuelle Charpentier

1. Author's interviews with Emmanuelle Charpentier. This chapter also draws from Uta Deffke, "An Artist in Gene Editing," *Max Planck Research Magazine*, Jan. 2016; "Interview with Emmanuelle Charpentier," *FEMS Microbiology Letters*, Feb. 1, 2018; Alison Abbott, "A CRISPR Vision," *Nature*, Apr. 28, 2016; Kevin Davies, "Finding Her Niche: An Interview with Emmanuelle Charpentier," *CRISPR Journal*, Feb. 21, 2019; Margaret Knox, "The Gene Genie," *Scientific American*, Dec. 2014; Jennifer Doudna, "Why Genome Editing Will Change Our Lives," *Financial Times*, Mar. 24, 2018; Martin Jinek, Krzysztof Chylinski, Ines Fonfara, Michael Hauer, Jennifer Doudna, and Emmanuelle Charpentier, "A Programmable Dual-RNA–Guided DNA Endonuclease in Adaptive Bacterial Immunity," *Science*, Aug. 17, 2012.
2. Author's interview with Emmanuelle Charpentier.
3. Author's interviews with Rodger Novak, Emmanuelle Charpentier; Rodger Novak, Emmanuelle Charpentier, Johann S. Braun, and Elaine Tuomanen, "Signal Transduction by a Death Signal Peptide Uncovering the Mechanism of Bacterial Killing by Penicillin," *Molecular Cell*, Jan. 1, 2000.
4. Emmanuelle Charpentier . . . Pamela Cowin, et al., "Plakoglobin Suppresses Epithelial Proliferation and Hair Growth in Vivo," *Journal of Cell Biology*, May 2000; Monika Mangold . . . Rodger Novak, Richard Novick, Emmanuelle Charpentier, et al., "Synthesis of Group A Streptococcal Virulence Factors Is Controlled by a Regulatory RNA Molecule," *Molecular Biology*, Aug. 3, 2004; Davies, "Finding Her Niche"; Philip Hemme, "Fireside Chat with Rodger Novak," *Refresh Berlin*, May 24, 2016, Labiotech.eu.
5. Author's interview with Emmanuelle Charpentier.
6. Elitza Deltcheva, Krzysztof Chylinski . . . Emmanuelle Charpentier, et al., "CRISPR RNA Maturation by Trans-encoded Small RNA and Host Factor RNase III," *Nature*, Mar. 31, 2011.
7. Author's interviews with Emmanuelle Charpentier, Jennifer Doudna, Erik Sontheimer; Doudna and Sternberg, *A Crack in Creation*, 71–73.
8. Author's interviews with Martin Jinek, Jennifer Doudna. See also Kevin Davies, interview with Martin Jinek, *CRISPR Journal*, Apr. 2020.

Chapter 17: CRISPR-Cas9

1. Author's interviews with Martin Jinek, Jennifer Doudna, Emmanuelle Charpentier.
2. Richard Asher, "An Interview with Krzysztof Chylinski," *Pioneers Zero21*, Oct. 2018.
3. Author's interviews with Jennifer Doudna, Emmanuelle Charpentier, Martin Jinek, Ross Wilson.
4. Author's interviews with Jennifer Doudna, Martin Jinek.
5. Author's interviews with Jennifer Doudna, Martin Jinek, Sam Sternberg, Rachel Haurwitz, Ross Wilson.

Chapter 18: Science, *2012*

1. Author's interviews with Jennifer Doudna, Emmanuelle Charpentier, and Martin Jinek.
2. Jinek et al., "A Programmable Dual-RNA–Guided DNA Endonuclease in Adaptive Bacterial Immunity."
3. Author's interview with Emmanuelle Charpentier.
4. Author's interviews with Emmanuelle Charpentier, Jennifer Doudna, Martin Jinek, and Sam Sternberg.

Chapter 19: Dueling Presentations

1. Author's interview with Virginijus Šikšnys.
2. Giedrius Gasiunas, Rodolphe Barrangou, Philippe Horvath, and Virginijus Šikšnys, "Cas9–crRNA Ribonucleoprotein Complex Mediates Specific DNA Cleavage for Adaptive Immunity in Bacteria," *PNAS*, Sept. 25, 2012 (received May 21, 2012; approved Aug. 1; published online Sept. 4).
3. Author's interview with Rodolphe Barrangou.
4. Author's interview with Eric Lander.
5. Author's interviews with Erik Lander, Jennifer Doudna.
6. Author's interview with Rodolphe Barrangou.
7. Virginijus Šikšnys et al., "RNA-Directed Cleavage by the Cas9-crRNA Complex," international patent application WO 2013/142578 Al, priority date Mar. 20, 2012, official filing Mar. 20, 2013, publication Sept. 26, 2013.
8. Author's interviews with Virginijus Šikšnys, Jennifer Doudna, Sam Sternberg, Emmanuelle Charpentier, and Martin Jinek.
9. Author's interviews with Sam Sternberg, Rodolph Barrangou, Erik Sontheimer, Virginijus Šikšnys, Jennifer Doudna, Martin Jinek, and Emmanuelle Charpentier.

Chapter 20: A Human Tool

1. Srinivasan Chandrasegaran and Dana Carroll, "Origins of Programmable Nucleases for Genome Engineering," *Journal of Molecular Biology*, Feb. 27, 2016.

Chapter 21: The Race

1. Author's interviews with Jennifer Doudna; Doudna and Sternberg, *A Crack in Creation*, 242.
2. Ferric C. Fang and Arturo Casadevall, "Is Competition Ruining Science?," *American Society for Microbiology*, Apr. 2015; Melissa Anderson . . . Brian Martinson, et al., "The Perverse Effects of Competition on Scientists' Work and Relationships," *Science Engineering Ethics*, Dec. 2007; Matt Ridley, "Two Cheers for Scientific Backbiting," *Wall Street Journal*, July 27, 2012.
3. Author's interview with Emmanuelle Charpentier.

Chapter 22: Feng Zhang

1. Author's interviews with Feng Zhang. This section also draws from Eric Topol, podcast interview with Feng Zhang, Medscape, Mar. 31, 2017; Michael Specter, "The Gene Hackers," *New Yorker*, Nov. 8, 2015; Sharon Begley, "Meet One of the World's Most Groundbreaking Scientists," *Stat*, Nov. 6, 2015.
2. Galen Johnson, "Gifted and Talented Education Grades K–12 Program Evaluation," Des Moines Public Schools, September 1996.

3. Edward Boyden, Feng Zhang, Ernst Bamberg, Georg Nagel, and Karl Deisseroth, "Millisecond-Timescale, Genetically Targeted Optical Control of Neural Activity," *Nature Neuroscience*, Aug. 14, 2005; Alexander Aravanis, Li-Ping Wang, Feng Zhang . . . and Karl Deisseroth, "An Optical Neural Interface: In vivo Control of Rodent Motor Cortex with Integrated Fiberoptic and Optogenetic Technology," *Journal of Neural Engineering*, Sept. 2007.
4. Feng Zhang, Le Cong, Simona Lodato, Sriram Kosuri, George M. Church, and Paola Arlotta, "Efficient Construction of Sequence-Specific TAL Effectors for Modulating Mammalian Transcription," *Nature Biotechnology*, Jan. 19, 2011.

Chapter 23: George Church

1. This section is based on author's interviews and visits with George Church and also Ben Mezrich, *Woolly* (Atria, 2017); Anna Azvolinsky, "Curious George," *The Scientist*, Oct. 1, 2016; Sharon Begley, "George Church Has a Wild Idea to Upend Evolution," *Stat*, May 16, 2016; Prashant Nair, "George Church," *PNAS*, July 24, 2012; Jeneen Interlandi, "The Church of George Church," *Popular Science*, May 27, 2015.
2. Mezrich, *Woolly*, 43.
3. George Church Oral History, National Human Genome Research Institute, July 26, 2017.
4. Nicholas Wade, "Regenerating a Mammoth for $10 Million," *New York Times*, Nov. 19, 2008; Nicholas Wade, "The Wooly Mammoth's Last Stand," *New York Times*, Mar. 2, 2017; Mezrich, *Woolly*.
5. Author's interviews with George Church and Jennifer Doudna.

Chapter 24: Zhang Tackles CRISPR

1. Josiane Garneau . . . Rodolphe Barrangou . . . Philippe Horvath, Alfonso H. Magadán, and Sylvain Moineau, "The CRISPR/Cas Bacterial Immune System Cleaves Bacteriophage and Plasmid DNA," *Nature*, Nov. 3, 2010.
2. Davies, *Editing Humanity*, 80; author's interview with Le Cong.
3. Author's interviews with Eric Lander, Feng Zhang; Begley, "George Church Has a Wild Idea . . ."; Michael Specter, "The Gene Hackers," *New Yorker*, Nov. 8, 2015; Davies, *Editing Humanity*, 82.
4. Feng Zhang, "Confidential Memorandum of Invention," Feb. 13, 2013.
5. David Altshuler, Chad Cowan, Feng Zhang, et al., Grant application 1R01DK097758-01, "Isogenic Human Pluripotent Stem Cell-Based Models of Human Disease Mutations," National Institutes of Health, Jan. 12, 2012.
6. Broad Opposition 3; UC reply 3.
7. Author's interviews with Luciano Marraffini and Erik Sontheimer; Marraffini and Sontheimer, "CRISPR Interference Limits Horizontal Gene Transfer in Staphylococci by Targeting DNA"; Sontheimer and Marraffini, "Target DNA Interference with crRNA," U.S. Provisional Patent Application; Kevin Davies, "Interview with Luciano Marraffini," *CRISPR Journal*, Feb. 2020.
8. Author's interviews with Luciano Marraffini and Feng Zhang; Zhang email to Marraffini, Jan. 2, 2012 (given to me by Marraffini).
9. Marraffini email to Zhang, Jan. 11, 2012.
10. Eric Lander, "The Heroes of CRISPR," *Cell*, Jan. 14, 2016.
11. Author's interviews with Feng Zhang.
12. Feng Zhang, "Declaration in Connection with U.S. Patent Application Serial 14 /0054,414," USPTO, Jan. 30, 2014.

13. Shuailiang Lin, "Summary of CRISPR Work during Oct. 2011–June 2012," Exhibit 14 to Neville Sanjana Declaration, July 23, 2015, UC et al. Reply 3, exhibit 1614, in *Broad v. UC*, Patent Interference 106,048.
14. Shuailiang Lin email to Jennifer Doudna, Feb. 28, 2015.
15. Antonio Regalado, "In CRISPR Fight, Co-Inventor Says Broad Institute Misled Patent Office," *MIT Technology Review*, Aug. 17, 2016.
16. Author's interviews with Dana Carroll; Dana Carroll, "Declaration in Support of Suggestion of Interference," University of California Exhibit 1476, Interference No. 106,048, Apr. 10, 2015.
17. Carroll, "Declaration"; Berkeley et al., "List of Intended Motions," Patent Interference No. 106,115, USPTO, July 30, 2019.
18. Author's interviews with Jennifer Doudna and Feng Zhang; Broad et al., "Contingent Responsive Motion 6" and "Constructive Reduction to Practice by Embodiment 17," USPTO, Patent Interference 106,048, June 22, 2016.
19. Author's interviews with Feng Zhang and Luciano Marraffini. See also Davies, "Interview with Luciano Marraffini."

Chapter 25: Doudna Joins the Race

1. Author's interviews with Martin Jinek and Jennifer Doudna.
2. Melissa Pandika, "Jennifer Doudna, CRISPR Code Killer," *Ozy*, Jan. 7, 2014.
3. Author's interviews with Jennifer Doudna and Martin Jinek.

Chapter 26: Photo Finish

1. Author's interviews with Feng Zhang; Fei Ann Ran, "CRISPR-Cas9," *NABC Report* 26, ed. Alan Eaglesham and Ralph Hardy, Oct. 8, 2014.
2. Le Cong, Fei Ann Ran, David Cox, Shuailiang Lin . . . Luciano Marraffini, and Feng Zhang, "Multiplex Genome Engineering Using CRISPR/Cas Systems," *Science*, Feb. 15, 2013 (received Oct. 5, 2012; accepted Dec. 12; published online Jan. 3, 2013).
3. Author's interviews with George Church, Eric Lander, and Feng Zhang.
4. Author's email interviews with Le Cong.
5. Author's interview with George Church.
6. Prashant Mali . . . George Church, et al., "RNA-Guided Human Genome Engineering via Cas9," *Science*, Feb. 15, 2013 (received Oct. 26, 2012; accepted Dec. 12, 2012; published online Jan. 3, 2013).

Chapter 27: Doudna's Final Sprint

1. Pandika, "Jennifer Doudna, CRISPR Code Killer."
2. Author's interviews with Jennifer Doudna and Martin Jinek.
3. Michael M. Cox, Jennifer Doudna, and Michael O'Donnell, *Molecular Biology: Principles and Practice* (W. H. Freeman, 2011). The first edition cost $195.
4. It was Detlef Weigel, at the Max Planck Institute for Developmental Biology.
5. Author's interviews with Emmanuelle Charpentier and Jennifer Doudna.
6. Detlef Weigel decision letter and Jennifer Doudna author response, *eLife*, Jan. 29, 2013.
7. Martin Jinek, Alexandra East, Aaron Cheng, Steven Lin, Enbo Ma, and Jennifer Doudna, "RNA-Programmed Genome Editing in Human Cells," *eLife*, Jan. 29, 2013 (received Dec. 15, 2012; accepted Jan. 3, 2013).
8. Jin-Soo Kim email to Jennifer Doudna, July 16, 2012; Seung Woo Cho, Sojung Kim, Jong Min Kim, and Jin-Soo Kim, "Targeted Genome Engineering in Human Cells

with the Cas9 RNA-Guided Endonuclease," *Nature Biotechnology*, Mar. 2013 (received Nov. 20, 2012; accepted Jan. 14, 2013; published online Jan. 29, 2013).

9. Woong Y. Hwang . . . Keith Joung, et al., "Efficient Genome Editing in Zebrafish Using a CRISPR-Cas System," *Nature Biotechnology*, Jan. 29, 2013.

Chapter 28: Forming Companies

1. Author's interviews with Andy May, Jennifer Doudna, and Rachel Haurwitz.
2. George Church interview, "Can Neanderthals Be Brought Back from the Dead?," *Spiegel*, Jan. 18, 2013; David Wagner, "How the Viral Neanderthal-Baby Story Turned Real Science into Junk Journalism," *The Atlantic*, Jan. 22, 2013.
3. Author's interview with Rodger Novak; Hemme, "Fireside Chat with Rodger Novak"; Jon Cohen, "Birth of CRISPR Inc.," *Science*, Feb. 17, 2017; author's interviews with Emmanuelle Charpentier.
4. Author's interviews with Jennifer Doudna, George Church, and Emmanuelle Charpentier.
5. Author's interviews with Rodger Novak and Emmanuelle Charpentier.
6. Author's interview with Andy May.
7. Hemme, "Fireside Chat with Rodger Novak."
8. Author's interviews with Jennifer Doudna.
9. Editas Medicine, SEC 10-K filing 2016 and 2019; John Carroll, "Biotech Pioneer in 'Gene Editing' Launches with $43M in VC Cash," *FierceBiotech*, Nov. 25, 2013.
10. Author's interviews with Jennifer Doudna, Rachel Haurwitz, Erik Sontheimer, and Luciano Marraffini.

Chapter 29: Mon Amie

1. Author's interviews with Jennifer Doudna, Emmanuelle Charpentier, and Martin Jinek; Martin Jinek . . . Samuel Sternberg . . . Kaihong Zhou . . . Emmanuelle Charpentier, Eva Nogales, Jennifer A. Doudna, et al., "Structures of Cas9 Endonucleases Reveal RNA-Mediated Conformational Activation," *Science*, Mar. 14, 2014.
2. Jennifer Doudna and Emmanuelle Charpentier, "The New Frontier of Genome Engineering with CRISPR-Cas9," *Science*, Nov. 28, 2014.
3. Author's interviews with Jennifer Doudna and Emmanuelle Charpentier.
4. Hemme, "Fireside Chat with Rodger Novak"; author's interview with Rodger Novak.
5. Author's interview with Rodolphe Barrangou.
6. Davies, *Editing Humanity*, 96.
7. Author's interview with Jennifer Doudna; "CRISPR Timeline," Broad Institute website, broadinstitute.org.
8. Author's interview with Eric Lander; Breakthrough Prize ceremony, Mar. 19, 2015.
9. Author's interviews with Jennifer Doudna, George Church; Gairdner Awards ceremony, Oct. 27, 2016.

Chapter 30: The Heroes of CRISPR

1. Author's interviews with Eric Lander and Emmanuelle Charpentier.
2. Lander, "The Heroes of CRISPR."
3. Michael Eisen, "The Villain of CRISPR," *It Is Not Junk*, Jan. 25, 2016.
4. "Heroes of CRISPR," eighty-four comments, PubPeer, https://pubpeer.com/publications/D400145518C0A557E9A79F7BB20294; Sharon Begley, "Controversial CRISPR History Set Off an Online Firestorm," *Stat*, Jan. 19, 2016.

5. Nathaniel Comfort, "A Whig History of CRISPR," *Genotopia*, Jan. 18, 2016; @nccom fort, "I made a hashtag that became a thing! #Landergate," Twitter, Jan. 27, 2016.
6. Antonio Regalado, "A Scientist's Contested History of CRISPR," *MIT Technology Review*, Jan. 19, 2016.
7. Ruth Reader, "These Women Helped Create CRISPR Gene Editing. So Why Are They Written Out of Its History?," *Mic*, Jan. 22, 2016; Joanna Rothkopf, "How One Man Tried to Write Women Out of CRISPR, the Biggest Biotech Innovation in Decades," *Jezebel*, Jan. 20, 2016.
8. Stephen Hall, "The Embarrassing, Destructive Fight over Biotech's Big Breakthrough," *Scientific American*, Feb. 4, 2016.
9. Tracy Vence, "'Heroes of CRISPR' Disputed," *The Scientist*, Jan. 19, 2016.
10. Author's interview with Jack Szostak.
11. Eric Lander, email to the Broad Institute staff, Jan. 28, 2016.
12. Joel Achenbach, "Eric Lander Talks CRISPR and the Infamous Nobel 'Rule of Three,'" *Washington Post*, Apr. 21, 2016.

Chapter 31: Patents

1. *Diamond v. Chakrabarty*, 447 U.S. 303, U.S. Supreme Court, 1980; Douglas Robinson and Nina Medlock, "*Diamond v. Chakrabarty*: A Retrospective on 25 Years of Biotech Patents," *Intellectual Property & Technology Law Journal*, Oct. 2005.
2. Michael Eisen, "Patents Are Destroying the Soul of Academic Science," *it is NOT junk* (blog), Feb. 20, 2017. See also Alfred Engelberg, "Taxpayers Are Entitled to Reasonable Prices on Federally Funded Drug Discoveries," *Modern Healthcare*, July 18, 2018.
3. Author's interview with Eldora Ellison.
4. Martin Jinek, Jennifer Doudna, Emmanuelle Charpentier, and Krzysztof Chylinski, U.S. Patent Application 61/652,086, "Methods and Compositions, for RNA-Directed Site-Specific DNA Modification," filed May 25, 2012; Jacob Sherkow, "Patent Protection for CRISPR," *Journal of Law and the Biosciences*, Dec. 7, 2017.
5. "CRISPR-Cas Systems and Methods for Altering Expressions of Gene Products," provisional application No. 61/736,527, filed on Dec. 12, 2012, which in 2014 resulted in U.S. Patent No. 8,697,359. This application, later revised, included Luciano Marraffini as well as Feng Zhang, Le Cong, and Shuailiang Lin as inventors.
6. The main patent application and related filings of Zhang/Broad can be found through the U.S. Patent Office as U.S. Provisional Patent Application No. 61/736,527. The Doudna/Charpentier/Berkeley filings are under U.S. Provisional Patent Application No. 61/652,086. A good guide to the patent issues is the work of Jacob Sherkow of New York Law School, which includes "Law, History and Lessons in the CRISPR Patent Conflict," *Nature Biotechnology*, Mar. 2015; "Patents in the Time of CRISPR," *Biochemist*, June 2016; "Inventive Steps: The CRISPR Patent Dispute and Scientific Progress," *EMBO Reports*, May 23, 2017; "Patent Protection for CRISPR."
7. Author's interviews with George Church, Jennifer Doudna, Erik Lander, and Feng Zhang.
8. "CRISPR-Cas Systems and Methods for Altering Expressions of Gene Products," provisional application No. 61/736,527.
9. Author's interviews with Luciano Marraffini.
10. Author's interviews with Feng Zhang and Eric Lander; Lander, "Heroes of CRISPR."
11. U.S. Patent No. 8,697,359.

12. Author's interviews with Andy May and Jennifer Doudna.
13. Provisional patent application U.S. 2012/61652086P and published patent application U.S. 2014/0068797A1 of Doudna et al.; Provisional patent application U.S. 2012 /61736527P (Dec. 12, 2012) and granted patent US 8,697,359 B1 (Apr. 15, 2014) of Zhang et al.
14. "Suggestion of Interference" and "Declaration of Dana Carroll, PhD, in Support of Suggestion of Interference," in re Patent Application of Jennifer Doudna et al., serial no. 2013/842859, U.S. Patent and Trademark Office, Apr. 10 and 13, 2015; Mark Summerfield, "CRISPR—Will This Be the Last Great US Patent Interference?," *Patentology*, July 11, 2015; Jacob Sherkow, "The CRISPR Patent Interference Showdown Is On," Stanford Law School blog, Dec. 29, 2015; Antonio Regalado, "CRISPR Patent Fight Now a Winner-Take-All Match," *MIT Technology Review*, Apr. 15, 2015.
15. Feng Zhang, "Declaration," in re Patent Application of Feng Zhang, Serial no. 2014 /054,414, Jan. 30, 2014, provided privately to the author.
16. *In re Dow Chemical Co.*, 837 F.2d 469, 473 (Fed. Cir. 1988).
17 Jacob Sherkow, "Inventive Steps: The CRISPR Patent Dispute and Scientific Progress," *EMBO Reports*, May 23, 2017; Broad et al. contingent responsive motion 6 for benefit of Broad et al. Application 61/736,527, USPTO, June 22, 2016; University of California et al., Opposition motion 2, Patent Interference case 106,048, USPTO, Aug. 15, 2016 (Opposing Broad's Allegations of No Interference-in-Fact).
18. Alessandra Potenza, "Who Owns CRISPR?," *The Verge*, Dec. 6, 2016; Jacob Sherkow, "Biotech Trial of the Century Could Determine Who Owns CRISPR," *MIT Technology Review*, Dec. 7, 2016; Sharon Begley, "CRISPR Court Hearing Puts University of California on the Defensive," *Stat*, Dec. 6, 2016.
19. Transcript of oral arguments before the patent trial board, Dec. 6, 2016, Patent Interference Case 106,048, U.S. Patent and Trademark Office.
20. Jennifer Doudna interview, *Catalyst*, UC Berkeley College of Chemistry, July 10, 2014.
21. Berkeley substantive motion 4, Patent Interference Case 106,048, May 23, 2016. See also Broad substantive motions 2, 3, and 5.
22. Patent Trial Board Judgment and Decision on Motions, Patent Interference Case 106,048, Feb. 15, 2017.
23. Judge Kimberly Moore, decision, Patent Interference Case 106,048, United States Court of Appeals for the Federal Circuit, Sept. 10, 2018.
24. Author's interviews with Eldora Ellison.
25. Patent Interference No. 106,115, Patent Trial and Appeal Board, June 24, 2019.
26. Oral argument, Patent Interference No. 106,115, Patent Trial and Appeal Board, May 18, 2020.
27. "Methods and Compositions for RNA-Directed Target DNA Modification," European Patent Office, patent EP2800811, granted Apr. 7, 2017; Jef Akst, "UC Berkeley Receives CRISPR Patent in Europe," *The Scientist*, Mar. 24, 2017; Sherkow, "Inventive Steps."
28. Author's interviews with Luciano Marraffini; "Engineering of Systems, Methods, and Optimized Guide Compositions for Sequence Manipulation," European Patent Office, patent EP2771468; Kelly Servick, "Broad Institute Takes a Hit in European CRISPR Patent Struggle," *Science*, Jan. 18, 2018; Rory O'Neill, "EPO Revokes Broad's CRISPR Patent," *Life Sciences Intellectual Property Review*, Jan. 16, 2020.
29. Author's interview with Andy May.

Chapter 32: Therapies

1. Rob Stein, "In a First, Doctors in U.S. Use CRISPR Tool to Treat Patient with Genetic Disorder," *Morning Edition*, NPR, July 29, 2019; Rob Stein, "A Young Mississippi Woman's Journey through a Pioneering Gene-Editing Experiment," *All Things Considered*, NPR, Dec. 25, 2019.
2. "CRISPR Therapeutics and Vertex Announce New Clinical Data," CRISPR Therapeutics, June 12, 2020.
3. Rob Stein, "A Year In, 1st Patient to Get Gene-Editing for Sickle Cell Disease Is Thriving," *Morning Edition*, NPR, June 23, 2020.
4. Author's interview with Emmanuelle Charpentier.
5. Author's interview with Jennifer Doudna.
6. "Proposal for an IGI Sickle Cell Initiative," Innovative Genomics Institute, February 2020.
7. Preetika Rana, Amy Dockser Marcus, and Wenxin Fan, "China, Unhampered by Rules, Races Ahead in Gene-Editing Trials," *Wall Street Journal*, Jan. 21, 2018.
8. David Cyranoski, "CRISPR Gene-Editing Tested in a Person for the First Time," *Nature*, Nov. 15, 2016.
9. Jennifer Hamilton and Jennifer Doudna, "Knocking Out Barriers to Engineered Cell Activity," *Science*, Feb. 6, 2020; Edward Stadtmauer . . . Carl June, et al., "CRISPR-Engineered T Cells in Patients with Refractory Cancer," *Science*, Feb. 6, 2020.
10. "CRISPR Diagnostics in Cancer Treatments," Mammoth Biosciences website, June 11, 2019.
11. "Single Ascending Dose Study in Participants with LCA10," ClinicalTrials.gov, Mar. 13, 2019, identifier: NCT03872479; Morgan Maeder . . . and Haiyan Jiang, "Development of a Gene-Editing Approach to Restore Vision Loss in Leber Congenital Amaurosis Type 10," *Nature*, Jan. 21, 2019.
12. Marilynn Marchione, "Doctors Try 1st CRISPR Editing in the Body for Blindness," AP, Mar. 4, 2020.
13. Sharon Begley, "CRISPR Babies' Lab Asked U.S. Scientist for Help to Disable Cholesterol Gene in Human Embryos," *Stat*, Dec. 4, 2018; Anthony King, "A CRISPR Edit for Heart Disease," *Nature*, Mar. 7, 2018.
14. Matthew Porteus, "A New Class of Medicines through DNA Editing," *New England Journal of Medicine*, Mar. 7, 2019; Sharon Begley, "CRISPR Trackr: Latest Advances," *Stat Plus*.

Chapter 33: Biohacking

1. Josiah Zayner, "DIY Human CRISPR Myostatin Knock-Out," YouTube, Oct. 6, 2017; Sarah Zhang, "Biohacker Regrets Injecting Himself with CRISPR on Live TV," *The Atlantic*, Feb. 20, 2018; Stephanie Lee, "This Guy Says He's the First Person to Attempt Editing His DNA with CRISPR," *BuzzFeed*, Oct. 14, 2017.
2. Kate McLean and Mario Furloni, "Gut Hack," *New York Times* op-doc, Apr. 11, 2017; Arielle Duhaime-Ross, "A Bitter Pill," *The Verge*, May 4, 2016.
3. "About us," The Odin, https://www.the-odin.com/about-us/; author's interviews with Josiah Zayner.
4. Author's interviews with Josiah Zayner and Kevin Doxzen.
5. Author's interview with Josiah Zayner. See also Josiah Zayner, "CRISPR Babies Scientist He Jiankui Should Not Be Villainized," *Stat*, Jan. 2, 2020.

Chapter 34: DARPA and anti-CRISPR

1. Heidi Ledford, "CRISPR, the Disruptor," *Nature*, June 3, 2015. Danilo Maddalo . . . and Andrea Ventura, "In vivo Engineering of Oncogenic Chromosomal Rearrangements with the CRISPR/Cas9 System," *Nature*, Oct. 22, 2014; Sidi Chen, Neville E. Sanjana . . . Feng Zhang, and Phillip A. Sharp, "Genome-wide CRISPR Screen in a Mouse Model of Tumor Growth and Metastasis," *Cell*, Mar. 12, 2015.
2. James Clapper, "Threat Assessment of the U.S. Intelligence Community," Feb. 9, 2016; Antonio Regalado, "The Search for the Kryptonite That Can Stop CRISPR," *MIT Technology Review*, May 2, 2019; Robert Sanders, "Defense Department Pours $65 Million into Making CRISPR Safer," *Berkeley News*, July 19, 2017.
3. Defense Advanced Research Projects Agency, "Building the Safe Genes Toolkit," July 19, 2017.
4. Author's interview with Jennifer Doudna.
5. Author's interview with Joe Bondy-Denomy; Joe Bondy-Denomy, April Pawluk . . . Alan R. Davidson, et al., "Bacteriophage Genes That Inactivate the CRISPR/Cas Bacterial Immune System," *Nature*, Jan. 17, 2013; Elie Dolgin, "Kill Switch for CRISPR Could Make Gene Editing Safer," *Nature*, Jan. 15, 2020.
6. Jiyung Shin . . . Joseph Bondy-Denomy, and Jennifer Doudna, "Disabling Cas9 by an Anti-CRISPR DNA Mimic," *Science Advances*, July 12, 2017.
7. Nicole D. Marino . . . and Joseph Bondy-Denomy, "Anti-CRISPR Protein Applications: Natural Brakes for CRISPR-Cas Technologies," *Nature Methods*, Mar. 16, 2020.
8. Author's interview with Fyodor Urnov; Emily Mullin, "The Defense Department Plans to Build Radiation-Proof CRISPR Soldiers," *One Zero*, Sept. 27, 2019.
9. Author's interviews with Jennifer Doudna and Gavin Knott.
10. Author's interviews with Josiah Zayner.

Chapter 35: Rules of the Road

1. Robert Sinsheimer, "The Prospect of Designed Genetic Change," *Engineering and Science*, Caltech, Apr. 1969.
2. Bentley Glass, Presidential Address to the AAAS, Dec. 28, 1970, *Science*, Jan. 8, 1971.
3. John Fletcher, *The Ethics of Genetic Control: Ending Reproductive Roulette* (Doubleday, 1974), 158.
4. Paul Ramsey, *Fabricated Man* (Yale, 1970), 138.
5. Ted Howard and Jeremy Rifkin, *Who Should Play God?* (Delacorte, 1977), 14; Dick Thompson, "The Most Hated Man in Science," *Time*, Dec. 4, 1989.
6. Shane Crotty, *Ahead of the Curve* (University of California, 2003), 93; Mukherjee, *The Gene*, 225.
7. Paul Berg et al., "Potential Biohazards of Recombinant DNA Molecules, " *Science*, July 26, 1974.
8. Author's interview with David Baltimore; Michael Rogers, "The Pandora's Box Conference," *Rolling Stone*, June 19, 1975; Michael Rogers, *Biohazard* (Random House, 1977); Crotty, *Ahead of the Curve*, 104–8; Mukherjee, *The Gene*, 226–30; Donald S. Fredrickson, "Asilomar and Recombinant DNA: The End of the Beginning," in *Biomedical Politics* (National Academies Press, 1991); Richard Hindmarsh and Herbert Gottweis, "Recombinant Regulation: The Asilomar Legacy 30 Years On," *Science as Culture*, Fall 2005; Daniel Gregorowius, Nikola Biller-Andorno, and Anna Deplazes-Zemp, "The Role of Scientific Self-Regulation for the Control of Genome Editing

in the Human Germline," *EMBO Reports*, Feb. 20, 2017; Jim Kozubek, *Modern Prometheus* (Cambridge, 2016), 124.

9. Author's interviews with James Watson and David Baltimore.
10. Paul Berg et al., "Summary Statement of the Asilomar Conference on Recombinant DNA Molecules," *PNAS*, June 1975.
11. Paul Berg, "Asilomar and Recombinant DNA," *The Scientist*, Mar. 18, 2002.
12. Hindmarsh and Gottweis, "Recombinant Regulation," 301.
13. Claire Randall, Rabbi Bernard Mandelbaum, and Bishop Thomas Kelly, "Message from Three General Secretaries to President Jimmy Carter," June 20, 1980.
14. Morris Abram et al., *Splicing Life*, President's Commission for the Study of Ethical Problems in Medicine and Biomedical and Behavioral Research, Nov. 16, 1982.
15. Alan Handyside et al., "Birth of a Normal Girl after in vitro Fertilization and Preimplantation Diagnostic Testing for Cystic Fibrosis," *New England Journal of Medicine*, Sept. 1992.
16. Roger Ebert, *Gattaca* review, Oct. 24, 1997, rogerebert.com.
17. Gregory Stock and John Campbell, *Engineering the Human Germline* (Oxford, 2000), 73–95; author's interviews with James Watson; Gina Kolata, "Scientists Brace for Changes in Path of Human Evolution," *New York Times*, Mar. 21, 1998.
18. Steve Connor, "Nobel Scientist Happy to 'Play God' with DNA," *The Independent*, May 17, 2000.
19. Lee Silver, *Remaking Eden* (Avon, 1997), 4.
20. Lee Silver, "Reprogenetics: Third Millennium Speculation," *EMBO Reports*, Nov. 15, 2000.
21. Gregory Stock, *Redesigning Humans: Our Inevitable Genetic Future* (Houghton Mifflin, 2002), 170.
22. Council of Europe, "Oviedo Convention and Its Protocols," April 4, 1997.
23. Sheryl Gay Stolberg, "The Biotech Death of Jesse Gelsinger," *New York Times*, Nov. 28, 1999.
24. Meir Rinde, "The Death of Jesse Gelsinger," *Science History Institute*, June 4, 2019.
25. Harvey Flaumenhaft, "The Career of Leon Kass," *Journal of Contemporary Health Law & Policy*, 2004; "Leon Kass," Conversations with Bill Kristol, Dec. 2015, https://conversationswithbillkristol.org/video/leon-kass/.
26. Leon Kass, "What Price the Perfect Baby?," *Science*, July 9, 1971; Leon Kass, "Review of *Fabricated Man* by Paul Ramsey," *Theology Today*, Apr. 1, 1971; Leon Kass, "Making Babies: the New Biology and the Old Morality," *Public Interest*, Winter 1972.
27. Michael Sandel, "The Case against Perfection," *The Atlantic*, Apr. 2004; Michael Sandel, *The Case Against Perfection* (Harvard, 2007).
28. Francis Fukuyama, *Our Posthuman Future* (Farrar, Straus and Giroux, 2000), 10.
29. Leon Kass et al., *Beyond Therapy: Biotechnology and the Pursuit of Happiness*, report of the President's Council on Bioethics, October 2003.

Chapter 36: Doudna Steps In

1. Doudna and Sternberg, *A Crack in Creation*, 198; Michael Specter, "Humans 2.0," *New Yorker*, Nov. 16, 2015; author's interview with Jennifer Doudna.
2. Author's interviews with Sam Sternberg and Lauren Buchman.
3. Author's interviews with George Church and Lauren Buchman.
4. Doudna and Sternberg, *A Crack in Creation*, 199–220; author's interviews with Jennifer Doudna and Sam Sternberg.

5. Author's interviews with David Baltimore, Jennifer Doudna, Sam Sternberg, and Dana Carroll.
6. David Baltimore, et al., "A Prudent Path Forward for Genomic Engineering and Germline Gene Modification," *Science*, Apr. 3, 2015 (published online Mar. 19).
7. Nicholas Wade, "Scientists Seek Ban on Method of Editing the Human Genome," *New York Times*, Mar. 19, 2015.
8. See, for example, Edward Lanphier, Fyodor Urnov, et al., "Don't Edit the Human Germ Line," *Nature*, Mar. 12, 2015.
9. Author's interviews with Jennifer Doudna, Sam Sternberg; Doudna and Sternberg, *A Crack in Creation*, 214ff.
10. Puping Liang . . . Junjiu Huang, et al., "CRISPR/Cas9-Mediated Gene Editing in Human Tripronuclear Zygotes," *Protein & Cell*, May 2015 (published online Apr. 18).
11. Rob Stein, "Critics Lash Out at Chinese Scientists Who Edited DNA in Human Embryos," *Morning Edition*, NPR, April 23, 2015.
12. Author's interviews with Ting Wu, George Church, Jennifer Doudna; Johnny Kung, "Increasing Policymaker's Interest in Genetics," pgEd briefing paper, Dec. 1, 2015.
13. Jennifer Doudna, "Embryo Editing Needs Scrutiny," *Nature*, Dec. 3, 2015.
14. George Church, "Encourage the Innovators," *Nature*, Dec. 3, 2015.
15. Steven Pinker, "A Moral Imperative for Bioethics," *Boston Globe*, Aug. 1, 2015; Paul Knoepfler, Steven Pinker interview, *The Niche*, Aug. 10, 2015.
16. Author's interviews with Jennifer Doudna, David Baltimore, and George Church; *International Summit on Human Gene Editing, Dec. 1–3, 2015* (National Academies Press, 2015); Jef Akst, "Let's Talk Human Engineering," *The Scientist*, Dec. 3, 2015.
17. R. Alto Charo, Richard Hynes, et al., "Human Genome Editing: Scientific, Medical, and Ethical Considerations," report of the National Academies of Sciences, Engineering, Medicine, 2017.
18. Françoise Baylis, *Altered Inheritance: CRISPR and the Ethics of Human Genome Editing* (Harvard, 2019); Jocelyn Kaiser, "U.S. Panel Gives Yellow Light to Human Embryo Editing," *Science*, Feb. 14, 2017; Kelsey Montgomery, "Behind the Scenes of the National Academy of Sciences' Report on Human Genome Editing," *Medical Press*, Feb. 27, 2017.
19. "Genome Editing and Human Reproduction," Nuffield Council on Bioethics, July 2018; Ian Sample, "Genetically Modified Babies Given Go Ahead by UK Ethics Body," *Guardian*, July 17, 2018; Clive Cookson, "Human Gene Editing Morally Permissible, Says Ethics Study," *Financial Times*, July 17, 2018; Donna Dickenson and Marcy Darnovsky, "Did a Permissive Scientific Culture Encourage the 'CRISPR Babies' Experiment?," *Nature Biotechnology*, Mar. 15, 2019.
20. Consolidated Appropriations Act of 2016, Public Law 114-113, Section 749, Dec. 18, 2015; Francis Collins, "Statement on NIH Funding of Research Using Gene-Editing Technologies in Human Embryos," Apr. 28, 2015; John Holdren, "A Note on Genome Editing," May 26, 2015.
21. "Putin said scientists could create Universal Soldier-style supermen," YouTube, Oct. 24, 2017, youtube.com/watch?v=9v3TNGmbArs; "Russia's Parliament Seeks to Create Gene-Edited Babies," *EU Observer*, Sept. 3, 2019; Christina Daumann, "'New Type of Society'," *Asgardia*, Sept. 4, 2019.
22. Achim Rosemann, Li Jiang, and Xinqing Zhang, "The Regulatory and Legal Situation of Human Embryo, Gamete and Germ Line Gene Editing Research and Clinical Applications in the People's Republic of China," Nuffield Council on Bioethics, May

2017; Jing-ru Li, et. al., "Experiments That Led to the First Gene-Edited Babies," *Journal of Zhejiang University Science B*, Jan. 2019.

Chapter 37: He Jiankui

1. This section is based on Xi Xin and Xu Yue, "The Life Track of He Jiankui," *Jiemian News*, Nov. 27, 2018; Jon Cohen, "The Untold Story of the 'Circle of Trust' behind the World's First Gene-Edited Babies," *Science*, Aug. 1, 2019; Sharon Begley and Andrew Joseph, "The CRISPR Shocker," *Stat*, Dec. 17, 2018; Zach Coleman, "The Businesses behind the Doctor Who Manipulated Baby DNA," *Nikkei Asian Review*, Nov. 27, 2018; Zoe Low, "China's Gene Editing Frankenstein," *South China Morning Post*, Nov. 27, 2018; Yangyang Cheng, "Brave New World with Chinese Characteristics," *Bulletin of the Atomic Scientists*, Jan. 13, 2019; He Jiankui, "Draft Ethical Principles," YouTube, Nov. 25, 2018, youtube.com/watch?v=MyNHpMoPkIg; Antonio Regalado, "Chinese Scientists Are Creating CRISPR Babies," *MIT Technology Review*, Nov. 25, 2018; Marilynn Marchione, "Chinese Researcher Claims First Gene-Edited Babies," AP, Nov. 26, 2018; Christina Larson, "Gene-Editing Chinese Scientist Kept Much of His Work Secret," AP, Nov. 27, 2018; Davies, *Editing Humanity*.
2. Jiankui He and Michael W. Deem, "Heterogeneous Diversity of Spacers within CRISPR," *Physical Review Letters*, Sept. 14, 2010.
3. Mike Williams, "He's on a Hot Streak," *Rice News*, Nov. 17, 2010.
4. Cohen, "The Untold Story"; Coleman, "The Businesses behind the Doctor."
5. Davies, *Editing Humanity*, 209.
6. Yuan Yuan, "The Talent Magnet," *Beijing Review*, May 31, 2018.
7. Luyang Zhao . . . Jiankui He, et al., "Resequencing the *Escherichia coli* Genome by GenoCare Single Molecule," bioRxiv, posted online July 13, 2017.
8. Teng Jing Xuan, "CCTV's Glowing 2017 Coverage of Gene-Editing Pariah He Jiankui," *Caixan Global*, Nov. 30, 2018; Rob Schmitz, "Gene-Editing Scientist's Actions Are a Product of Modern China," *All Things Considered*, NPR, Feb. 5, 2019.
9. "Welcome to the Jiankui He Lab," http://sustc-genome.org.cn/people.html (site no longer active); Regalado, "Chinese Scientists Are Creating CRISPR Babies."
10. He Jiankui, "CRISPR Gene Editing Meeting," blog post (in Chinese), Aug. 24, 2016, http://blog.sciencenet.cn/home.php?mod=space&uid=514529&do=blog&id=998292.
11. Cohen, "The Untold Story"; Begley and Joseph, "The CRISPR Shocker"; author's interviews with Jennifer Doudna; Jennifer Doudna and William Hurlbut, "The Challenge and Opportunity of Gene Editing," Templeton Foundation grant 217,398.
12. Davies, *Editing Humanity*, 221; George Church, "Future, Human, Nature: Reading, Writing, Revolution," Innovative Genomics Institute, January 26, 2017, innovativegenomics.org/multimedia-library/george-church-lecture/.
13. He Jiankui, "The Safety of Gene-Editing of Human Embryos to Be Resolved," blog post (in Chinese), Feb. 19, 2017, blog.sciencenet.cn/home.php?mod=space&uid=514529&do=blog&id=1034671.
14. Author's interview with Jennifer Doudna.
15. He Jiankui, "Evaluating the Safety of Germline Genome Editing in Human, Monkey, and Mouse Embryos," Cold Spring Harbor Lab Symposium, July 29, 2017, youtube.com/watch?v=llxNRGMxyCc&t=3s; Regalado, "Chinese Scientists Are Creating CRISPR Babies."
16. Medical Ethics Approval Application Form, HarMoniCare Shenzhen Women's and Children's Hospital, March 7, 2017, theregreview.org/wp-content/uploads/2019/05

/He-Jiankui-Documents-3.pdf; Cohen, "The Untold Story"; Kathy Young, Marilynn Marchione, Emily Wang, et al., "First Gene-Edited Babies Reported in China," YouTube, Nov. 25, 2018, https://www.youtube.com/watch?v=C9V3mqswbv0; Gerry Shih and Carolyn Johnson, "Chinese Genomics Scientist Defends His Gene-Editing Research," *Washington Post*, Nov. 28, 2018.

17. Jiankui He, "Informed Consent, Version: Female 3.0," Mar. 2017, theregreview.org/wp-content/uploads/2019/05/He-Jiankui-Documents-3.pdf; Cohen, "The Untold Story"; Marilynn Marchione, "Chinese Researcher Claims First Gene-Edited Babies," AP, Nov. 26, 2018; Larson, "Gene-Editing Chinese Scientist Kept Much of His Work Secret."
18. Kiran Musunuru, *The Crispr Generation* (BookBaby, 2019).
19. Begley and Joseph, "The CRISPR Shocker." See also Pam Belluck, "How to Stop Rogue Gene-Editing of Human Embryos?," *New York Times*, Jan. 23, 2019; Preetika Rana, "How a Chinese Scientist Broke the Rules to Create the First Gene-Edited Babies," *Wall Street Journal*, May 10, 2019.
20. Author's interviews with Matthew Porteus.
21. Cohen, "The Untold Story"; Begley and Joseph, "The CRISPR Shocker"; Marilyn Marchione and Christina Larson, "Could Anyone Have Stopped Gene-Edited Babies Experiment?," AP, Dec. 2, 2018.
22. Pam Belluck, "Gene-Edited Babies: What a Chinese Scientist Told an American Mentor," *New York Times*, Apr. 14, 2019; "Statement on Fact-Finding Review related to Dr. Jiankui He," *Stanford News*, Apr. 16, 2019. Belluck was the first to publish the emails between He and Quake.
23. He Jiankui, question-and-answer session, the Second International Summit on Human Genome Editing, Hong Kong, Nov. 28, 2018; Cohen, "The Untold Story"; Marchione and Larson, "Could Anyone Have Stopped Gene-Edited Babies Experiment?"; Marchione, "Chinese Researcher Claims First Gene-Edited Babies"; Jane Qiu, "American Scientist Played More Active Role in 'CRISPR Babies' Project Than Previously Known," *Stat*, Jan. 31, 2019; Todd Ackerman, "Lawyers Say Rice Professor Not Involved in Controversial Gene-Edited Babies Research," *Houston Chronicle*, Dec. 13, 2018; decommissioned web page: Rice University, Faculty, https://profiles.rice.edu/faculty/michael-deem; see Michael Deem search on Rice website: https://search.rice.edu/?q=michael+deem&tab=Search.
24. Cohen, "The Untold Story."
25. He Jiankui, Ryan Ferrell, Chen Yuanlin, Qin Jinzhou, and Chen Yangran, "Draft Ethical Principles for Therapeutic Assisted Reproductive Technologies," *CRISPR Journal*, originally published Nov. 26, 2019, but later retracted and removed from the website. See also Henry Greeley, "CRISPR'd Babies," *Journal of Law and the Biosciences*, Aug. 13, 2019.
26. Allen Buchanan, *Better Than Human* (Oxford, 2011), 40, 101.
27. He Jiankui, "Draft Ethical Principles for Therapeutic Assisted Reproductive Technologies."
28. He Jiankui, "Designer Baby Is an Epithet" and "Why We Chose HIV and *CCR5* First," The He Lab, YouTube, Nov. 25, 2018.
29. He Jiankui, "HIV Immune Gene CCR5 Gene Editing in Human Embryos," Chinese Clinical Trial Registry, ChiCTR1800019378, Nov. 8, 2018.
30. Jinzhou Qin . . . Michael W. Deem, Jiankui He, et al., "Birth of Twins after Genome Editing for HIV Resistance," submitted to *Nature* Nov. 2019 (never published; I was given

a copy by an American researcher who received it from He Jiankui); Qiu, "American Scientist Played More Active Role in 'CRISPR Babies' Project Than Previously Known."

31. Greely, "CRISPR'd Babies"; Musunuru, *The Crispr Generation*; author's interview with Dana Carroll.
32. Regalado, "Chinese Scientists Are Creating CRISPR Babies."
33. Marchione, "Chinese Researcher Claims First Gene-Edited Babies"; Larson, "Gene-Editing Chinese Scientist Kept Much of His Work Secret."
34. He Jiankui, "About Lulu and Nana," YouTube, Nov. 25, 2018.

Chapter 38: The Hong Kong Summit

1. Author's interview with Jennifer Doudna.
2. Author's interview with David Baltimore.
3. Cohen, "The Untold Story."
4. Author's interviews with Victor Dzau, David Baltimore, Jennifer Doudna.
5. Author's interviews with Duanqing Pei.
6. Author's interviews with Jennifer Doudna; Robin Lovell-Badge, "CRISPR Babies," *Development*, Feb. 6, 2019.
7. Cached story deleted from the *China's People's Daily*, Nov. 26, 2018, ithome.com/html/discovery/396899.htm.
8. Author's interviews with Duanqing Pei, Jennifer Doudna.
9. Author's interviews with Jennifer Doudna, Victor Dzau.
10. Second International Summit on Genome Editing, University of Hong Kong, Nov. 27–29, 2018.
11. He Jiankui session, the Second International Summit on Human Genome Editing, Hong Kong, Nov. 28, 2018.
12. Davies, *Editing Humanity*, 235.
13. Author's interview with David Baltimore.
14. Author's interview with Matthew Porteus.
15. Author's interviews with Jennifer Doudna.
16. Author's interview with Duanqing Pei.
17. Author's interviews with Jennifer Doudna, David Baltimore.
18. Author's interviews with Matthew Porteus, David Baltimore.
19. Mary Louise Kelly, "Harvard Medical School Dean Weighs In on Ethics of Gene Editing," *All Things Considered*, NPR, Nov. 29, 2018. See also Baylis, *Altered Inheritance*, 140; George Daley, Robin Lovell-Badge, and Julie Steffann, "After the Storm—A Responsible Path for Genome Editing," and R. Alta Charo, "Rogues and Regulation of Germline Editing," *New England Journal of Medicine*, Mar. 7, 2019; David Cyranoski and Heidi Ledford, "How the Genome-Edited Babies Revelation Will Affect Research," *Nature*, Nov. 27, 2018.
20. David Baltimore, et al., "Statement by the Organizing Committee of the Second International Summit on Human Genome Editing," Nov. 29, 2018.

Chapter 39: Acceptance

1. Author's interview with Josiah Zayner.
2. Zayner, "CRISPR Babies Scientist He Jiankui Should Not Be Villainized."
3. Author's interview with Josiah Zayner.
4. Author's interview with Jennifer Doudna and dinner with her and Andrew Doudna Cate.

5. Author's interviews with Jennifer Doudna, Bill Cassidy.
6. Author's interview with Margaret Hamburg and Victor Dzau; Walter Isaacson, "Should the Rich Be Allowed to Buy the Best Genes?," *Air Mail*, July 27, 2019.
7. Belluck, "How to Stop Rogue Gene-Editing of Human Embryos?"
8. Eric S. Lander, et. al., "Adopt a Moratorium on Heritable Genome Editing," *Nature*, Mar. 13, 2019.
9. Ian Sample, "Scientists Call for Global Moratorium on Gene Editing of Embryos," *Guardian*, Mar. 13, 2019; Joel Achenbach, "NIH and Top Scientists Call for Moratorium on Gene-Edited Babies," *Washington Post*, Mar. 13, 2019; Jon Cohen, "New Call to Ban Gene-Edited Babies Divides Biologists," *Science*, Mar. 13, 2019; Francis Collins, "NIH Supports International Moratorium on Clinical Application of Germline Editing," National Institutes of Health statement, Mar. 13, 2019.
10. Author's interview with Margaret Hamburg. See also Sara Reardon, "World Health Organization Panel Weighs In on CRISPR-Babies Debate," *Nature*, Mar. 19, 2019.
11. Author's interview with Jennifer Doudna. For a strong critique of Doudna's argument, see Baylis, *Altered Inheritance*, 163–66.
12. Kay Davies, Richard Lifton, et al., "Heritable Human Genome Editing," International Commission on the Clinical Use of Human Germline Genome Editing, Sept. 3, 2020.
13. "He Jiankui Jailed for Illegal Human Embryo Gene-Editing," Xinhua news agency, Dec. 30, 2019.
14. Philip Wen and Amy Dockser Marcus, "Chinese Scientist Who Gene-Edited Babies Is Sent to Prison," *Wall Street Journal*, Dec. 30, 2019.

Chapter 40: Red Lines

1. This chapter draws on a wealth of writing about the ethics of genetic engineering. These include Françoise Baylis, Michael Sandel, Leon Kass, Francis Fukuyama, Nathaniel Comfort, Jason Scott Robert, Eric Cohen, Bill McKibben, Marcy Darnovsky, Erik Parens, Josephine Johnston, Rosemarie Garland-Thomson, Robert Sparrow, Ronald Dworkin, Jürgen Habermas, Michael Hauskeller, Jonathan Glover, Gregory Stock, John Harris, Maxwell Mehlman, Guy Kahane, Jamie Metzl, Allen Buchanan, Julian Savulescu, Lee Silver, Nick Bostrom, John Harris, Ronald Green, Nicholas Agar, Arthur Caplan, and Hank Greeley. I also drew on the work of the Hastings Center, the Center for Genetics and Society, the Oxford Uehiro Centre for Practical Ethics, and the Nuffield Council on Bioethics.
2. Sandel, *The Case against Perfection*; Robert Sparrow, "Genetically Engineering Humans," *Pharmaceutical Journal*, Sept. 24, 2015; Jamie Metzl, *Hacking Darwin* (Sourcebooks, 2019); Julian Savulescu, Ruud ter Meulen, and Guy Kahane, *Enhancing Human Capacities* (Wiley, 2011).
3. Gert de Graaf, Frank Buckley, and Brian Skotko, "Estimates of the Live Births, Natural Losses, and Elective Terminations with Down Syndrome in the United States," *American Journal of Medical Genetics*, Apr. 2015.
4. Steve Boggan, Glenda Cooper, and Charles Arthur, "Nobel Winner Backs Abortion 'for Any Reason,'" *The Independent*, Feb. 17, 1997.

Chapter 41: Thought Experiments

1. Matt Ridley, *Genome* (Harper Collins, 2000), chapter 4, powerfully describes Huntington's and the work of Nancy Wexler in researching it.
2. Baylis, *Altered Inheritance*, 30; Tina Rulli, "The Ethics of Procreation and Adoption," *Philosophy Compass*, June 6, 2012.

3. Adam Bolt, director, and Elliot Kirschner, executive producer, *Human Nature*, documentary, the Wonder Collaborative, 2019.
4. My questions to David Sanchez and his responses were relayed through the *Human Nature* producer, Meredith DeSalazar.
5. Rosemarie Garland-Thomson, "Welcoming the Unexpected," in Erik Parens and Josephine Johnston, *Human Flourishing in an Age of Gene Editing* (Oxford, 2019); Rosemarie Garland-Thomson, "Human Biodiversity Conservation," *American Journal of Bioethics*, Jan. 2015. See also Ethan Weiss, "Should 'Broken' Genes Be Fixed?" *Stat*, Feb. 21, 2020.
6. Jory Fleming, *How to Be Human* (Simon & Schuster, 2021).
7. Liza Mundy, "A World of Their Own," *Washington Post*, Mar. 31, 2002; Sandel, *The Case against Perfection*; Marion Andrea Schmidt, *Eradicating Deafness?* (Manchester University Press, 2020).
8. Craig Pickering and John Kiely, "ACTN#: More Than Just a Gene for Speed," *Frontiers in Physiology*, Dec. 18, 2017; David Epstein, *The Sports Gene* (Current, 2013); Haran Sivapalan, "Genetics of Marathon Runners," *Fitness Genes*, Sept. 26, 2018.
9. The Americans with Disabilities Act defines a disability as "a physical or mental impairment that substantially limits one or more major life activity."
10. Fred Hirsch, *Social Limits to Growth* (Routledge, 1977); Glenn Cohen, "What (If Anything) Is Wrong with Human Enhancement? What (If Anything) Is Right with It?," *Tulsa Law Review*, Apr. 21, 2014.
11. Nancy Andreasen, "The Relationship between Creativity and Mood Disorders," *Dialogues in Clinical Psychology*, June 2018; Neel Burton, "Hide and Seek: Bipolar Disorder and Creativity," *Psychology Today*, Mar. 19, 2012; Nathaniel Comfort, "Better Babies," *Aeon*, Nov. 17, 2015.
12. Robert Nozick, *Anarchy, State, and Utopia* (Basic Books, 1974).
13. See Erik Parens and Josephine Johnston, eds., *Human Flourishing in an Age of Gene Editing* (Oxford, 2019).
14. Jinping Liu . . . Yan Wu, et al., "The Role of NMDA Receptors in Alzheimer's Disease," *Frontiers in Neuroscience*, Feb. 8, 2019.

Chapter 42: Who Should Decide?

1. National Academy of Sciences, "How Does Human Gene Editing Work?" 2019, https://thesciencebehindit.org/how-does-human-gene-editing-work/, page removed; Marilynn Marchione, "Group Pulls Video That Stirred Talk of Designer Babies," AP, Oct. 2, 2019.
2. Twitter thread, @FrancoiseBaylis, @pknoepfler, @UrnovFyodor, @theNASAcademies, and others, Oct. 1, 2019.
3. John Rawls, *A Theory of Justice* (Harvard, 1971), 266, 92.
4. Nozick, *Anarchy, State and Utopia*, 315n.
5. Colin Gavaghan, *Defending the Genetic Supermarket* (Routledge-Cavendish, 2007); Peter Singer, "Shopping at the Genetic Supermarket," in John Rasko, ed., *The Ethics of Inheritable Genetic Modification* (Cambridge, 2006); Chris Gyngell and Thomas Douglas, "Stocking the Genetic Supermarket," *Bioethics*, May 2015.
6. Fukuyama, *Our Posthuman Future*, chapter 1; George Orwell, *1984* (Harcourt, 1949); Aldous Huxley, *Brave New World* (Harper, 1932).
7. Aldous Huxley, *Brave New World Revisited* (Harper, 1958), 120.
8. Aldous Huxley, *Island* (Harper, 1962), 232; Derek So, "The Use and Misuse of Brave New World in the CRISPR Debate," *CRISPR Journal*, Oct. 2019.

9. Nathaniel Comfort, "Can We Cure Genetic Diseases without Slipping into Eugenics?," *The Nation*, Aug. 3, 2015; Nathaniel Comfort, *The Science of Human Perfection* (Yale, 2012); Mark Frankel, "Inheritable Genetic Modification and a Brave New World," *Hastings Center Report*, Mar. 6, 2012; Arthur Caplan, "What Should the Rules Be?," *Time*, Jan. 14, 2001; Françoise Baylis and Jason Scott Robert, "The Inevitability of Genetic Enhancement Technologies," *Bioethics*, Feb. 2004; Daniel Kevles, "If You Could Design Your Baby's Genes, Would You?," *Politico*, Dec. 9, 2015; Lee M. Silver, "How Reprogenetics Will Transform the American Family," *Hofstra Law Review*, Fall 1999; Jürgen Habermas, *The Future of Human Nature* (Polity, 2003).
10. Author's interview with George Church, and similarly quoted in Rachel Cocker, "We Should Not Fear 'Editing' Embryos to Enhance Human Intelligence," *The Telegraph*, Mar. 16, 2019; Lee Silver, *Remaking Eden* (Morrow, 1997); John Harris, *Enhancing Evolution* (Princeton, 2011); Ronald Green, *Babies by Design* (Yale, 2008).
11. Julian Savulescu, "Procreative Beneficence: Why We Should Select the Best Children," *Bioethics*, Nov. 2001.
12. Antonio Regalado, "The World's First Gattaca Baby Tests Are Finally Here," *MIT Technology Review*, Nov. 8, 2019; Genomic Prediction company website, "Frequently Asked Questions," retrieved July 6, 2020; Hannah Devlin, "IVF Couples Could Be Able to Choose the 'Smartest' Embryo," *Guardian*, May 24, 2019; Nathan Treff . . . and Laurent Tellier, "Preimplantation Genetic Testing for Polygenic Disease Relative Risk Reduction," *Genes*, June 12, 2020; Louis Lello . . . and Stephen Hsu, "Genomic Prediction of 16 Complex Disease Risks," *Nature*, Oct. 25, 2019. In November 2019, *Nature* issued a conflict-of-interest correction saying that some of the authors did not disclose that they were affiliated with the company Genomic Prediction.
13. In addition to the sources cited above, see Laura Hercher, "Designer Babies Aren't Futuristic. They're Already Here," *MIT Technology Review*, Oct. 22, 2018; Ilya Somin, "In Defense of Designer Babies," *Reason*, Nov. 11, 2018.
14. Francis Fukuyama, "Gene Regime," *Foreign Policy*, Mar. 2002.
15. Francis Collins in Patrick Skerrett, "Experts Debate: Are We Playing with Fire When We Edit Human Genes?," *Stat*, Nov. 17, 2016.
16. Russell Powell and Allen Buchanan, "Breaking Evolution's Chains," *Journal of Medical Philosophy*, Feb. 2011; Allen Buchanan, *Better Than Human* (Oxford, 2011); Charles Darwin to J. D. Hooker, July 13, 1856.
17. Sandel, *The Case against Perfection*; Leon Kass, "Ageless Bodies, Happy Souls," *The New Atlantis*, Jan. 2003; Michael Hauskeller, "Human Enhancement and the Giftedness of Life," *Philosophical Papers*, Feb. 26, 2011.

Chapter 43: Doudna's Ethical Journey

1. Author's interviews with Jennifer Doudna; Doudna and Sternberg, *A Crack in Creation*, 222–40; Hannah Devlin, "Jennifer Doudna: 'I Have to Be True to Who I Am as a Scientist,'" *The Observer*, July 2, 2017.

Chapter 44: Quebec

1. Sanne Klompe . . . Samuel Sternberg, et al., "Transposon-Encoded CRISPR-Cas Systems Direct RNA-Guided DNA Integration," *Nature*, July 11, 2019 (received Mar. 15, 2019; accepted June 4; published online June 12); Jonathan Strecker . . . Eugene Koonin, Feng Zhang, et al., "RNA-Guided DNA Insertion with CRISPR-Associated Transposases," *Science*, July 5, 2019 (received May 4, 2019; accepted May 29; published online June 6).

2. Author's interviews with Sam Sternberg, Martin Jinek, Jennifer Doudna, Joe Bondy-Denomy.
3. Author's interviews with Feng Zhang.

Chapter 45: I Learn to Edit

1. Author's interviews with Gavin Knott.
2. "Alt-R CRISPR-Cas9 System: Delivery of Ribonucleoprotein Complexes into HEK-293 Cells Using the Amaxa Nucleofector System," IDTDNA.com; "CRISPR Gene-Editing Tools," GeneCopoeia.com.
3. Author's interviews with Jennifer Hamilton.

Chapter 46: Watson Revisited

1. Author's interviews with James Watson, Jennifer Doudna; "The CRISPR/Cas Revolution," Cold Spring Harbor Laboratory meeting, Sept. 24–27, 2015.
2. David Dugan, producer, *DNA*, documentary, Windfall Films for WNET/PBS and BBC4, 2003; Shaoni Bhattacharya, "Stupidity Should Be Cured, Says DNA Discoverer," *The New Scientist*, Feb. 28, 2003. See also Tom Abate, "Nobel Winner's Theories Raise Uproar in Berkeley," *San Francisco Chronicle*, Nov. 13, 2000.
3. Michael Sandel, "The Case against Perfection," *The Atlantic*, Apr. 2004.
4. Charlotte Hunt-Grubbe, "The Elementary DNA of Dr Watson," *Sunday Times* (London), Oct. 14, 2007; author's interviews with James Watson.
5. Author's interviews with James Watson; Roxanne Khamsi, "James Watson Retires amidst Race Controversy," *The New Scientist*, Oct. 25, 2007.
6. Author's interview with Eric Lander; Sharon Begley, "As Twitter Explodes, Eric Lander Apologizes for Toasting James Watson," *Stat*, May 14, 2018.
7. Author's interviews with James Watson.
8. "Decoding Watson."
9. Amy Harmon, "James Watson Had a Chance to Salvage His Reputation on Race. He Made Things Worse," *New York Times*, Jan. 1, 2019.
10. Harmon, "James Watson Had a Chance to Salvage His Reputation on Race."
11. "Decoding Watson"; Harmon, "James Watson Had a Chance to Salvage His Reputation on Race"; author's interviews with James Watson.
12. James Watson, "An Appreciation of Linus Pauling," *Time* magazine seventy-fifth anniversary dinner, Mar. 3, 1998.
13. Author's interviews with James Watson. I used some of these quotes, as well as other passages, in a piece I wrote, "Should the Rich Be Allowed to Buy the Best Genes?"
14. "Decoding Watson."
15. Author's meetings with James Watson, Rufus Watson, Elizabeth Watson.
16. Malcolm Ritter, "Lab Revokes Honors for Controversial DNA Scientist Watson," AP, Jan. 11, 2019.

Chapter 47: Doudna Pays a Visit

1. Author's visit with James Watson and Jennifer Doudna. The conference book was designed by Megan Hochstrasser, who works in Doudna's lab.
2. Author's interviews with Jennifer Doudna.

Chapter 48: Call to Arms

1. Robert Sanders, "New DNA-Editing Technology Spawns Bold UC Initiative," *Berkeley News*, Mar. 18, 2014; "About Us," Innovative Genomics Institute website, https://

innovativegenomics.org/about-us/. It was relaunched in January 2017 as the Innovative Genomics Institute.

2. Author's interview with Dave Savage; Benjamin Oakes . . . Jennifer Doudna, David Savage, et al., "CRISPR-Cas9 Circular Permutants as Programmable Scaffolds for Genome Modification," *Cell*, Jan 10, 2019.
3. Author's interviews with Dave Savage, Gavin Knott, and Jennifer Doudna.
4. Jonathan Corum and Carl Zimmer, "Bad News Wrapped in Protein: Inside the Coronavirus Genome," *New York Times*, Apr. 3, 2020; GenBank, National Institutes of Health, SARS-CoV-2 Sequences, updated Apr. 14, 2020.
5. Alexander Walls . . . David Veesler, et al., "Structure, Function, and Antigenicity of the SARS-CoV-2 Spike Glycoprotein," *Cell*, Mar. 9, 2020; Qihui Wang . . . and Jianxun Qi, "Structural and Functional Basis of SARS-CoV-2 Entry by Using Human ACE2," *Cell*, May 14, 2020; Francis Collins, "Antibody Points to Possible Weak Spot on Novel Coronavirus," NIH, Apr. 14, 2020; Bonnie Berkowitz, Aaron Steckelberg, and John Muyskens, "What the Structure of the Coronavirus Can Tell Us," *Washington Post*, Mar. 23, 2020.
6. Author's interviews with Megan Hochstrasser, Jennifer Doudna, Dave Savage, and Fyodor Urnov.

Chapter 49: Testing

1. Shawn Boburg, Robert O'Harrow Jr., Neena Satija, and Amy Goldstein, "Inside the Coronavirus Testing Failure," *Washington Post*, Apr. 3, 2020; Robert Baird, "What Went Wrong with Coronavirus Testing in the U.S.," *New Yorker*, Mar. 16, 2020; Michael Shear, Abby Goodnough, Sheila Kaplan, Sheri Fink, Katie Thomas, and Noah Weiland, "The Lost Month: How a Failure to Test Blinded the U.S. to COVID-19," *New York Times*, Mar. 28, 2020.
2. Kary Mullis, "The Unusual Origin of the Polymerase Chain Reaction," *Scientific American*, Apr. 1990.
3. Boburg et al., "Inside the Coronavirus Testing Failure"; David Willman, "Contamination at CDC Lab Delayed Rollout of Coronavirus Tests," *Washington Post*, Apr. 18, 2020.
4. JoNel Aleccia, "How Intrepid Lab Sleuths Ramped Up Tests as Coronavirus Closed In," *Kaiser Health News*, Mar. 16, 2020.
5. Julia Ioffe, "The Infuriating Story of How the Government Stalled Coronavirus Testing," *GQ*, Mar. 16, 2020; Boburg et al., "Inside the Coronavirus Testing Failure." Greninger's email to a friend is in the excellent *Washington Post* reconstruction.
6. Boburg et al., "Inside the Coronavirus Testing Failure"; Patrick Boyle, "Coronavirus Testing: How Academic Medical Labs Are Stepping Up to Fill a Void," *AAMC*, Mar. 12, 2020.
7. Author's interview with Eric Lander; Leah Eisenstadt, "How Broad Institute Converted a Clinical Processing Lab into a Large-Scale COVID-19 Testing Facility in a Matter of Days," *Broad Communications*, Mar. 27, 2020.

Chapter 50: The Berkeley Lab

1. IGI COVID-19 Rapid Response Research meeting, Mar. 13, 2020. I was allowed to attend the meetings of the rapid-response team and its working groups, most of which took place on Zoom with discussion in Slack channels.
2. Author's interviews with Fyodor Urnov. Dmitry Urnov became a professor at Adelphi

University in New York. He is an accomplished horseman who once accompanied three horses on a sea voyage when Nikita Khrushchev wanted to give them as a gift to the American industrialist Cyrus Eaton. He and his wife, Julia Palievsky, wrote *A Kindred Writer: Dickens in Russia*. They are also scholars of William Faulkner.

3. Author's interviews with Jennifer Hamilton; Jennifer Hamilton, "Building a COVID-19 Pop-Up Testing Lab," *CRISPR Journal*, June 2020.
4. Author's interviews with Enrique Lin Shiao.
5. Author's interviews with Fyodor Urnov, Jennifer Doudna, Jennifer Hamilton, Enrique Lin Shiao; Hope Henderson, "IGI Launches Major Automated COVID-19 Diagnostic Testing Initiative," *IGI News*, Mar. 30, 2020; Megan Molteni and Gregory Barber, "How a Crispr Lab Became a Pop-Up COVID Testing Center," *Wired*, Apr. 2, 2020.
6. Innovative Genomics Institute SARS-CoV-2 Testing Consortium, Dirk Hockemeyer, Fyodor Urnov, and Jennifer A. Doudna, "Blueprint for a Pop-up SARS-CoV-2 Testing Lab," *medRxiv*, Apr. 12, 2020.
7. Author's interviews with Fyodor Urnov, Jennifer Hamilton, and Enrique Lin Shiao.

Chapter 51: Mammoth and Sherlock

1. Author's interview with Lucas Harrington and Janice Chen.
2. Janice Chen . . . Lucas B. Harrington . . . Jennifer A. Doudna, et al., "CRISPR-Cas12a Target Binding Unleashes Indiscriminate Single-Stranded DNase Activity," *Science*, Apr. 27, 2018 (received Nov. 29, 2017; accepted Feb. 5, 2018; published online Feb. 15); John Carroll, "CRISPR Legend Jennifer Doudna Helps Some Recent College Grads Launch a Diagnostics Up-start," *Endpoints*, Apr. 26, 2018.
3. Sergey Shmakov, Omar Abudayyeh, Kira S. Makarova . . . Konstantin Severinov, Feng Zhang, and Eugene V. Koonin, "Discovery and Functional Characterization of Diverse Class 2 CRISPR-Cas Systems," *Molecular Cell*, Nov. 5, 2015 (published online Oct. 22, 2015); Omar Abudayyeh, Jonathan Gootenberg . . . Eric Lander, Eugene Koonin, and Feng Zhang, "C2c2 Is a Single-Component Programmable RNA-Guided RNA-Targeting CRISPR Effector," *Science*, Aug. 5, 2016 (published online June 2, 2016).
4. Author's interviews with Feng Zhang.
5. Alexandra East-Seletsky . . . Jamie Cate, Robert Tjian, and Jennifer Doudna, "Two Distinct RNase Activities of CRISPR-C2c2 Enable Guide-RNA Processing and RNA Detection," *Nature*, Oct. 13, 2016. CRISPR-C2c2 was renamed CRISPER-Cas13a.
6. Jonathan Gootenberg, Omar Abudayyeh . . . Cameron Myhrvold . . . Eugene Koonin . . . Feng Zhang et al., "Nucleic Acid Detection with CRISPR-Cas13a/C2c2," *Science*, Apr. 28, 2017.
7. Jonathan Gootenberg, Omar Abudayyeh . . . Feng Zhang, et al., "Multiplexed and Portable Nucleic Acid Detection Platform with Cas13, Cas12a, and Csm6," *Science*, Apr. 27, 2018. See also Abudayyeh et al., "C2c2 Is a Single Component Programmable RNA-Guided RNA-Targeting CRISPR Effector."
8. Author's interview with Feng Zhang; Carey Goldberg, "CRISPR Comes to COVID," WBUR, July 10, 2020.
9. Emily Mullin, "CRISPR Could Be the Future of Disease Diagnosis," *OneZero*, July 25, 2019; Emily Mullin, "CRISPR Pioneer Jennifer Doudna on the Future of Disease Detection," *OneZero*, July 30, 2019; Daniel Chertow, "Next-Generation Diagnostics with CRISPR," *Science*, Apr. 27, 2018; Ann Gronowski "Who or What Is SHERLOCK?," *EJIFCC*, Nov. 2018.

Chapter 52: Coronavirus Tests

1. Author's interviews with Feng Zhang.
2. Feng Zhang, Omar Abudayyeh, and Jonathan Gootenberg, "A Protocol for Detection of COVID-19 Using CRISPR Diagnostics," Broad Institute website, posted Feb. 14, 2020; Carl Zimmer, "With Crispr, a Possible Quick Test for the Coronavirus," *New York Times*, May 5, 2020.
3. Goldberg, "CRISPR Comes to COVID"; "Sherlock Biosciences and Binx Health Announce Global Partnership to Develop First CRISPR-Based Point-of-Care Test for COVID-19," *PR Newswire*, July 1, 2020.
4. Author's interviews with Janice Chen and Lucas Harrington; Jim Daley, "CRISPR Gene Editing May Help Scale Up Coronavirus Testing," *Scientific American*, Apr. 23, 2020; John Cumbers, "With Its Coronavirus Rapid Paper Test Strip, This CRISPR Startup Wants to Help Halt a Pandemic," *Forbes*, Mar. 14, 2020; Lauren Martz, "CRISPR-Based Diagnostics Are Poised to Make an Early Debut amid COVID-19 Outbreak," *Biocentury*, Feb. 28, 2020.
5. James Broughton . . . Charles Chiu, Janice Chen, et al., "A Protocol for Rapid Detection of the 2019 Novel Coronavirus SARS-CoV-2 Using CRISPR Diagnostics: SARS-CoV-2 DETECTR," Mammoth Biosciences website, posted Feb. 15, 2020. The full Mammoth paper with patient data and other details is James Broughton . . . Janice Chen, and Charles Chiu, "CRISPR–Cas12-Based Detection of SARS-CoV-2," *Nature Biotechnology*, Apr. 16, 2020 (received Mar. 5, 2020). See also Eelke Brandsma . . . and Emile van den Akker, "Rapid, Sensitive and Specific SARS Coronavirus-2 Detection: A Multi-center Comparison between Standard qRT-PCR and CRISPR Based DETECTR," *medRxiv*, July 27, 2020.
6. Julia Joung . . . Jonathan S. Gootenberg, Omar O. Abudayyeh, and Feng Zhang, "Point-of-Care Testing for COVID-19 Using SHERLOCK Diagnostics," *medRxiv*, May 5, 2020.
7. Author's interview with Feng Zhang.
8. Author's interview with Janice Chen.

Chapter 53: Vaccines

1. Ochsner Health System, phase 2/3 study by Pfizer Inc. and BioNTech SE of investigational vaccine, BNT162b2, against SARS-CoV-2, beginning July 2020.
2. Author's interview with Jennifer Doudna.
3. Simantini Dey, "Meet Sarah Gilbert," *News18*, July 21, 2020; Stephanie Baker, "Covid Vaccine Front-Runner Is Months Ahead of Her Competition," *Bloomberg Business-Week*, July 14, 2020; Clive Cookson, "Sarah Gilbert, the Researcher Leading the Race to a Covid-19 Vaccine," *Financial Times*, July 24, 2020.
4. Author's interviews with Ross Wilson, Alex Marson; IGI white paper seeking funding for DNA vaccine delivery systems, Mar. 2020; Ross Wilson report at IGI COVID-response meeting, June 11, 2020.
5. "A Trial Investigating the Safety and Effects of Four BNT162 Vaccines against COVID-2019 in Healthy Adults," ClinicalTrials.gov, May 2020, identifier: NCT04380701; "BNT162 SARS-CoV-2 Vaccine," *Precision Vaccinations*, Aug. 14, 2020; Mark J. Mulligan . . . Uğur Şahin, Kathrin Jansen, et. al., "Phase 1/2 Study of COVID-19 RNA Vaccine BNT162b1 in Adults," *Nature*, Aug. 12, 2020.
6. Joe Miller, "The Immunologist Racing to Find a Vaccine," *Financial Times*, Mar. 20, 2020.
7. Author's interview with Phil Dormitzer; Matthew Herper, "In the Race for a

COVID-19 Vaccine, Pfizer Turns to a Scientist with a History of Defying Skeptics," *Stat*, Aug. 24, 2020.

8. Author's interviews with Noubar Afeyan, Christine Heenan.
9. Author's interview and emails with Josiah Zayner; Kristen Brown, "One Biohacker's Improbable Bid to Make a DIY Covid-19 Vaccine," *Bloomberg Business Week*, June 25, 2020; Josiah Zayner videos, www.youtube.com/josiahzayner.
10. Jingyou Yu . . . and Dan H. Barouch, "DNA Vaccine Protection against SARS-CoV-2 in Rhesus Macaques," *Science*, May 20, 2020.
11. Author's interviews with Josiah Zayner; Kristen Brown, "Home-Made Vaccine Appeared to Work, but Questions Remain," *Bloomberg BusinessWeek*, Oct. 10, 2020.
12. The Ochsner Health system clinical trial of Pfizer/BioNTech vaccine BNT162b2, led by Julia Garcia-Diaz, director of Clinical Infectious Diseases Research, and Leonardo Seoane, chief academic officer.
13. Author's interview with Francis Collins; "Bioethics Consultation Service Consultation Report," Department of Bioethics, NIH Clinical Center, July 31, 2020.
14. Sharon LaFraniere, Katie Thomas, Noah Weiland, David Gelles, Sheryl Gay Stolberg and Denise Grady, "Politics, Science and the Remarkable Race for a Coronavirus Vaccine," *New York Times*, Nov. 21, 2020; author's interviews with Noubar Afeyan, Moncef Slaoui, Philip Dormitzer, Christine Heenan.

Chapter 54: CRISPR Cures

1. David Dorward . . . and Christopher Lucas, "Tissue-Specific Tolerance in Fatal COVID-19," *medRxiv*, July 2, 2020; Bicheng Zhag . . . and Jun Wan, "Clinical Characteristics of 82 Cases of Death from COVID-19," *Plos One*, July 9, 2020.
2. Ed Yong, "Immunology Is Where Intuition Goes to Die," *The Atlantic*, Aug. 5, 2020.
3. Author's interview with Cameron Myhrvold.
4. Jonathan Gootenberg, Omar Abudayyeh . . . Cameron Myhrvold . . . Eugene Koonin . . . Pardis Sabeti . . . and Feng Zhang, "Nucleic Acid Detection with CRISPR-Cas13a/C2c2," *Science*, Apr. 28, 2017.
5. Cameron Myhrvold, Catherine Freije, Jonathan Gootenberg, Omar Abudayyeh . . . Feng Zhang, and Pardis Sabeti, "Field-Deployable Viral Diagnostics Using CRISPR-Cas13," *Science*, Apr. 27, 2018.
6. Author's interview with Cameron Myhrvold.
7. Cameron Myhrvold to Pardis Sabeti, Dec. 22, 2016.
8. Defense Advanced Research Projects Agency (DARPA) grant D18AC00006.
9. Susanna Hamilton, "CRISPR-Cas13 Developed as Combination Antiviral and Diagnostic System," *Broad Communications*, Oct. 11, 2019.
10. Catherine Freije, Cameron Myhrvold . . . Omar Abudayyeh, Jonathan Gootenberg . . . Feng Zhang, and Pardis Sabeti, "Programmable Inhibition and Detection of RNA Viruses Using Cas13," *Molecular Cell*, Dec. 5, 2019 (received Apr. 16, 2019; revised July 18, 2019; accepted Sept. 6, 2019; published online Oct. 10, 2019); Tanya Lewis, "Scientists Program CRISPR to Fight Viruses in Human Cells," *Scientific American*, Oct. 23, 2019.
11. Cheri Ackerman, Cameron Myhrvold . . . and Pardis C. Sabeti, "Massively Multiplexed Nucleic Acid Detection with Cas13m," *Nature*, Apr. 29, 2020 (received Mar. 20, 2020; accepted Apr. 20, 2020).
12. Jon Arizti-Sanz, Catherine Freije . . . Pardis Sabeti, and Cameron Myhrvold, "Integrated Sample Inactivation, Amplification, and Cas13-Based Detection of SARS-CoV-2," *bioRxiv*, May 28, 2020.

13. Author's interviews with Stanley Qi.
14. Silvana Konermann . . . and Patrick Hsu, "Transcriptome Engineering with RNA-Targeting Type VI-D CRISPR Effectors," *Cell*, Mar. 15, 2018.
15. Steven Levy, "Could CRISPR Be Humanity's Next Virus Killer?," *Wired*, Mar. 10, 2020.
16. Timothy Abbott . . . and Lei [Stanley] Qi, "Development of CRISPR as a Prophylactic Strategy to Combat Novel Coronavirus and Influenza," *bioRxiv*, Mar. 14, 2020.
17. Author's interview with Stanley Qi.
18. IGI weekly Zoom meeting, Mar. 22, 2020; author's interviews with Stanley Qi and Jennifer Doudna.
19. Stanley Qi, Jennifer Doudna, and Ross Wilson, "A White Paper for the Development of Novel COVID-19 Prophylactic and Therapeutics Using CRISPR Technology," unpublished, Apr. 2020.
20. Author's interviews with Ross Wilson; Ross Wilson, "Engineered CRISPR RNPs as Targeted Effectors for Genome Editing of Immune and Stem Cells In Vivo," unpublished, Apr. 2020.
21. Theresa Duque, "Cellular Delivery System Could Be Missing Link in Battle against SARS-CoV-2," *Berkeley Lab News*, June 4, 2020.

Chapter 55: Cold Spring Harbor Virtual

1. Kevin Bishop and others gave me permission to quote them from the meeting.
2. Andrew Anzalone . . . David Liu, et al., "Search-and-Replace Genome Editing without Double-Strand Breaks or Donor DNA," *Nature*, Dec. 5, 2019 (received Aug. 26; accepted Oct. 10; published online Oct. 21).
3. Megan Molteni, "A New Crispr Technique Could Fix Almost All Genetic Diseases," *Wired*, Oct. 21, 2019; Sharon Begley, "New CRISPR Tool Has the Potential to Correct Almost All Disease-Causing DNA Glitches," *Stat*, Oct. 21, 2019; Sharon Begley, "You Had Questions for David Liu," *Stat*, Nov. 6, 2019.
4. Beverly Mok . . . David Liu, et al., "A Bacterial Cytidine Deaminase Toxin Enables CRISPR-Free Mitochondrial Base Editing," *Nature*, July 8, 2020.
5. Jonathan Hsu . . . David Liu, Keith Joung, Lucan Pinello, et al., "PrimeDesign Software for Rapid and Simplified Design of Prime Editing Guide RNAs," *bioRxiv*, May 4, 2020.
6. Audrone Lapinaite, Gavin Knott . . . David Liu, and Jennifer A. Doudna, "DNA Capture by a CRISPR-Cas9–Guided Adenine Base Editor," *Science*, July 31, 2020.

Chapter 56: The Nobel Prize

1. Author's interviews with Heidi Ledford, Jennifer Doudna, Emmanuelle Charpentier.
2. Jennifer Doudna, "How COVID-19 Is Spurring Science to Accelerate," *The Economist*, June 5, 2020. See also Jane Metcalfe, "COVID-19 Is Accelerating Human Transformation—Let's Not Waste It," *Wired*, July 5, 2020.
3. Michael Eisen, "Patents Are Destroying the Soul of Academic Science," *it is NOT junk* (blog), Feb. 20, 2017.
4. "SARS-CoV-2 Sequence Read Archive Submissions," National Center for Biotechnology Information, https://www.ncbi.nlm.nih.gov/sars-cov-2/, n.d.
5. Simine Vazire, "Peer-Reviewed Scientific Journals Don't Really Do Their Job," *Wired*, June 25, 2020.
6. Author's interview with George Church.
7. Author's interview with Emmanuelle Charpentier.

INDEX

Page numbers in *italics* refer to photographs.

Abbott, Tim, 455
Abbott Labs, 430
abortion, 293, 337
Abudayyeh, Omar, 425, *426*, 428, 451
ACE2 protein, 404
Achenbach, Joel, 229
ACTN3 gene, 349
adenine, 26, *27*
adenoviruses, 438, 456
Advanced Photon Source, 85–86
Aeschylus, 243
Afeyan, Noubar, 442, 447
African Americans
  COVID vaccines and, 461
  as researchers, 461
  sickle-cell anemia and, *see* sickle-cell anemia
  and skin color as disadvantage, 347–48
  Watson's comments on race, 37, 49, 386–92, 394
Agee, James, 265
AIDS and HIV, 166, 305, 390, 431
  He Jiankui's removal of HIV receptor in embryos, xv–xvi, 251, 303–13, 316–24, 369, 404
  sperm-washing and, 305, 322
Alda, Alan, 221
Alexander, Lamar, 328
Alliance of Chief Executives, 116
Altman, Sidney, 44–45, 53
Alzheimer's, 250, 251, 324, 339, 365
American Association for the Advancement of Science, 268
American Cancer Society, 465
American Masters: Decoding Watson (PBS documentary), *384*, 386, 389–90, 392, 393
American Society for Microbiology conference (2011), 119, 127–28
American Telephone and Telegraph Company, 90
Anarchy, State, and Utopia (Nozick), 357–58
Anderson, French, 276
Andreasen, Nancy, 352
Andromeda Strain, The (Crichton), 271
antibiotics
  bacterial resistance to, 121–22
  penicillin, 56, 122
antigen tests, 430–31
APOE4 gene, 251
Apple, 358, 463
  "Think Different" ad, 326, 371
Apollo program, 417
archaea, 71–73, 155
Argonne National Laboratory, 85–86
Aristotle, 464
Ashkenazi Jews, 389
Asilomar, *266*, 269–73, 278, 286–88, 330, 331
Asimov, Isaac, 13
Aspen Institute, 228
Associated Press (AP), 308–10, 319, 321
AstraZeneda, 439
Atlantic, 281
atom bomb, xvii, 89, 117, 224, 265, 338
atoms, 8
autism, 346
Avery, Oswald, 17
Avoid Boring People (Watson), 386–87
Ax, Emanuel, 388
Azar, Alex, 407–9

babies, designer, 355–56
  see also CRISPR babies
bacteria, 34, 71–73, 155, 216
  antibiotic resistance in, 121–22
  cloning genes from, 34–35
  E. coli, 71, 72, 74, 98
  fruiting bodies formed by, 33
  oil spills and, 232
  in ponds, 402
  Streptococcus pyogenes, 122, 139, 182
  Streptococcus thermophilus, 91, 182
  in yogurt starter cultures, xviii, 90–92, 133, 143
bacteriophages (viruses that attack bacteria), *see* phages

Baltimore, David, 270, 272, *282,* 287–88, 293, 330
at International Summit on Human Genome Editing, 315, 321, 323, 324
Bancel, Stéphane, 442
Banfield, Jillian, *78,* 79, 97, 402, 472
Doudna invited to work on CRISPR by, xviii, 77, 79–80, 81, 93, 471
Bannister, Roger, 326
Bardeen, John, 90
Barnabé, Duilio, 395
Barrangou, Rodolphe, *88,* 90–93, 143–46, 148, 149, 177, 178, 213, 217, 220, 223, 309, 470
Bayh-Dole Act, 232
BBC, 386
Beagle, HMS, 11–12
Beam Therapeutics, 461
Beckinsale, Kate, 219
Beijing Review, 301
Bell Labs, 89–90, 201
Belluck, Pam, 307
Berg, Paul, 98–100, 153, 232, 287–88, 330
at Asilomar, *266,* 269–73
Watson and, 271
Berkeley, University of California at, 64–65, 114, 115, 117, 397
COVID and, xiii-xv
Doudna at, xiii–xv, xvii, 64–65, 83, 98, 101, 107
Doudna and Charpentier's patent with, 207–8, 210, 219, 224, 233–40
Doudna's lab at, 103–11, *104,* 117
Isaacson learns to edit at, xiv, *378,* 379–83
entrepreneurship and, 115–16
Haas School of Business, 115
Innovative Genomics Institute, 248, 261, 380–82, 401–3, 456
COVID and, 401–5, 413–18, *419*
Berkeley Fire Department, *406,* 418
Berkeley National Laboratory, 64, 87, 117, 456–57
Better Than Human (Buchanan), 310
Biden, Joe, 465
Bill and Melinda Gates Foundation, 118, 248, 439
biochemistry, 18, 29–35, 51, 80, 183, 201, 478
bioconservatives, 267–69, 280
Biogen, 206
biohacking, 253–57, 263, 285, 288, 443–44
biology, 8, 133, 183, 474
bringing into the home, 432–33, 445, 460
central dogma of, 44, 47, 270
public interest in, 460, 473
structural, 18, 51–53, 80, 166
BioNTech, 435, 441, 445, 447
Bio-Revolution and Its Implications for Army Combat Capabilities, The (conference), 262–63
biotechnology, 64, 98–100, 113–15, 153, 211, 232, 461, 477
Asilomar and, 269–73
commercial development of, 288
crowdsourcing and, 256–57, 285
moral concerns about, *see* moral questions
patents in, 232
regulation of, 270, 278, 281
utopians vs. bioconservatives and, 267–69
see also genetic engineering
bipolar disorder, 177, 327, 352–53
Bishop, Kevin, 461
bits (binary digits), xvii, 28
Bitterman, Kevin, 209–10
Black Death, 447
black markets, 291
blindness, xviii, 250
Blue Room meeting, 206, 207
Bohr, Niels, 224
Bondy-Denomy, Joseph, *258,* 260–61
Borgia, Cesare, 259
Boston Globe, 291
Bourla, Albert, 447
Boyer, Herbert, *96,* 98–100, 153, 232, *266,* 270
Bragg, Lawrence, 20, 23, 26
brain, 167, 207
experience machine and, 353
Brannigan, David, 418
Brave New World (Huxley), xv, 267, 269, 280, 281, 353, 358–59
Breakthrough Prize in Life Sciences, 219–21
breeding, 11, 12, 14
Brenner, Sydney, *266,* 272
Brigham and Women's Hospital, 410
Brin, Sergey, 219
Britain, 278, 294
Medical Research Council in, 26
Nuffield Council in, 293, 294
Parliamentary and Scientific Committee in, 333
Royal Society in, 292
Broad, Eli and Edythe, 175
Broad Institute of MIT and Harvard, xv, 64, 144, 160, 175–78, 187, 189, 192, 206, 330, 388, 425, 450, 453, 461
COVID and, 410–11, 429
founding of, 175
Lander as director of, 64, 144, 160, 175, 216, 220, 223, 225–28
Zhang at, xv, 159, 161, 177, 179, 184–86, 189, 374, 421, 423, 424, 450–51, 454
Zhang's patent with, 192, 207–8, 210–11, 219, 224–27, 233–40
Brouns, Stan, 84
Brown, Louise (first test-tube baby), 274, 313, 328
Buchanan, Allen, 309–10
Buchman, Lauren, 284–86
Burger, Warren, 232
Bush, George W., 280
Bush, Vannevar, 89, 117

calculus, 159
California Institute for Quantitative Biosciences (QB3), 116
Caltech, 270
Cambridge University Cavendish Laboratory, 20–23, 26
Camus, Albert, 399

cancer, xviii, 100, 177, 259, 339, 365, 441, 442
  CRISPR and, xviii, 249–51
Canseco, José, 349
Carey, Mariah, 352
Caribou Biosciences, 113–18, 203, 207–8, 213
  Intellia Therapeutics, 213
CARMEN, 453, 460
Carroll, Dana, 185
Carter, Jimmy, 273
CARVER, 452–53, 455–57
"Case Against Perfection, The" (Rawls), 281
"Case for Developing America's Talent, A" (study), 164
Cas enzymes, *see* CRISPR-associated enzymes
Casey Eye Institute, 250
Cassidy, Bill, 328–29
Cate, Andrew, xiii, *62,* 63, 97, 101, 114, 132, 199, 209, 328, 368, 402, 465, *468,* 470
Cate, Jamie, xiii, 56, 57, 60, *62,* 63–64, 97, 101, 114, 199, 381, 402, 422, 424, *468,* 470, 472
  Jennifer's marriage to, 63
Cavendish Laboratory, 20–23, 26
CDC, 407–10
CDKN1C gene, 349–50
Cech, Thomas, 44–45, 49, 53–56
Celera, 39, 40
Cell, 143, 146, 376
  Lander's "Heroes of CRISPR" piece in, 182, 223–29, 388
Cell Reports, 143
Centers for Disease Control (CDC), 407–10
Chan, Priscilla, 472
character, 345–46
Chargaff, Erwin, 273
Charité, 124
Charles, Prince, 100
Charo, Alta, 288, 316
Charpentier, Emmanuelle, 119–28, *120,* 129–32, *130, 138,* 147, *204, 214,* 283, 292, 330, 385, 477
  Breakthrough Prize in Life Sciences awarded to, 219–21
  Church and, 206
  Church's email to Doudna and, 173, 197
  company formations and, 205–7, 209, 213
  CRISPR Therapeutics, 207, 213, 245, 465
  CRISPR paper of (2011), 125–27, 131, 147, 179–80, 185, 218
  CRISPR paper written by Doudna and (2012), 126, 137–41, 144–46, 148, 149, 156, 172–73, 179, 180, 182–86, 191, 192, 194–95, 202, 227, 233, 237–38, 374–75
  CRISPR paper written by Doudna and (2014), 215
  CRISPR patent of Doudna and, 207–8, 210, 219, 224, 233–40
  Doudna's collaboration with, 135, 140, 160, 183, 199–200, 206, 215–21, 464–67, 475
  Doudna's meeting of, 119, 122, 127–28, 200
  Doudna's rift with, 215–21, 464, 466
  Gray and, 247
  Jinek and, 128
  Kavli Prize awarded to, 221
  Lander and, 220, 223–24
  Lander's "Heroes of CRISPR" and, 223–29, 388
  media and, 216
  Nobel Prize awarded to Doudna and, xix, *468,* 469–73
  Novak and, 122, 205–7, 213
  proprietary feelings about CRISPR-Cas9, 216
  tracrRNA work of, 124–27, 131, 217–18, 225
Chase, Leah, 479
chemistry, 8, 19
  biochemistry, 18, 29–35, 51, 80, 183, 201, 478
  Doudna's study of, 31–34
Chen, Janice, *420,* 421–24, 429–32, 451, 460
Chernobyl disaster, 262
Chez Boulay, 376
Chez Panisse, 140–41, 145
chimpanzees, 438
China, 294–95, 300, 427
  Academy of Sciences in, 292, 319
  cancer treatment in, 249
  COVID in, 263, 403, 452
  editing of nonviable embryos in, 288, 290
  genetic engineering in, 301
  germline editing prohibited in, 292, 317
  SARS outbreaks in, 65, 403
  South University of Science and Technology in, 300–301
cholesterol, 251
Choudhary, Amit, 260
Church, Gaylord, 170
Church, George, 40–41, 48, *158,* 160, 167, *168,* 169–74, 211, 212, 228, 262, 291, 360, 385, 470, 472, 474
  Charpentier and, 206
  childhood of, 170–71
  company formations and, 203–6, 208, 209, 213
  in competition to adapt CRISPR into tool for gene editing in humans, 64, 160, 173–74, 192–95, 197–99, 201, 202
  Cong and, 177, 178, 192–94
  CRISPR paper of, 192, 194–95, 197, 199, 201, 202
  CRISPR paper sent to Doudna by, 197, 234
  DARPA grant received by, 260
  Doudna on, 172–74
  at Duke, 171–72
  email to Doudna and Charpentier, 173, 197
  germline editing and, 291, 292, 303
  Happy Healthy Baby and, 286
  at Harvard, 172
  Lander and, 172, 227–28
  Lander's "Heroes of CRISPR" and, 225–26
  Neanderthal idea of, 205
  Personal Genetics Education Project of, 220–21
  prizes and, 220–21
  woolly mammoth project of, 160, 169, 172
  Zayner and, 255
  Zhang and, 167, 174, 175, 178, 192–94

Chylinski, Krzysztof, 125, 128, 129–31, *130*, 135, *142*
CRISPR conference presentation of, 141, 147–48
CRISPR paper and, 137, 140
citizen involvement in science, 263, 445
Clapper, James, 259
Clinton, Bill, 39–40, 309
cloning, 34–35, 98, 153, 280, 313, 415–16
codons, 190, 191–92, 194
cognitive skills, 354, 376
intelligence, 354, 360–62, 376
Watson's comments on, 385–87, 389–90, 392, 397
memory, 354
Cohen, Jon, 316
Cohen, Stanley, 98, 153, 232
Cold Spring Harbor Laboratory, 37–38, 48–49, 53, 271, *298*, 385–91, *458*
Blackford Bar at, 462–63
CRISPR conferences at, 302, 304, 385, 391, 393, 397, 415, *458*, 459–67
founding of, 37, 463
Watson as director of, 37
Watson disassociated from, 386, 387, 390
Watson portrait at, 302, *384*, 386, 390, *396*
Collins, Francis, *36*, 39–40, 294, *325*, 329–31, 364, 390, 446
Columbia University, 98
Comfort, Nathaniel, 226–27
computers, xvii, 117, 163–64, 256, 358–59, 432
hacking of, 263
internet, xvii, 117, 262, 359
software, 432
open-source, 163, 256
Cong, Le, 177–78, 184, 192–94
Congress, U.S., 290–91, 294, *325*, 328–29
Constitution, U.S., 231
contact tracing, 257
Conti, Elena, 84–85
continuum conundrum, 337–38
Copenhagen (Frayn), 224
Coppola, Francis Ford, 352
coronaviruses, 47, 90, 270, 339, 365, 451
CRISPR systems to detect, 118
name of, 404
PAC-MAN and, 454–57, 460
RNA of, 65, 403–4
RNA interference and, 66–67
SARS, 65, 403, 441
SARS-CoV-2, 403–4
see also COVID-19 pandemic
coronary heart disease, 251
Costolo, Dick, 219
Council of Europe, 278
Count Me In, 177
COVID-19 pandemic, xiii, xv, xvii, 251, 257, 262, 335, 373, 399–475, 477
Berkeley and, xiii–xv
Innovative Genomics Institute, 401–5, 413–18, *419*
China origin of, 263, 403, 452
Cold Spring Harbor conference and, 459–62
CRISPR cures and, xviii, 449–57
PAC-MAN, 454–57, 460
deaths in, 404, 449
legacies of, 473–75
SARS-CoV-2 in, 403–4
testing and, xiv–xv, 405, *406*, 407–11, 413–18, 427–33
antigen tests, 430–31
Broad Institute and, 410–11, 429
CDC and, 407–10
CRISPR and, 413, 417, 421–25, 427–33, 453
Fauci and, 410
FDA and, 407–10
first person in U.S. to test positive, 407
Mammoth Biosciences and, 423–25, 429–32
PCR tests, 408, 413, 416, 421, 425, 428, 430
Sherlock Biosciences and, 428, 430
Trump administration and, 411
University of Washington and, 409–10, 415
Zhang and, 421, 427–32
vaccines and, *434*, 435–47, 449
clinical trials of, 435–36, 440–41, 445–46, 461
Fauci and, 442
FDA and, 446
Isaacson in trial of, 435–36, 440–41, 445–46
cowpox, 436
Crichton, Michael, 271
Crick, Francis, 20–28, 47, 166, 389, 475
at Asilomar, 269
on central dogma of biology, 44
in DNA double-helix structure discovery, xviii, 7, 11, 26–28, 29–31, 46, 51, 58, 159, 423, 470
with DNA model, *16*
Franklin's work and, 25, 26
Nobel Prize awarded to, 28, 470
Watson's meeting of, 20
CRISPR (clustered regularly interspaced short palindromic repeats), 69–149, 449–50, 477, 480
anti-CRISPRs, 260–61
as bacterial immune system, xiv, xviii, xix, 67, 86, 87
Banfield's enlistment of Doudna to work on, xviii, 77, 79–80, 81, 93, 471
Barrangou and Horvath's work on, 90–93, 143–44, 177, 178, 220, 223, 470
diagram of mechanism of, *133*
discovery of, 71–77
DNA targeting in, xiv, xviii, 67, 76–77, 86, 87, 93–94, 106, 111, 131, 132, *133*, 146, 217
Doudna's creation of team for studying, 103–11
enzymes in, *see* CRISPR-associated enzymes
in vitro vs. *in vivo* studies of, 80, 94–95, 183, 191, 202
Koonin's study of, 76–77, 80, 92
and linear model of innovation, 90
Mojica's study of, 71–76, 80, 91, 92, 220, 223, 224, 470, 479
naming of, 73
RNA in, 79, 108, *133*
Cas13 targeting of RNA, 423–25, 429, 451, 452, 454–57

CRISPR RNA (crRNA), 86, 106, 124–26, 131–35, 137–39, 143, 146, 147, 180, 184, 186, 217
RNA interference, 77, 79, 93
single-guide RNA (sgRNA), 134–35, 139, 186, 190, 191, 195, 198
trans-activating CRISPR RNA (tracrRNA), 124–27, 131, 134, 135, 139, 146–47, 179–80, 182, 184–86, 217–18, 225
viruses' disabling of, 260–61
yogurt starter cultures and, xviii, 90–92, 133, 143
CRISPR applications, 243–63
affordability of, 247–48
biohacking, 253–57, 263, 285, 288, 443–44
coronavirus testing, 413, 417
cures for viral infections, 449–57
CARVER, 452–53, 455–57
PAC-MAN, 454–57, 460
defense against terrorists, 259–63
diagnostic tests for cancers, 250
diagnostic tests for viruses, 115, 118, 421–25, 431–32, 451
at-home tests, 430–32, 460
coronavirus, 413, 417, 421–25, 427–33, 453
HPV, 422–23
ex vivo and *in vivo*, 246, 250
germline (inheritable) editing, *see* human germline editing
somatic editing, 246, 276–77, 288, 329, 336, 338, 341
therapies, 233, 245–51
blindness, xviii, 250
cancer, xviii, 249–51
safe viruses as delivery method in, 456
sickle cell, xviii, 245–49, 329, *340*
CRISPR-associated (Cas) enzymes, 73, 76, 84, 86, 92
Cas1, 86–87, 106, 128
Cas6, 106–7, 113, 128
as diagnostic tool, 115, 118
Cas9, 86, 92, 119, 124–25, 127, 128, 129–36, *133*, 422–24, 440
purchasing of, 380
see also CRISPR-Cas9 gene editing
Cas12, 86, 422–24, 429
Cas13, 86, 379, 423–25, 429, 451, 452, 454–57
CASCADE array of, 111
fluorescence microscopy and, 107–8
Zhang's work with multiple proteins, 181
CRISPR babies, 245–47, 297–332, 325–32, 335, 337
birth of, 311–13, 315, 318
He Jiankui's conviction and sentencing for his work, 332
He Jiankui's removal of HIV receptor in, xv–xvi, 251, 303–13, 316–24, 369, 404
CRISPR-Cas9 gene editing, xiii, xiv, xvii–xix, 60, 90, 94, 97–98, 132–36, 151–241, 272
applications of, *see* CRISPR applications
codons and, 190, 191–92, 194
companies formed to commercialize applications of, 203–13, 430
CRISPR Therapeutics, 207, 213, 245, 465
Editas Medicine, 209–13, 250, 454
Intellia Therapeutics, 213
competition to prove human application of, xv, 64, 143–49, 155–56, 157–60, 201–2, 374
Church in, 64, 160, 173–74, 192–95, 197–99, 201, 202
Doudna in, xv, 64, 160, 173–74, 179, 187–90, 197–202
Lander in, 64, 160
Zhang in, xv, 64, 159–60, 174, 175–86, 191–95, 197, 199–202, 227
Doudna and Charpentier's 2012 paper on, 126, 137–41, 144–46, 148, 149, 156, 172–73, 179, 180, 182–86, 191, 192, 194–95, 202, 227, 233, 237–38, 374–75
in eukaryotic cells, 148–49, 184, 237–38
as inevitable, 202
in vitro vs. *in vivo*, 184, 382
Isaacson's learning of, xiv, *378*, 379–83
licensing of, 208, 211
moral questions concerning, *see* moral questions
patents for, 92, 94, 115, 135–36, 144, 156, 232–41
of Doudna, Charpentier, and Berkeley, 207–8, 210, 219, 224, 233–40
Doudna's battle with Zhang over, 184, 194, 208, 224, 234–40, 425, 470
pool idea for, 207–8
of Zhang and the Broad Institute, 192, 207–8, 210–11, 219, 224–27, 233–40
regulation of, 255, 361
Šikšnys's paper and, 144
transposons and, 374, 375
virus infections and, xv, xix
CRISPR conferences, 93, 129
in Berkeley (2012), 140, 141, 145–49
at Cold Spring Harbor, 302, 304, 385, 391, 393, 397, 415, *458*, 459–67
in Quebec (2019), 373–77
CRISPR Journal, 85, 309, 319
CRISPR Therapeutics, 207, 213, 245, 465
crowdsourcing, 256–57, 285, 464
cryocooling, 57, 63
crystallography, 84, 187
Cas1 and, 86–87
DNA and, 19–23, 51
"photograph 51," 25, *25*, 395
phase problem in, 57
RNA and, 53, 55–57, 66, 171, 181
Cumberbatch, Benedict, 219
Cure Sickle Cell Initiative, 248
Curie, Marie, 8, 470
curiosity, 478–79
of Doudna, xix, 4, 5, 8, 31, 46, 51, 479
science driven by, xix, 89, 90, 457, 473, 479
Current Contents, 72
cystic fibrosis, 248
cytosine, 26, *27*

Daley, George, *282*, 316, 324, 369
Danisco, 90–92, 143, 148, 177, 220

Dantseva, Dariia, *434,* 443–45
DARPA, 259–62, 351, 452, 454
  Safe Genes project of, 260, 262, 380
Darwin, Charles, *10,* 11–14, 28, 159, 292, 302, 364, 475
Davenport, Charles, 37
Davies, Kevin, 85, 309, 320
Davis, Miles, 345
Davos, World Economic Forum in, 368
deafness, 346–48
Decoding Watson (PBS documentary), *384,* 386, 389–90, 392, 393
Deem, Michael, *298,* 300, 301, 307–9
Defense Advanced Research Projects Agency (DARPA), 259–62, 351, 452, 454
  Safe Genes project of, 260, 262, 380
Defense Department, U.S., 4, 259, 262
Defoe, Daniel, 414
Deisseroth, Karl, 167
Delbrück, Max, 37
Deltcheva, Elitza, 125–26
Dengue virus, 424
depression, 166, 352–53
  bipolar disorder, 177, 327, 352–53
Derain, André, 395
designer babies, 355–56
  see also CRISPR babies
Desmond-Hellmann, Sue, 100
DETECTR (DNA endonuclease targeted CRISPR trans reporter), 422–23, 429, 460
Dhanda, Rahul, 428
diabetes, 99
Diana, Princess, 100
Diaz, Cameron, 219
Dicer enzyme, 66, 79
Didion, Joan, 7
digital technology, xvii, 28, 114
disabilities, 345–48
diversity, 362, 376, 480
DNA (deoxyribonucleic acid), xvii, xviii, 17–28, 43–47
  bases in, 26, *27,* 37, 71, 382, 480
  in central dogma of biology, 44, 47, 270
  cloning, 34–35, 98, 153, 280, 313, 415–16
  CRISPR sequences of, *see* CRISPR
  crystallography and, 19–23, 51
  discovery of double-helix structure of, xviii, 7, 11, 26–28, 29–31, 46, 51, 58, 159, 386, 390, 395, 397, 423, 470
  editing of, *see* gene editing
  Franklin's work on, xviii, 7, 22–26, 28, 31, 46, 51, 53, 227, 389, 459, 470
  "photograph 51," 25, *25,* 395
  Human Genome Project and, 37–41, 43, 45, 74, 172, 175, 280, 330, 352, 388, 390, 391
  "junk," 416
  model of, *16,* 27
  mutations in, 41, 302
  Pauling's work on structure of, 21–25, 51, 159, 391
  "photograph 51" of, 25, *25,* 395
  race to discover structure of, 21–28, 159, 390, 391
  reverse transcription and, 270, 408
  sequencing of, 71–72, 74, 79, 80, 83, 172
  vaccines, 439–41, 444–45, 456
  in viruses, 65
  in yeast, 35, 45
Dolly the sheep, 280, 313, 415–16
Double Helix, The (Watson), xix, 6–8, *9,* 20, 21, 27, 29–31, 50, 157, 328, 391, 397, 398, 414, 459, 477–78
Doudna, Dorothy (mother), *2,* 4, 5, 32, 98, 472–73
Doudna, Ellen (sister), *2,* 472–73
Doudna, Jennifer, *2, 62, 130, 138, 158, 188, 204, 214, 282, 298*
  as bench scientist, 103
  at Berkeley, xiii–xv, xvii, 64–65, 83, 98, 101, 107
  Berkeley lab of, 103–11, *104,* 117
  Isaacson learns to edit at, xiv, *378,* 379–83
  birth of, 4
  business school considered by, 98
  Caribou Biosciences, 113–18, 203, 207–8, 213
  Intellia Therapeutics, 213
  chemistry studied by, 31–34, 478
  childhood of, xix, *2,* 3–8, 31, 123
  Charpentier's rift with, 215–21, 464, 466
  Church's email to, 173, 197
  coauthors on papers of, 48
  at conferences, 287–90, 292, 303, 332, 367, 373, 375, 385, 459–61
  at congressional hearing, *325,* 328–29
  CRISPR work of
  Banfield calls and invites her to collaborate, xviii, 77, 79–80, 81, 93, 471
  Breakthrough Prize in Life Sciences awarded for, 219–21
  Church's paper and, 197, 234
  collaboration with Charpentier, 135, 140, 160, 183, 199–200, 206, 215–21, 464–67, 475
  company formations and, 203–13
  competition to adapt CRISPR into tool for gene editing in humans, xv, 64, 160, 173–74, 179, 187–90, 197–202
  creation of lab team, 103–11
  defense applications and, 259–61
  diagnostic tests and, 422–23
  Editas Medicine and, 209–13
  ethical considerations of, 367–70
  Kavli Prize awarded for, 221
  Lander's "Heroes of CRISPR" and, 223–29, 388
  medical applications and, 247–49, 251
  meeting Charpentier, 119, 122, 127–28, 200
  Nobel Prize awarded for, xix, *468,* 469–73
  paper for *eLife,* 198–201, 215, 225, 237
  paper written with Charpentier (2012), 126, 137–41, 144–46, 148, 149, 156, 172–73, 179, 180, 182–86, 191, 192, 194–95, 202, 227, 233, 237–38, 374–75
  patent for, 207–8, 210, 219, 224, 233–40
  patent battle with Zhang, 184, 194, 208, 224, 234–40, 425, 470

possible partnership with Zhang, 207–8
publicity for, 216–17, 220, 223
review article written with Charpentier (2014), 215
Zhang's emails and, 200–201, 207
see also CRISPR-Cas9 gene editing
curiosity of, xix, 4, 5, 8, 31, 46, 51, 479
The Double Helix read by, xix, 6–8, 29–31, 157, 328, 391, 397, 398, 414, 459, 477
first marriage of, 54–56, 63
Genentech and, 97–102, 103, 105, 107, 113, 114
on gene therapy, 279
germline editing views of, 331–32, 367–70
at Harvard, 34–35, 40, 64, 208
He Jiankui and, *298,* 302–4, 315–17, 319, 321–24, 328, 332
in high school, 31
Hitler nightmare of, 283, 286
at Hong Kong summit (International Summit on Human Gene Editing, 2015), *282,* 292–93, *366*
husband of, *see* Cate, Jamie
Innovative Genomics Institute, 248, 261, 380–82, 401–3
COVID and, 401–5, 413–18, *419*
Mammoth Biosciences and, xv, 249–50, 423
medical school considered by, 98
midlife crisis of, 97, 101
Napa Valley conference organized by, 287–90, 292, 303, 332, 367
in National Academy video, 355
at Pomona College, *30,* 31–34
Qi and, 453–56
at Quebec conference, 373, 375
RNA work of, xviii, 45–49, 51–61, 63, 65–67, 134, 181, 220, 435–36, 446
son of, *see* Cate, Andrew
Watson and, xviii, 8, 29, 48–50, *396*
Watson visited by, 395–98
at Yale, *52,* 57, 60–61, 63
Doudna, Martin (father), *2,* 3–8, 32, 34, 58–60, 397–98, 472–73
cancer of, 58–59
death of, 59–60
Watson's *The Double Helix* given to Jennifer by, xix, 6–8, 477
Doudna, Sarah (sister), *2,* 3, 472–73
Down's syndrome, 337
Doxzen, Kevin, 255
dragonflies, 170
Dr. John, 479
drugs, 233, 451
insulin, 99
Duchesneau, Sharon, 346–47
Duke University, 171–72
Durbin, Richard, *325,* 328
Dzau, Victor, 316, 329

East-Seletsky, Alexandra, 189–90, 424
Ebert, Roger, 276
Ebola, 425
E. coli, 71, 72, 74, 98
Economist, 473
Editas Medicine, 209–13, 250, 454
Education, U.S. Department of, 164
Einstein, Albert, xvii, 4, 90, 201, 262, 479
Eisen, Michael, 226, 232, 473
electroporation, 439–40, 444, 456
elephants, 251
eLife, 198
Doudna's paper in, 198–201, 215, 225, 237
Ellingsen, Heidi Ruud, 221
Ellison, Eldora, *230,* 239
embryos, animal, 284, 289
embryos, human
genetic diagnosis of, 274–76, 285, 302, 327, 337, 342, 361–62
genetic editing of
Happy Healthy Baby and, 284–86
in nonviable embryos, 288, 290, 304
see also CRISPR babies; human germline editing
in Huxley's *Brave New World*, 359
prenatal testing of, 337
purchase of, 326
Emerson, Ralph Waldo, 6
empathy, 365, 480
Engineering the Human Germline conference, 276–78
Enlightenment, 357
entrepreneurs, 114–16, 285
enzymes, 44, 85, 86, 98, 164, 177
Cas, *see* CRISPR-associated enzymes
Dicer, 66, 79
FokI, 155
functions of, 86
nucleases, 155, 177
TALENs (transcription activator-like effector nucleases), 155, 167, 177–79, 187, 190
ZFNs (zinc-finger nucleases), 155, 178, 179
ribozymes, 44–45, 47, 53, 60
"Essay on the Principle of Population, An" (Malthus), 12–13
Esvelt, Kevin, 260
ethical questions, *see* moral questions
Ethics of Genetic Control, The (Fletcher), 268
eugenics, 37, 268, 359, 386
free-market, 360–63
eukaryotic cells, 148–49, 184, 237–38
evolution, 12–13, 17, 159, 277, 302, 360, 364, 481
experience machine, 353

Fabricated Man (Ramsey), 268–69
Fauci, Anthony, 410, 442
Faulkner, William, 414
fecal transplants, 254–55
FBI, 23
FDA (Food and Drug Administration), 294, 363, 446
Ferrell, Ryan, 308–10, 312
Fire, Andrew, 66
Fisher, Carrie, 352

Flagship Ventures, 209
Fleming, Alexander, 56
Fleming, Jory, 346
Fletcher, Joseph, 268
flu, 118, 300, 431, 433, 438, 441, 447, 452, 454
fluorescence microscopy, 107–8
fluorescent proteins, 165–67, 382–83
flytrap plants, 171
FokI enzyme, 155
Fonfara, Ines, 137
Food and Drug Administration (FDA), 294, 363, 446
  coronavirus tests and, 407–10
Foy, Shaun, 205–7
Frangoul, Haydar, *244*
Frankenstein (Shelley), 267, 269, 283, 286, 297
Franklin, Benjamin, 479
Franklin, Rosalind, 21–26, *25,* 459–60, 462, 465, 470
  death of, 28, 470
  DNA work of, xviii, 7, 22–26, 28, 31, 46, 51, 53, 227, 389, 459, 470
  "photograph 51," 25, *25,* 395
  Watson and, 7–8, 22–26, 50, 389, 395
  Wilkins and, 21–24
Frayne, Michael, 224
free market, 331, 357–59, 363
  eugenics and, 360–63
Free Speech Movement Café, 80, 93, 471
frog genetic engineering kit, 253, 255, 348
fruit flies, 216
Fukuyama, Francis, 281

Gabriel, Stacey, 410
Gairdner Award, 220
Garland-Thomson, Rosemarie, 345–46
Gates, Bill, 373, 439
Gates Foundation, 118, 248, 439
Gattaca, xv, 275–76, 337, 361
Gelsinger, Jesse, 279, 304
GeneCopoeia, 380
gene drives, 260
gene editing, xviii, 46, 47, 60, 61, 478
  base editing, 462
  double-strand break in, 154, 374, 382
  homology-directed repair of, 143, 195, 201
  nonhomologous end-joining of, 143
  invention of, 154–55
  moral concerns about, *see* moral questions
  prime editing, 462
  TALENs (transcription activator-like effector nucleases) in, 155, 167, 177–79, 187, 190
  transposons in, 373–76
  in yeast, 35, 148–49
  ZFNs (zinc-finger nucleases) in, 155, 178, 179
  Zhang's research on, 167, 177–79
  see also CRISPR-Cas9 gene editing
Genentech, 97–102, 113–14
  Doudna and, 97–102, 103, 105, 107, 113, 114
genes, 11–15, 17–19, 46, 478
  jumping, 49, 373–76
  mutations in, 12, 13, 41
  see also DNA
Genesis, Book of, 1
gene therapies, 153–54, 274
  Gelsinger and, 279, 304
genetically modified foods, 329, 478
genetic engineering, 64, 98, 100, 113, 153, 167, 172, 268, 478
  American vs. European attitudes toward, 278
  Asilomar and, 269–73, 330, 331
  bioconservatives and, 267–69, 280
  in China, 301
  Kass Commission and, 280–81
  moral questions concerning, *see* moral questions
  moratorium proposed for, 270, 272
  patents for, 392
  recombinant DNA, 98–100, 153, 232, 269–71, 274, 287, 330, 331
  regulation of, 270, 278, 281
  utopians and, 267–69
  see also biotechnology; CRISPR-Cas9 gene editing; gene editing
genomes, 74
  human, 74
  Human Genome Project, 37–41, 43, 45, 74, 172, 175, 280, 330, 352, 388, 390, 391
  non-coding regions of, 416
Genomic Prediction, 361–62
George, Robert, 281
Germany, 278
  Nazi, 356, 359
germline (inheritable) editing, *see* human germline editing
Gilbert, Sarah, 438–39
Gilbert, Walter, 47, 172
Glass, Bentley, 268, 280
GlaxoSmithKline, 431
Glendon, Mary Ann, 281
Global Synthetic Biology Summit, 253–55
Google, 114
Gootenberg, Jonathan, 425, *426,* 428, 451
Gore, Al, 40
Gould, John, 12
Graham, John, 395
Graham, Lindsey, 328
Graves, Joseph, 389
Gray, Victoria, *244,* 245–47, 343
Greene, Eric, 107, 108
Greene, Graham, 352
Greninger, Alex, 409–10
Griffin, Tom, 54–56
guanine, 26, *27*
Guardian, 293
Gut Hack, 255

hacking, 256, 259
  computer, 263
  biohacking, 253–57, 23, 285, 288, 443–44

Hall, Stephen, 227–28
Hamburg, Margaret, 329, 331
Hamilton, Jennifer, xiv, 381–83, *412,* 415, 456
Hamilton STARlet, 417
Hapai, Marlene, 6
Happy Healthy Baby, 284–86
Harald V, King, 221
Harmon, Amy, 390
Harmonicare Women and Children's Hospital, 304
Harrington, Lucas, 110, *420,* 422, 424, 429–30, 451, 460
Harrison, Brian, 410
Harvard University, 35, 114, 208, 228, 235n, 324, 450
  Church at, 172
  Doudna at, 34–35, 40, 64, 208
  Medical School, 167, 172, 175, 193, 369
  vaccines and, 442–44
  Zhang at, 166, 167, 175
  see also Broad Institute of MIT and Harvard
Hastings Center, 310
Hauer, Michael, 137
Haurwitz, Rachel, 101, *104,* 105–6, 108, *112,* 113–15, 136, 205, 208, 213, 239
  Caribou Biosciences, 113–18, 203, 207–8, 213
  Intellia Therapeutics, 213
Hawaii, xix, 98
  Hilo, *2,* 3–6, 8, 58, 59, 123
Health and Human Services, U.S. Department of, 407
heart disease, 251
Heenan, Christine, 228–29
height, 349–51, 361, 376
Heisenberg, Werner, 224
He Jiankui, 299–313, 481
  Deem and, 300, 301, 307–9
  Doudna and, *298,* 302–4, 315–17, 319, 321–24, 328, 332
  early life of, 299–300
  editing of PCSK9 gene studied by, 251
  human embryo editing principles of, 309, 323
  Hurlbut and, 306, 307
  at International Summit on Human Genome Editing, *314,* 315–24
  Porteus and, 306–7, *314,* 321, 323
  PR campaign of, 308–11
  Quake and, 300, 301, 307
  removal of HIV receptor in embryos by, xv–xvi, 251, 303–13, 316–24, 369, 404
  birth of babies, 311–13, 315, 318
  conviction and sentencing for, 332
Hemingway, Ernest, 352, 353
Hemmes, Don, *2,* 5–6, 32–33, 59
hemoglobin, 246, 247
hepatitis B, 437
hepatitis C, 118
heredity, 11–15, 17, 18, 20, 28
Hernandez, Lisa, 417–18
"Heroes of CRISPR, The" (Lander), 182, 223–29, 388
Hewlett-Packard, 114
Hilo, Hawaii, *2,* 3–6, 8, 58, 59, 123
Hinckley, John, 352
Hitler, Adolf, 283, 286
HIV and AIDS, 166, 305, 390, 431
  He Jiankui's removal of HIV receptor in embryos, xv–xvi, 251, 303–13, 316–24, 369, 404
  sperm-washing and, 305, 322
Hockemeyer, Dirk, *406,* 418
Hockney, David, 395
Holbrooke, Richard, 224
Holdren, John, 294
Homebrew Computer Club, 256, 373
Hong Kong summit (International Summit on Human Genome Editing, 2018), 295, 312, *314,* 315–24, *366*
Hoover, Jessica, 115
Horvath, Philippe, *88,* 90–93, 143–44, 177, 178, 220, 223, 470
Hospital for Sick Children, 369
Hotchkiss v. Greenwood, 231
How Innovation Works (Ridley), 90
How to Be Human (Fleming), 346
HPV, 422–23
Hsu, Patrick, *420,* 454, 460
hubris, 268, 281
  playing God, 269, 273, 333, 363–65, 480–81
HUDSON, 453
Human Genome Project, 37–41, 43, 45, 74, 172, 175, 280, 330, 352, 388, 390, 391
human germline (inheritable) editing, xv, 245–47, 276–78, 303, 324, 376
  absence of laws against, 294
  accessibility of technology for, 326, 380, 383
  broad societal consensus on, 292–93
  congressional hearing on, *325,* 328–29
  CRISPR babies, 245–47, 297–332, 325–32, 335, 337
  birth of, 311–13, 315, 318
  He Jiankui's conviction and sentencing for his work, 332
  HIV receptor removal in, xv–xvi, 251, 303–13, 316–24, 369, 404
  criteria for allowing, 293
  Doudna's views on, 331–32, 367–70
  He Jiankui's principles of, 309, 323
  international commissions on, 329–30
  International Summit on Human Gene Editing (2015), *282,* 292–93, 315, 317
  medical tourism and, 291, 329
  moral issues with, 245–47, 276–77, 287–93, 332, 341
  He Jiankui and, 304
  Napa Valley conference and, 287–90, 292, 332
  as red line, 276–77, 322, 324, 336–38
  see also moral questions
  moratorium issue and, 289, 292, 293, 323, 324, 330–32, 335
  National Academy of Sciences and, 292–94, 309, 329
  video and, 355–56

human germline (inheritable) editing (*cont.*)
Nuffield Council and, 293, 294
"prudent path forward" goal for, 272, 287, 289, 292, 331, 332, 370
regulations on, 361
global, 294–95
restraints on, 288–89, 292, 293, 323–24
unintended consequences of, 277, 290, 364
Human Nature, 344
human papillomavirus (HPV, 422–23
Hung, Deborah, 410
Hunt-Grubbe, Charlotte, 387
Huntington's disease, 41, 329, 337, 341–42, 361, 365, 368–69, 376, 392
Hurlbut, William, 306, 307
Huxley, Aldous
Brave New World, xv, 267, 269, 280, 281, 353, 358–59
Island, 359
Huxley, Julian, 352

IDT, 380
Illumina, 301
IMAGe syndrome, 349–50
immune system, 449–50, 455, 457
Indiana University, 18–19
individual choice, 331, 358, 359, 369
community vs., 356–59
free-market eugenics and, 360–63
inequality, 274, 288, 358, 360, 362–63, 369, 376–77, 391–92
influenza, 118, 300, 431, 433, 438, 441, 447, 452, 454
information technology, 358, 359
inheritable (germline) editing, *see* human germline editing
innovation, 432
government-business-university partnerships and, 117–18, 232–33
institutions and, 405
linear model of, 89–90
military spending and, 259
see also science; technology
Innovative Genomics Institute (IGI), 248, 261, 380–82, 401–3, 456
COVID and, 401–5, 413–18, *419*
Inovio Pharmaceuticals, 439–40
insulin, 99
Intel, 241
intellectual property, 405, 473–74
licensing
COVID and, 405
of CRISPR technology, 208, 211
patents, *see* patents
intelligence, 354, 360–62, 376
Watson's comments on, 385–87, 389–90, 392, 397
Intellia Therapeutics, 213
Intel Science Search, 166
International Summit on Human Gene Editing (2015, Washington D.C.), *282*, 292–93
International Summit on Human Genome Editing (2018, Hong Kong), 295, 312, *314*, 315–24, *366*
internet, xvii, 117, 262, 359
introns, 44–45, 49, 53, 55, 66
in vitro fertilization (IVF), 274, 280, 284–85, 327, 328, 337, 361
Louise Brown (first test-tube baby), 274, 313, 328
Ioffe, Julia, 410
IQ, *see* intelligence
Isaacson, Walter
in COVID vaccine trial, 435–36, 440–41, 445–46
CRISPR editing learned by, xiv, *378*, 379–83
Watson visited by, 391–94
Ishee, David, *434*, 443–45
Ishino, Yoshizumi, 71, 72
Island (Huxley), 359

Jansen, Kathrin, 441
Jansen, Ruud, 73
Jasin, Maria, 465
Jefferson Conundrum, 390–91
jellyfish, 165–66
Jenner, Edward, 436
Jenner Institute, 438
Jerome, Keith, 409
Jews, 388–89
Jezebel, 227
Jinek, Martin, 84–85, 87, *104*, 105, 106, 108–10, 128, 129–32, *130*, 134–36, *142*, 237, 287, 374, 380–81, 465, 470
Charpentier and, 128
in competition to adapt CRISPR into tool for gene editing in humans, 187–90, 197–99, 201
CRISPR conference presentation of, 141, 147–48
CRISPR papers and, 108, 137, 140, 215
Wiedenheft and, 85
Jobs, Steve, 4, 103, 358, 373, 390–91, 463, 479
"Think Different" ad, 326, 371
Johannsen, Wilhelm, 15
Johnson & Johnson, 438
Joung, Keith, 202, 209
Journal of Bacteriology, 33–34
Journal of Molecular Evolution, 76
Journal of the Plague Year, A (Defoe), 414
J.P. Morgan, 210
jumping genes (transposons), 49, 373–76
June, Carl, 249
Jurassic Park, xviii, 164

Kass, Leon, 280–81
Kass Commission, 280–81
Katz, Deborah, 238
Kavli Prize, 221
Ketchens, Doreen, 480
Kilby, Jack, 240–41
Kim, Jin-Soo, 202

Kingsley, David, 385
Kissinger, Henry, 4
Klee, Paul, 395
Klompe, Sanne, 375
Knoepfler, Paul, 356
Knott, Gavin, 260, *378,* 379–81, 402–3, 462
Kolter, Roberto, 34–35
Koonin, Eugene, 76–77, 80, 92, 423
Krauthammer, Charles, 281

Lam, Wifredo, 395
Lander, Eric, 39, 175–77, 193, 207, 209, 211, 212, 216, *222,* 374, 388
as Broad Institute director, 64, 144, 160, 175, 216, 220, 223, 225–28
Charpentier and, 220, 223–24
Church and, 172, 227–28
in competition to adapt CRISPR into tool for gene editing in humans, 64, 160
COVID and, 410–11
Doudna and, 144–45
Doudna and Charpentier's Nobel and, 471–72
"The Heroes of CRISPR," 182, 223–29, 388
moratorium issue and, 330–32
prizes and, 219–20
Zhang and, 182–83, 220
Larson, Christina, 310
Lasky, Larry, 115
Lassa fever, 425
Lawrence Berkeley National Laboratory, 64, 87, 117, 456–57
Leber congenital amaurosis, 250
Leibniz, Gottfried Wilhelm, 159
Leonardo da Vinci, xix, 4, 259, 338, 375, 457, 478, 479
Lessing, Doris, 7
Levy, John, 165–66
Levy, Steven, 454
licensing
COVID and, 405
of CRISPR technology, 208, 211
life, origin of, xvii, 45–49
Lifton, Richard, 47
light-sensitive proteins, 167
Lin, Shuailiang, 184, 192
Lincoln, Abraham, 64–65
linear model of innovation, 89–90
Lin Shiao, Enrique, *412,* 415–18
Linux, 163, 256
lipid nanoparticles, 441
lipitoids, 456–57
Litton Industries, 114
Locke, John, 357
Lord of the Rings, The (Tolkien), 415
Lovell-Badge, Robin, *314,* 316–22
Liu, David, 209, 321–22, 461–62, 472
Luria, Salvador, 19, 37

MacFarlane, Seth, 219
Mahler, Gustav, 352
Makarova, Kira, 376
malaria, 343, 390
Malthus, Thomas, 12–13
mammoth, woolly, 160, 169, 172
Mammoth Biosciences, xv, 249–50, 423–25, 429–32, 460
Manson, Charles, 352
Mäntyranta, Eero, 348
Marchione, Marilynn, 310, 313
Mardi Gras, *476,* 480
Marletta, Michael, 101
Marraffini, Luciano, *70,* 93–94, *176,* 213, 220, 376, 470
Doudna's meeting of, 180–81
patent and, 234–35, 240
Zhang and, 180–82, 186, 192, 200, 213, 234–35, 240
Marson, Alex, 440, 456
Max Planck Institute, 123–24, 140
May, Andy, 203–7, 235–36, 240
Mayo Clinic, 410
McCarthyism, 23
McClintock, Barbara, 49, 463
McCullough, Candy, 346–47
McGovern Institute, 428
McGwire, Mark, 349
Medical Research Council, 26
medical tourism, 291, 329
Mello, Craig, 66
memory, 354
Mendel, Gregor, *10,* 11, 13–15, 292, 475
mental illness, 166–67, 177, 352–54
bipolar disorder, 177, 327, 352–53
depression, 166, 352–53
schizophrenia, 177, 352–53
of Rufus Watson, 38, 277, 352, 386, 389, 393
MERS (Middle East respiratory syndrome), 409, 438, 441
Mezrich, Ben, 171
Mic, 227
mice, 74, 122, 167, 259, 260, 348, 354
microchips, xvii, xix, 117, 201, 432, 473
patent rights for, 240–41
microfluids, 460
Microsoft, 206, 450
Middle East respiratory syndrome (MERS), 409, 438, 441
military defense, 260–63
Mill, John Stuart, 357
Milner, Yuri, 219
Miró, Joan, 395
MIT, 63, 64, 89, 117, 235n, 269
McGovern Institute, 428
see also Broad Institute of MIT and Harvard
MIT Technology Review, 227, 312
Moderna, 441–42, 447
Moineau, Sylvain, 148, 220
Mojica, Francisco, *70,* 71–76, 79, 80, 81, 90–93, 220, 223, 224, 470, 479
molecules, 7, 8, 29, 34, 51, 473

Moore, Henry, 395
moral questions, xvi, 268–69, 273, 333–70
continuum conundrum and, 337–38
diversity, 362, 376, 480
decision making, 355–65
Doudna and, 367–70
Engineering the Human Germline conference and, 276–78
enhancements, 276, 281, 302, 303, 326–27, 355–56, 360, 365, 376
absolute vs. positional improvements, 350–51
attractiveness, 386
cognitive skills, 354, 376
height, 349–51, 361, 376
intelligence, 354, 360–62, 376
memory, 354
muscle mass and athletic ability, 253–55, 348–49, 361, 376
super-enhancements, 339, 351–52
treatments vs., 338–39, 341, 350, 369–70, 376
free market, 331, 357–59, 363
eugenics, 360–63
germline editing, 245–47, 276–77, 287–93, 332, 341
He Jiankui and, 304
Napa Valley conference and, 287–90, 292, 332, 367
as red line, 276–77, 322, 324, 336–38
individual choice, 331, 358, 359, 369
community vs., 356–59
free-market eugenics and, 360–63
inequality, 274, 288, 358, 360, 362–63, 369, 376–77, 391–92
National Academy of Sciences and, 355–56
playing God, 269, 273, 333, 363–65, 480–81
preventions, 339
red lines, 335–39
germline as, 276–77, 322, 324, 336–38
slippery slope, xvi, 277, 289, 361, 481
societal discussion about, 355–56
Splicing Life report and, 273–74
thought experiments, 339, 341–54, 361, 362
character, 345–46
cognitive skills, 354
deafness, 346–48
disabilities, 345–48
Huntington's disease, 341–42, 361, 365
psychological disorders, 352–54
sexual orientation, 347–48, 361
sickle cell, 343–45, 361
skin color, 347–48
unintended consequences, 277, 290, 364
Moreau, Sylvain, 93
Morrill Land-Grant Act, 65
mosquitoes, 122, 260, 261
Mount Sinai Medical Center, 381
Moviegoer, The (Percy), 345
MSTN gene, 348
Mukherjee, Siddhartha, 20–21
Muller, Hermann, 19
Mullis, Kary, 408
muscles and sports, 253–55, 348–49, 361, 376
Musunuru, Kiran, 312
Myhrvold, Cameron, *448,* 450–54, 457, 460
Myhrvold, Nathan, *448,* 450
myostatin, 253

Napa Valley conference, 287–90, 292, 303, 332, 367
Napolitano, Janet, 417
NASA, 254
Apollo program, 417
Nash, John, 352
National Academy of Medicine, 363
National Academy of Sciences, 143, 262, 269, 292–94, 309, 329
video of, 355–56
National Center for Biotechnology Information, 376
National Council of Churches, 273
National Human Genome Research Institute, 309
National Institutes of Health (NIH), 39, 56, 94, 117–18, 248, 294, 329, 331, 364, 390, 410, 423, 442, 446, 461
National Science Foundation, 89
National Transportation Systems Center, 206
Natural Science Society, 14–15
natural selection, 12, 13, 302
nature, xviii, xix, 5, 7, 8, 12, 17, 29, 33, 46, 51, 69, 75, 157, 280, 294, 310, 336, 342, 343, 364–65, 367, 375, 447, 457, 481
Nature, 47, 75, 108, 111, 198, 290, 291, 330, 374, 376, 469, 470, 474
Charpentier's 2011 paper in, 125–27, 131, 147, 179–80, 185, 218
He Jiankui and, 309, 311, 312, 315, 316
Nazi Germany, 356, 359
Neanderthal, 205
nematodes, 66, 187
neurons, 167
Newton, Isaac, 159
New York Times, 40, 172, 289, 304, 307, 390, 427
New York University, 122
1984 (Orwell), 358–59
NMDA receptors, 354
Nobel Prize, 19, 20, 21, 44, 49, 53, 66, 100, 219, 221, 226, 270, 305, 470, 472
awarded to Doudna and Charpentier, xix, *468,* 469–73
awarded to Watson, Crick, and Wilkins, 28, 470
Nogales, Eva, 111
Novak, Rodger, 122, *204,* 205–7, 213, 216
Novartis, 206
Noyce, Robert, 240–41
Nozick, Robert, 353, 357–58
NPR, 245, 247, 290
nuclear location signal or sequence (NLS), 182, 189–90, 191, 194, 380, 440

nuclear power, xvii
atom bomb, xvii, 89, 117, 224, 265, 338
attacks or disasters, 262, 351
nucleases, 155, 177
TALENs (transcription activator-like effector nucleases), 155, 167, 177–79, 187, 190
ZFNs (zinc-finger nucleases), 155, 178, 179
nucleic acids, 17
see also DNA; RNA
nucleofection, 382
Nuffield Council, 293, 294

Obama, Barack, 294
Ochsner Hospital, 445
ODIN, The, 255
Office of Scientific Research and Development, 89
oil spills, 232
On the Origin of Species (Darwin), 11, 28
open-source software, 163, 256
optogenetics, 167
Orwell, George, 358–59
osmium hexamine, 57
Our Posthuman Future (Fukuyama), 281
Oviedo Convention, 278
Oxford University, 438, 439

P53 gene, 250–51
PAC-MAN (prophylactic antiviral CRISPR in human cells), 454–57, 460
Palievsky, Julia, 414
Panasenko, Sharon, 33–34
pandemics, 256–57, 263, 357, 405, 447
Black Death, 447
COVID, *see* COVID-19
Pasteur, Louis, 123, 231–32
Pasteur Institute, 119–22, 471
patents, xv, 39, 76, 98, 99, 114, 231–41
on biological processes, 231–32
for CRISPR systems, 92, 94, 115, 135–36, 144, 156, 232–41
of Doudna and Charpentier at Berkeley, 207–8, 210, 219, 224, 233–40
Doudna's battle with Zhang over, 184, 194, 208, 224, 234–40, 425, 470
pool idea for, 207–8
of Zhang and the Broad Institute, 192, 207–8, 210–11, 219, 224–27, 233–40
effects on collaboration and competition, 219, 234, 375
for genetic engineering techniques, 392
for microchips, 240–41
of Pasteur, 231–32
universities and, 232, 473
Pauley, Jane, 352
Pauling, Linus, 21–25, 51, 159, 391
Pauling, Peter, 23–24
Pawluk, April, 261, 376
PBS, *384,* 386, 389–90, 392, 393
PCR (polymerase chain reaction), 408, 413, 416, 421, 425, 428, 430
PCSK9 gene, 251

PD-1 protein, 249
Peacock Initiative, 301
Pei, Duanqing, 295, 317–19
penicillin, 56, 122
Penrose, Roger, 469–70
People's Daily, 318
Percy, Walker, 345
Personal Genetics Education Project, 220–21
Pfizer, 435, 441, 442, 445, 447
Phage Group, 19, 479
phages (bacteriophages; viruses that attack bacteria), 19, 37, 75
bacteria's use of CRISPR against, xiv, xviii, 67, 76–77, 86, 87, 93–94, 106, 111
see also CRISPR
number of, 75
philanthropic foundations, 118
Phillips Academy, 171
physics, xvii, xix, 89
Pinker, Steven, 291–92
Pisano, Gary, 114
Pixar, 463
Plague, The (Camus), 399
Planck, Max, 471
Plath, Sylvia, 352
playing God, 269, 273, 333, 363–65, 480–81
PNAS, 143–44
pneumonia, 121
pneumonic plague, 159
Poe, Edgar Allan, 352
Poincaré, Henri, 69
Polaris Partners, 209
polio, 345, 437
polymerase chain reaction (PCR), 408, 413, 416, 421, 425, 428, 430
Pomona College, *30,* 31–34
Porteus, Matthew, 344
He Jiankui and, 306–7, *314,* 321, 323
preimplantation genetic diagnosis, 274–76, 285, 327, 337, 342, 361–62
prizes, *see* scientific prizes
procreative beneficence, 360
Project Lightspeed, 441
Project McAfee, 443
Prometheus, 268, 286, 363, 365
Prometheus Bound (Aeschylus), 243
Protein & Cell, 290
proteins, 17, 43, 44, 45, 47, 51, 59, 61, 65
enzymes, *see* enzymes
fluorescent, 165–67, 382–83
light-sensitive, 167
Pauling's work on, 21
psychological disorders, *see* mental illness
PubPeer, 226
Putin, Vladimir, 294

Quake, Stephen, 300, 301, 307
Quebec conference, 373–77
Qi, Stanley, *448,* 453–57, 460
Quiz Kids, 18

race
African Americans
COVID vaccines and, 461
as researchers, 461
and skin color as disadvantage, 347–48
Watson's comments on, 37, 49, 386–92, 394
radiation exposure, 260, 351
Rakeman, Jennifer, 409
Ramsey, Paul, 268–69
Rawls, John, 281, 357–58
Raytheon, 89, 117
Reader, Ruth, 227
Reagan, Ronald, 309
Rebennack, Mac, 479
Redesigning Humans (Stock), 278
Reed, Jack, 328
Reed, Lou, 352
Regalado, Antonio, 227, 312–13
religious organizations, 273
Remaking Eden (Silver), 277
reprogenetics, 277
reverse transcription, 270, 408
ribozymes, 44–45, 47, 53, 60
Rice University, 300, 307–9
Ridley, Matt, 90
crystallography and, 19–23, 51
Rifken, Jeremy, 269
RNA (ribonucleic acid), xvii, xviii, 17, 43–50, 85, 105, 385, 435–36, 446–47
in central dogma of biology, 44, 47, 270
of coronavirus, 65, 403–4
in CRISPR, 79, 108, *133*
Cas13 targeting of RNA, 423–25, 429, 451, 452, 454–57
CRISPR RNA (crRNA), 86, 106, 124–26, 131–35, 137–39, 143, 146, 147, 180, 184, 186, 217
RNA interference, 77, 79, 93
single-guide RNA (sgRNA), 134–35, 139, 186, 190, 191, 195, 198
trans-activating CRISPR RNA (tracrRNA), 124–27, 131, 134, 135, 139, 146–47, 179–80, 182, 184–86, 217–18, 225
crystallography and, 53, 55–57, 66, 171, 181
Doudna's work on, xviii, 45–49, 51–61, 63, 65–67, 134, 181, 220, 435–36, 446
first, origin of, 47
interference, 65–67, 77, 79, 85, 87, 93, 107
introns, 44–45, 49, 53, 55, 66
messenger (mRNA), 43–44, 65, 66, 77, 85, 439–42, 445
and origin of life, 45–47
replication in, 45, 47, 49, 51, 55
reverse transcription and, 270, 408
ribozymes, 44–45, 47, 53
structure of, xviii, 48, 51–61, 63, 65, 66, 134, 181, 220
self-splicing, 45, 47–49, 53–55, 60, 66
vaccines, 439, 440–42, 445–47, 449
in viruses, 65, 451–53
Rockefeller University, 17, 121, 181, 200, 205, 234–35, 270
Rogers, Michael, 270
Rolling Stone, 270
Roosevelt, Franklin, 345, 346, 386
Rossant, Janet, 369
Rothkopf, Joanna, 227
Royal Swedish Academy, 470
Russia, 294, 414

Sabeti, Pardis, 450–52
Sabin, Albert, 345, 437
Safe Genes, 260, 262, 380
Şahin, Uğur, 441
St. Jude Children's Research Hospital, 122
Salk, Jonas, 345, 437
Salk Institute, 454
salt-loving organisms, 72, 74, 76, 79
Sanchez, David, *340*, 343–45
Sandel, Michael, 281, 347, 365, 376, 386
Sanders, Bernie, 386
San Francisco Examiner, 100
Sangamo Therapeutics, 414
Sanofi, 205
Sarah Cannon Research Institute, *244*
SARS (severe acute respiratory syndrome), 65, 403, 441
SARS-CoV-2, 403–4
see also COVID-19 pandemic
Savage, Dave, 402, 403
Savulescu, Julian, 360
schizophrenia, 177, 352–53
of Rufus Watson, 38, 277, 352, 386, 393
Schrödinger, Erwin, 18
Schubert, Franz, 352
science, xix, 34, 35, 54, 124, 443, 473
citizen involvement in, 263, 445
collaboration across generations in, 475
commerce and, 114
competition in, 144–45, 157–59
curiosity-driven, xix, 89, 90, 457, 473, 479
in daily life, 474
effects of COVID pandemic on, 473–74
effects of prizes and patents on, 219, 234, 375
government-business-university partnerships and, 117–18, 232–33
institutions and, 405
invention and, 90
and linear model of innovation, 89–90
military spending and, 259
revolutions in, xvii, 28
teamwork in, 103
women in, xix, 8, 22, 31, 227, 459–61, 470, 471
Lander's "Heroes of CRISPR" and, 227
self-promotion and, 110–11
Science, 48, 63, 66–67, 198, 233, 249, 280, 289, 290, 316, 376, 451, 474
Barrangou and Horvath's paper in, 92
Church's paper in, 192, 194–95, 197, 199, 201, 202

Doudna, Chen, and Harrington's paper on CRISPR as detection tool in, 422–23, 451
Doudna and Charpentier's 2012 paper in, 126, 137–41, 144–46, 148, 149, 156, 172–73, 179, 180, 182–86, 191, 192, 194–95, 202, 227, 233, 237–38, 374–75
Doudna and Charpentier's 2014 review article in, 215
Sternberg's paper in, 374
Zhang's 2013 CRISPR paper in, 192, 195, 197, 199, 201, 202, 225, 233–34
Zhang's transposons paper in, 374–76
Science and Method (Poincaré), 69
Science Foo Camp, 193
Scientific American, 227
scientific prizes, 217, 218–21
Breakthrough Prize in Life Sciences, 219–21
effects on collaboration and competition, 219, 375
Gairdner Award, 220
Kavli Prize, 221
Nobel Prize, 19, 20, 21, 44, 49, 53, 66, 100, 219, 221, 226, 270, 305, 470, 472
awarded to Doudna and Charpentier, xix, *468*, 469–73
awarded to Watson, Crick, and Wilkins, 28, 470
Scientist, 228
sexual orientation, 347–48, 361
Shakespeare, William, 151, 414
Shannon, Claude, 28
Shelley, Mary, *Frankenstein*, 267, 269, 283, 286, 297
SHERLOCK (specific high sensitivity enzymatic reporter unlocking), 424–25, 428–31, 451, 453
STOP, 431, 460
Sherlock Biosciences, 425, 428, 430
Shibasaburō, Kitasato, 159
SHINE, 453
Shockley, William, 90
sickle-cell anemia, 41, 337, 343–45, 361, 376, 390, 392, 414
CRISPR therapy for, xviii, 245–49, 329, *340*
Šikšnys, Virginijus, *142*, 143–49, 179–80, 220, 221, 225, 375, 470
Silicon Valley, 114, 116, 219, 301
Silver, Lee, 277–78
Sinovac, 437
Sinsheimer, Robert, 268
skin color, 347–48
Slack, xv, 405, 459, 463
SLC24A5 gene, 347–48
smallpox, 436
social contract theories, 357, 358
social inequality, 274, 288, 358, 360, 362–63, 369, 376–77, 391–92
software, 432
open-source, 163, 256
somatic editing, 246, 276–77, 288, 329, 336, 338, 341
Sontheimer, Erik, *70*, 93–94, 139, 148–49, 180, 187, 213, 220, 376, 377, 470
Spiegel, 205
Splicing Life report, 273–74
sports and muscle enhancement, 253–55, 348–49, 361, 376
Stanford University, 114, 115, 232, 410
Children's Hospital, 344
Zhang at, 167
Startup in a Box, 116
Stat, 306, 308, 326
State Department, U.S., 23
Stein, Rob, 245
Steitz, Joan, 56
Steitz, Tom, 56–57, 63–64
Stimson, Henry, 338
STING program, 165
Stock, Gregory, 276, 278
Streptococcus pyogenes, 122, 139, 182
Streptococcus thermophilus, 91, 182
Sternberg, Sam, 107–8, 136, 146–47, 218, 287, *372*, 374–75
Happy Healthy Baby and, 284–86
transposons paper of, 374
Sunday Times, 26, 387, 392
Sun Microsystems, 114
Supreme Court, U.S., 231, 232, 239
Swanson, Robert A., *96*, 99
Synagogue Council of America, 273
Szostak, Jack, 35, 37, 40, *42*, 45–49, 53, 208, 228, 436, 442, 472

tadpoles, 171, 172
TALENs (transcription activator-like effector nucleases), 155, 167, 177–79, 187, 190
Tay-Sachs disease, 41, 329, 337
technology, xvii
biotechnology, *see* biotechnology
digital, xvii, 28, 114
and linear model of innovation, 89–90
Tempest, The (Shakespeare), 151
Templeton Foundation, 303
Terman, Frederick, 114
terrorists, 259–63
test-tube baby, 274, 313, 328
Texas Instruments, 240
Theory of Justice, A (Rawls), 357–58
therapies, 233, 245–51
blindness, xviii, 250
cancer, xviii, 249–51
safe viruses as delivery method in, 456
sickle cell, xviii, 245–49, 329, *340*
Third Rock Ventures, 209, 211
This Week in Virology, 415
Thoreau, Henry David, 6
thought experiments, 339, 341–54, 361, 362
character, 345–46
cognitive skills, 354
deafness, 346–48
disabilities, 345–48
Huntington's disease, 341–42, 361, 365
psychological disorders, 352–54

thought experiments (*cont.*)
sexual orientation, 347–48, 361
sickle cell, 343–45, 361
skin color, 347–48
Thousand Talents Recruitment Program, 301
Thurstone, Louis Leon, 392
thymine, 26, *27*
Tieu, Dori, *406,* 418
Time, 40, 100, 265, 269, 391
Time Machine, The (Wells), 267
Times (London), 26, 387, 392
Tjian, Robert, 405
Tolkien, J. R. R., 415
Torvalds, Linus, 256
transposons, 49, 373–76
transhumanism, 351–52
transistors, xvii, xix, 89–90, 117, 201
Truman, Harry, 89
Trump administration, 411
Tuomanen, Elaine, 121–22
Türeci, Özlem, 441
Turing, Alan, 28
Tuskegee experiments, 461
Twigg-Smith, Lisa Hinkley, 5, 59
Twitter, 226, 227, 388, 416
National Academy's video and, 355–56
Tyler, Anne, 7

unintended consequences, 277, 290, 364
universities
corporate involvement in research, 274
government-business-university partnerships, 117–18, 232–33
patents and, 232, 473
University of Alicante, 71, 72, 74, 76
University of California, 116
Berkeley, *see* Berkeley, University of California at
University of California, San Francisco (UCSF), 98, 261
Innovative Genomics Institute, 248, 261, 380–82, 401–3
COVID and, 401–5, 413–18, *419*
University of Colorado, 53, 55
University of Pennsylvania, 249, 279, 416
University of Vienna, 122, 233, 235n
University of Washington, 409–10, 415
University of Zurich, 380
Urnov, Dmitry, 414
Urnov, Fyodor, 261–62, *406,* 414–15, 417–18, *419,* 421, 459–60
U.S. Catholic Conference, 273
utilitarianism, 357
utopians, 267–69

vaccines, 357, 365, 435–47, 455
COVID, *434,* 435–47, 449
clinical trials of, 435–36, 440–41, 445–46, 461
Fauci and, 442
FDA and, 446
Isaacson in trial of, 435–36, 440–41, 445–46
DNA, 439–41, 444–45, 456
genetic, 437–39
RNA, 439, 440–42, 445–47, 449
traditional, 436–37, 449
for polio, 345, 437
van der Oost, John, 84, 127
van Gogh, Vincent, 352, 353
Vanity Fair, 219
Varian Associates, 114
Vence, Tracy, 228
Venter, Craig, *36,* 39–40
venture capitalists, 98, 99, 113–17, 205, 285, 301
viruses, xvi, 19, 65, 403, 449, 474
coronaviruses, *see* coronaviruses
CRISPR as tool for detecting, 115, 118, 421–25, 431–32, 451
at-home tests, 430–32, 460
coronavirus, 413, 417, 421–25, 427–33, 453
HPV, 422–23
CRISPR cures for, 449–57
CARVER, 452–53, 455–57
PAC-MAN, 454–57, 460
CRISPR used by bacteria against, *see* CRISPR
human vulnerability to, 350, 351
CRISPR gene editing and, xv, xix
pandemics, *see* pandemics
phage, *see* phages
RNA in, 65, 451–53
RNA interference and, 66–67

Wade, Nicholas, 40, 172, 289
Wallace, Alfred Russel, 12–13, 159
Wall Street Journal, 332
Wang, Emily, 310
Warren, Elizabeth, 290–91
Washington Post, 229, 347, 408
Waters, Alice, 140–41
Watson, Duncan, 397
Watson, Elizabeth, 38–39, 386, 388, 395, 397
Watson, James, 18–28, 37–39, 166, *266,* 302, 333, 352, 385–94, 475, 479
on abortion, 337
art collection of, 395
at Asilomar, 270–72
Avoid Boring People, 386–87
Berg and, 271
childhood of, 18
as Cold Spring Harbor director, 37
Cold Spring Harbor's cutting of ties with, 386, 387, 390
Crick's meeting of, 20
in DNA double-helix structure discovery, xviii, 7, 11, 26–28, 29–31, 46, 51, 58, 159, 386, 390, 395, 397, 423, 470
with DNA model, *16*
The Double Helix, xix, 6–8, *9,* 20, 21, 27, 29–31, 50, 157, 328, 391, 397, 398, 414, 459, 477–78
Doudna and, xviii, 8, 29, 48–50, *396*
Doudna's visit with, 395–98
Franklin and, 7–8, 22–26, 50, 389, 395

at gene-editing conference, 276–77
Human Genome Project and, 37–40, 388, 390, 391
on intelligence, 285–87, 389–90, 397
Isaacson's visit to, 391–94
Jefferson Conundrum and, 390–91
ninetieth birthday of, 388–89
Nobel Prize awarded to, 28, 470
oil portrait of, 302, *384,* 386, 390, *396*
PBS documentary on, *384,* 386, 389–90, 392, 393
on racial differences, 37, 49, 386–92, 394
Wilkins and, 19–21
Watson, Rufus, 37–39, 393–94, 395
in PBS documentary, *384*
schizophrenia of, 38, 277, 352, 386, 389, 393
Watters, Kyle, 260
Wegrzyn, Renee, 262
Wells, H. G., 267
West Coast Computer Faire, 373
West Nile Virus, 305, 390
What Is Life? (Schrödinger), 18
Whitehead Institute, 206
Who Should Play God? (Howard and Rifkin), 269
Wiedenheft, Blake, 81–87, *82,* 93, 95, *104,* 105, 106, 108, 111, 113, 284
Jinek and, 85
Wiley, Don, 166
Wilkins, Maurice, 19–25
Nobel Prize awarded to, 28, 470
Wilson, James Q., 281
Wilson, Ross, 109–10, 402, 440, 456
Wired, 454
wisdom, 354, 365
Wojcicki, Anne, 219
women in science, xix, 8, 22, 31, 227, 459–61, 470, 471
Lander's "Heroes of CRISPR" and, 227
self-promotion and, 110–11
Women in Science group, 381
Woods, Tiger, 351
woolly mammoth, 160, 169, 172
World Economic Forum in Davos, 368
World Health Organization (WHO), 329, 331, 409
World's Fair (1964), 171
World War II, 20, 114
atom bomb in, xvii, 89, 117, 224, 265, 338
Worldwide Threat Assessment, 259
Wozniak, Steve, 256
Wu, Ting, 220, 221

X-ray crystallography, *see* crystallography

Yale University, *52,* 57, 60–61, 63
yeast, 35, 45, 148–49, 232
Yellowstone National Park, 81–84
Yersin, Alexandre, 159
yogurt, xviii, 90–92, 133, 143, 478
Young, Mark, 83

Zayner, Josiah, *252,* 253–57, 262–63, 325–28, 348, 481
COVID vaccine and, 443–45
CRISPR babies as viewed by, 235–26
coronavirus vaccine and, *434*
zebrafish, 202
Zhang, Feng, *158,* 161–67, *162,* 259, 292, 302, 330, 376, 382, 385, 416, *420, 426,* 454, 459, 460, 472, 473, 477
at Broad Institute, xv, 159, 161, 177, 179, 184–86, 189, 374, 421, 423, 424, 450–51, 454
Cas13 and, 423–25, 454
childhood of, 161–64, 432
Church and, 167, 174, 175, 178, 192–94
company formations and, 205, 206, 208–10
Editas, 210–13, 250, 454
in competition to adapt CRISPR into tool for gene editing in humans, xv, 64, 159–60, 174, 175–86, 191–95, 197, 199–202, 227
Cong and, 177–78, 184, 192–94
and CRISPR as detection tool for coronavirus, 421, 427–32
home testing kits, 432–33, 460
SHERLOCK, 424–25, 428–31, 451, 453
Sherlock Biosciences, 425, 428, 430
STOP, 431, 460
as CRISPR-Cas9 inventor, 210–11
CRISPR-Cas9 paper of (2013), 192, 195, 197, 199, 201, 202, 225, 233–34
emails to Doudna, 200–201, 207
gene-editing research with TALENs, 167, 177–79
grant application of, 179–80
at Harvard, 166, 167, 175
in high school, 165–66
on *in vitro* vs. *in vivo* gene editing, 183, 382
Lander and, 182–83, 220
Lander's "Heroes of CRISPR" and, 182, 223–29
Levy and, 165–66
Lin and, 184, 192
Liu and, 461–62
Marraffini and, 180–82, 186, 192, 200, 213, 234–35, 240
patent of, with Broad Institute, 192, 207–8, 210–11, 219, 224–27, 233–40
in patent battle with Doudna, 184, 194, 208, 224, 234–40, 425, 470
possible partnership with Doudna, 207–8
at Quebec conference, 373, 375
Sabeti and, 450–51
at Stanford, 167
transposons paper of, 374–76
Zhou, Kaihong, *104*
Zika virus, 424
zinc-finger nucleases (ZFNs), 155, 178, 179
Zoom, xiv, xv, 404–5, 414, 418, *426,* 440, 459, 463, 464, 465, 472
Zuckerberg, Mark, 166, 219, 472

IMAGE CREDITS

By page number

Front and back endpaper: Carlos Chavarria/Redux
iv: Brittany Hosea-Small/UC Berkeley
xi (left and right): David Jacobs
xii: Jeff Gilbert/Alamy
2 :(clockwise): Courtesy of Jennifer Doudna; Leah Wyzykowski; courtesy of Jennifer Doudna
10: (left to right): George Richmond/Wikimedia/Public Domain; Wikimedia/Public Domain
16: A. Barrington Brown/Science Photo Library
25 (left to right): Universal History Archive/Universal Images Group/Getty Images; Courtesy Ava Helen and Linus Pauling Papers, Oregon State University Libraries
27: Historic Images/Alamy
30: Courtesy of Jennifer Doudna
36: Natl Human Genome Research Institute
42: Jim Harrison
52: YouTube
62: Courtesy of Jennifer Doudna
70 (top to bottom): Courtesy of BBVA Foundation; courtesy of Luciano Marraffini
78: The Royal Society / CC BY-SA (https://creativecommons.org/licenses/by-sa/3.0)
82: Mark Young
88: (top to bottom): Marc Hall/NC State courtesy of Rodolphe Barrangou; Franklin Institute /YouTube
96: Courtesy of Genetech
104: Roy Kaltschmidt/Lawrence Berkeley National Laboratory
112: Courtesy of Caribou Biosciences
120: Hallbauer & Fioretti/Wikimedia Commons
130: Berkeley Lab
133: MRS Bulletin
138: Miguel Riopa/AFP via Getty Images
142 (clockwise): Edgaras Kurauskas/Vilniaus universitetas; Heribert Corn/courtesy of Krzysztof Chylinski; Michael Tomes/ courtesy of Martin Jinek
152: Andriano_CZ/iStock by Getty Images
158: (top to bottom): Justin Knight/McGovern Institute; Seth Kroll / Wyss Institute at Harvard University; Thermal PR

162: Justin Knight/McGovern Institute
168: Seth Kroll / Wyss Institute at Harvard University
176: Wikimedia Commons
188: Anastasiia Sapon/The New York Times/Redux
196: Courtesy of Martin Jinek
204: Courtesy Rodger Novak
214: BBVA Foundation
222: Casey Atkins, courtesy Broad Institute
230: Courtesy of Sterne, Kessler, Goldstein & Fox P.L.L.C.
244: Amanda Stults, RN, Sarah Cannon Research Institute/The Children's Hos
252: Courtesy of The Odin
258: Susan Merrell/UCSF
266: (top to bottom): National Academy of Sciences, courtesy of Cold Spring Harbor Laboratory; Peter Breining/San Francisco Chronicle via Getty Images
282: Pam Risdom
298: (top to bottom): Courtesy He Jiankui; ABC News/YouTube
314: (top and bottom): Kin Cheung/AP/Shutterstock
325: Courtesy of UCDC
334: Tom & Dee Ann McCarthy/Getty Images
340: Wonder Collaborative
366: Isaac Lawrence/AFP/Getty Images
372: Nabor Godoy
384 (top to bottom): Lewis Miller; PBS
396: Courtesy of Jennifer Doudna
400: Irene Yi / UC Berekely
406: Fyodor Urnov
412: Courtesy of Innovative Genomics Institute
420 (top to bottom): Mammoth Biosciences; Justin Knight/McGovern Institute
426: Omar Abudayyeh
448 (top to bottom): Paul Sakuma; courtesy of Cameron Myhrvold
458 (top to bottom): Wikimedia Commons; Cold Spring Harbor Laboratory Archives
468: Brittany Hosea-Small/UC Berkeley E103
476: Gordon Russell

ABOUT THE AUTHOR

WALTER ISAACSON, a professor of history at Tulane, has been CEO of the Aspen Institute, chair of CNN, and editor of *Time*. He is the author of *Leonardo da Vinci*; *The Innovators*; *Steve Jobs*; *Einstein: His Life and Universe*; *Benjamin Franklin: An American Life*; and *Kissinger: A Biography*, and the coauthor of *The Wise Men: Six Friends and the World They Made*.